Unmanned System Technologies

Springer's Unmanned Systems Technologies (UST) book series publishes the latest developments in unmanned vehicles and platforms in a timely manner, with the highest of quality, and written and edited by leaders in the field. The aim is to provide an effective platform to global researchers in the field to exchange their research findings and ideas. The series covers all the main branches of unmanned systems and technologies, both theoretical and applied, including but not limited to:

- Unmanned aerial vehicles, unmanned ground vehicles and unmanned ships, and all unmanned systems related research in:
- Robotics Design
- Artificial Intelligence
- Guidance, Navigation and Control
- Signal Processing
- Circuit and Systems
- Mechatronics
- Big Data
- Intelligent Computing and Communication
- Advanced Materials and Engineering

The publication types of the series are monographs, professional books, graduate textbooks, and edited volumes.

More information about this series at http://www.springer.com/series/15608

Hussein A. Abbass • Robert A. Hunjet
Editors

Shepherding UxVs for Human-Swarm Teaming

An Artificial Intelligence Approach to Unmanned X Vehicles

Springer

Editors
Hussein A. Abbass
University of New South Wales
Canberra, ACT, Australia

Robert A. Hunjet
University of New South Wales
Canberra, ACT, Australia

ISSN 2523-3734 ISSN 2523-3742 (electronic)
Unmanned System Technologies
ISBN 978-3-030-60897-2 ISBN 978-3-030-60898-9 (eBook)
https://doi.org/10.1007/978-3-030-60898-9

This Springer imprint is published by the registered company Springer Nature Switzerland AG
The registered company address is: Gewerbestrasse 11, 6330 Cham, Switzerland

To a future where kindness, morality, and humanity shepherd humans and swarms. To a family I am blessed to have.

—HA

To Evie and Samuel, as much as I have guided you, you have guided me. Thank you.

—RH

Foreword

Research is the exploration of what is possible, and nature shows us that many things are possible. Our challenge as robotics researchers is to learn from nature to create new capabilities for man-made systems. I was intrigued by the request to write the foreword for this book *Shepherding UxVs for Human–Swarm Teaming*. As a research scientist and engineer with 30+ years of experience in robotics, artificial intelligence (AI), and control systems R&D, I have always been fascinated with the connection between natural and man-made systems and processes. With AI, we seek to recreate the capabilities found in humans (deemed intelligence) by creating explicit programs to mimic those functions in computers. In robotics, we utilise AI-based (and many non-AI-based) algorithms to create machines that can sense and interact with the physical world and replicate the functionality of humans at various tasks. Behavior-based approaches to control rely on the observation of biological systems in order to intuit their policies so that these policies can be programmed into artificial agents (physically manifested as robots or other machines). Swarm robotics lies at the intersection of bio-inspired and AI-based research with the use of bottom–up behavior-based approaches to the control of multi-agent systems, coupled with top–down AI-based approaches to coordination and planning of these systems, and with a need for humans to control the collective behavior of the swarm to accomplish a task.

This book focuses on a challenging problem in multi-robot systems: how to design (i.e., program) behaviors for a swarm of robots to perform a task observed in nature (shepherding). It draws inspiration from nature in several ways, including use of a biologically based group intelligence behavior (the shepherding behavior of a team of sheepdogs) versus the desire of the sheep (or adversarial agents) to escape, the need for individual behaviors or policies for each of the shepherding agents, and the control of the team of shepherding agents by a human. This challenge provides an extremely rich problem domain for exploring state-of-the-art techniques in AI, planning, sensing, machine learning, and inter-agent communication for teams and swarms of agents. This book explores these areas in depth and traces a cohesive path through the tools and techniques needed to design a man-made solution, including interfacing with and control of the swarm by a human operator.

Researchers in robotics and artificial intelligence often turn to nature for inspiration regarding what is possible, and to provide clues on how to create better designs or how to implement new capabilities in machines. As a robotics researcher and engineer, I often ask myself two questions: "What capabilities does a human have that I would like for my robot to have?" and, "How can I use algorithms and techniques from control theory, sensing and pattern recognition, planning, machine learning, etc. to implement these capabilities on my robot?" This book takes this thinking a step further to explore biologically inspired and analytical approaches to creating behaviors for swarms. The curious reader (whether roboticist, AI researcher, engineer, student, entrepreneur, or citizen scientist) will find much to appreciate in this book and, after reading it, will come away with a much deeper appreciation for the challenges and approaches to solutions for implementing real-world robotic swarms.

Donald Sofge
Swarm Roboticist
U.S. Naval Research Laboratory
Washington, D.C., USA

Preface

It is so often the case that scientists and engineers turn to nature when they are faced with complex or novel problems. The animal kingdom, the forest, the ocean, and the sky are classic sources of inspiration. Sometimes, those closest to us, the face in the mirror, our children, our family, or our co-workers, help us to learn and develop new understanding of cognition, intelligence, and communication. Society teaches us how trust others, and connectivity amongst people augments and expands the bounded capacity of a single individual with the near limitless capacity of the collective.

When taking inspiration from nature, some seek a seed for an idea and continue to be faithful to nature in every aspect of the solution as they grow the idea to a complete solution. This is normally the road that most scientists take to understand the limits, pros, and cons of a natural phenomenon and learn more about how it functions in nature. Others seek the inspiration for a seed and then deviate from nature to adopt and adapt the solution within the realm of what is technologically required to solve the problem at hand. This is normally the road of most engineers and technologists.

This book is an inquiry into both worlds: the world of scientists who collect data from nature and/or run basic simulations to understand fundamental phenomena and the world of engineers and technologists who seed solutions with ideas from nature but break new technological ground by deviating from nature, or even taking the idea out of its original context, applying it to more complex, yet practical settings.

In recent years, there has been much interest and advancement in the scientific community in the areas of swarm intelligence and swarm robotics. A swarm is a collective of individuals that display emergent global behaviour that is far more interesting than the individual behaviour each member of the swarm exhibits. It follows the ethos that the collective is more than the sum of its parts.

For the purpose of controlling multi-robot systems, the concepts brought by swarm intelligence are extremely important, especially from the perspective of scalability. As emergent global behaviours are realised through the local sensing (and perhaps communications) between robots, a swarm does not require a human operator to send control signals to each individual robot. Nor does it require an operator to define complex programs that prescribe the emergent behaviours. This begs further questions: *How does a human control a swarm? How do we interact and team with a swarm? How can we convey our intent to the swarm? How can we understand what the swarm is doing?* The concept of swarming is biologically inspired, and once again, we can turn to nature to elicit ideas for addressing these questions.

The context of this book is *shepherding*; how a farmer (shepherd) works together with one or more external smart actuators (sheepdogs) to control a large number of less capable and less intelligent individuals (sheep), but collectively complex groups (flock), to achieve the farmer's objective (herding). The biological behaviour alone is fascinating. The social dimensions of trust and friendship in the farmer–dog relationship, the lexicon used to command the dog, the interaction between the dog and the sheep, the information and behavioural cues displayed by the dog, the relationship between the dog's actions and its intent, and more, are all possible dimensions of scientific inquiry that could shed light on many fundamental scientific questions and inspire solutions for technological problems.

However, the majority of this book focuses more on the technological design factors that need to go into the smartness/cognition of a robotic sheepdog, and the artificial intelligence (AI) and multi-robot problems that need to be solved to create a robot as smart as a sheepdog. The book further explores how the shepherding concept can be re-positioned to inspire engineering solutions to control large swarm robotic setups with a fewer number of smarter robots in safety critical domains, such as unmanned aerial vehicle (UAV) traffic control systems.

Technologically, shepherding is the use of a few decision agents to guide a significantly larger number of, relatively less complex, decision agents. The shepherding concept provides a biologically inspired framework for multi-robot guidance and human–swarm teaming. The shepherd(s) interact with a manageable number of entities, which in turn exercise control over a large flock. As such, the cognitive load on a human operator is dramatically reduced from that required when explicitly controlling the individual members of a swarm. The human and animal ethics in the relationship, the social interaction, and trust between the agents are all dimensions worthy of further inquiry. Furthermore, the interactions between human and sheepdog, and sheepdog and sheep can be modelled using *force vectors* of attraction and repulsion. Herein lies the power of the shepherding abstraction.

First, force vectors are generic representations that could represent concepts in a variety of domains, from the movement of cars to the movement of information and ideas. The applications are near endless; first to mind, we may consider human operators controlling large numbers of drones, while only needing to interact with a few. Abstracting further, we may consider the mobility of software throughout a computer network or control of an optimisation meta-heuristic in a complex state space. Second, the representation encodes and transforms the sensorial information and action space in a swarm robotic system into vectors, matrices, and tensors. With advances in graphical processing units, tensor operations could be used to speed up processing by orders of magnitude over classic sequential processing of a complex piece of logic.

Third, the shepherding problem provides diversity not only in the application space but also affords us a rich playground for research in the areas of cognition, optimisation, and learning. *How can shepherding agents collaborate, learn, and dynamically optimise their performance to complete the shepherding task?* With applicability to real-world problems and relevance to a multitude of research fields such as AI and autonomous systems, we find the shepherding concept worthy of deeper investigation.

This book is not a comprehensive tome on shepherding multi-unmanned system swarms. Rather, it is the commencement of the journey we started some years back, investigating an exciting and rich problem space. Each chapter provides insight into an approach to tackling the problem, or how it could be applied, shedding light on current approaches and opening the door to new and exciting work.

We structure the book into four parts. The first part presents simulation models of abstract reactive shepherding. While reactive models such as BOIDS display complex dynamics on the group level, operating concepts such as shepherding to control a swarm and support a human operator begs for more complex AI models to monitor, understand, learn, act, recommend, and explain events in this complex environment. This is the context for the second part, which presents core machine learning and optimisation problems for shepherding.

The third part presents realistic and complex use of shepherding in the safety critical air domain and air-to-ground vehicle interaction. The fourth part focuses on the integration of shepherding with a human operator and presents on the need for transparency in the relationship, and human performance analysis and monitoring.

Chapter 1 defines our motivation for working in this space, proposing an architecture for contextually aware, AI-enabled shepherds and shepherding agents, as well as introducing a notation for consistency throughout the book and our future work.

Chapter 2 presents a hybrid shepherding approach for environmental observation in which certain waypoints are required to be examined in order. A classical shepherding approach is utilised to move the swarm towards the goal locations. A new behaviour is introduced in the event of swarm members straying from the *flock*, allowing these members to follow new trajectories to expedite goal location discovery.

Chapter 3 investigates how the use of swarming rules can enhance Strömbom's approach to sheepdog positioning, improving the performance of the shepherding task through the use of multiple shepherding agents in an arc formation.

Chapter 4 describes the dynamics of shepherding, paying particular attention to the concept of stall distance in Strömbom's model, examining the most appropriate setting for this distance to ensure a shepherding agent does not adversely affect flock cohesion upon approach.

Chapter 5 presents the decision-making processes within cognitive agents in the form of the optimisation sub-problems (task assignment, task allocation, path planning, and trajectory planning) that agents need to solve for mission planning in a shepherding context, with the agents presented as an unmanned x(any) vehicle (UxV) swarm.

Chapter 6 reviews previous learning approaches applied to shepherding, specifically identifying the benefit of applying prior knowledge to the design of such learning systems and introducing ontology-guided learning as a beneficial strategy to allow for symbolic prior knowledge to guide the non-symbolic machine learning system.

Chapter 7 investigates cognitive agents (shepherd, sheepdogs, and external observers) in the shepherding problem. As opposed to their reactive counterparts, these cognitive agents have the ability to reason; they are required to recognise the activity of other agents in the environment and infer intent, so as to provide a more accurate situational and contextual awareness, allowing for increased efficacy in the shepherding task.

Chapter 8 presents the use of shepherding to assist collective decision making within swarms. Individual and disparate swarm members gather information before they are herded by the shepherd, enabling fusion of the data that improves their confidence in the collective decision they need to make.

Chapter 9 presents the concept of a *sky shepherd*, an unmanned aerial vehicle (UAV), which takes the place of a sheepdog. Dorper sheep are exposed to the sky shepherd in order to understand the influence the UAV exhibits on the sheep.

Chapter 10 uses a new type of machine learning algorithm proposed by the authors, apprenticeship bootstrapping, to decompose the sky shepherding learning problem into sub-problems and then autonomously learn the combined skills needed by a sky shepherd to guide a group of unmanned ground vehicles operating in the Gazebo high-fidelity simulation environment.

Chapter 11 borrows shepherding to design a concept of operation for a swarm air traffic control system, where the authors modify the shepherding algorithm and combine it with a digital twin of the swarm traffic system to present a concept suitable for the safety critical nature of the operating environment.

Chapter 12 tackles the problem of transparency in human–autonomy teaming by first exploring human–swarm teaming and then proposing learning classifier systems to dynamically evolve the rules governing swarm dynamics.

The book concludes with Chap. 13, which focuses on the issues of task performance, human performance, situation awareness, and trust to synthesise the multi-dimensional set of indicators that will allow appropriate monitoring of the efficiency of human–swarm teams and balances the workload on the human and the swarm.

With a wide range of topics and applications covered, we hope that the reader finds inspiration and joins us in this research direction of shepherding for swarm control and human–swarm teaming.

Canberra, ACT, Australia Hussein A. Abbass

Canberra, ACT, Australia Robert A. Hunjet

July 2020

Acknowledgements

The editors wish to thank all authors for their contributions to this book and for their patience during the development of the book.

The first editor wishes to present his gratitude to his wife, Eleni, and children, Adam and Zac, who equally invested in this journey by allowing him the time and environment to complete this project and sacrificing their play time with daddy, so that daddy completes his book project.

The second editor wishes to thank his wife, Linda, for her support during this project, and his children, Evie and Samuel, his inspiration.

This project was initiated to synthesise the Australian work on shepherding as a first step for the work on a grant from the Office of Naval Research Global (ONRG), USA, in collaboration with the Defence Science and Technology Group (DST), Australia, and the University of New South Wales Canberra, Australia. What started as a first step took 1.5 years to complete. All of the chapters you will see are on-going research streams that have continued to advance and evolve as we have been compiling this book. We present this book as the start of a journey into what we see as an interdisciplinary field of inquiry, shepherding.

Contents

Contributors

Hussein A. Abbass School of Engineering and Information Technology, University of New South Wales, Canberra, ACT, Australia

Sreenatha Anavatti School of Engineering and Information Technology, University of New South Wales, Canberra, ACT, Australia

Michael Barlow School of Engineering and Information Technology, University of New South Wales, Canberra, ACT, Australia

Daniel Baxter School of Engineering and Information Technology, University of New South Wales, Canberra, ACT, Australia

Benjamin Campbell Advanced Vehicle Systems, Land Vehicles and Systems, Land Division, Defence Science and Technology Group, Edinburgh, SA, Australia

Essam Debie School of Engineering and Information Technology, University of New South Wales, Canberra, ACT, Australia

Heba El-Fiqi School of Engineering and Information Technology, University of New South Wales, Canberra, ACT, Australia

Justin Fidock Defence Science and Technology Group, Edinburgh, SA, Australia

Matthew Garratt School of Engineering and Information Technology, University of New South Wales, Canberra, ACT, Australia

Jack Healey School of Engineering and Information Technology, University of New South Wales, Canberra, ACT, Australia

Adam Hepworth School of Engineering and Information Technology, University of New South Wales, Canberra, ACT, Australia

Robert A. Hunjet Advanced Vehicle Systems, Land Vehicles and Systems, Land Division, Defence Science and Technology Group, Edinburgh, SA, Australia

Aya Hussein School of Engineering and Information Technology, University of New South Wales, Canberra, ACT, Australia

Jing Liu School of Engineering and Information Technology, University of New South Wales, Canberra, ACT, Australia

Nathan Long School of Engineering and Information Technology, University of New South Wales, Canberra, ACT, Australia

Nathan McIntyre School of Engineering and Information Technology, University of New South Wales, Canberra, ACT, Australia

Kathryn Kasmarik School of Engineering and Information Technology, University of New South Wales, Canberra, ACT, Australia

Hung Nguyen School of Engineering and Information Technology, University of New South Wales, Canberra, ACT, Australia

Jayden Park School of Engineering and Information Technology, University of New South Wales, Canberra, ACT, Australia

Anthony Perry Advanced Vehicle Systems, Land Vehicles and Systems, Land Division, Defence Science and Technology Group, Edinburgh, SA, Australia

Raul Fernandez Rojas Faculty of Science and Technology, University of Canberra, Canberra, ACT, Australia

Karl Sammut College of Science and Engineering at Flinders University, Bedford Park, SA, Australia

Daniel Sgarioto Maritime Division, Defence Science and Technology Group, Port Melbourne, VIC, Australia

Kate J. Yaxley School of Engineering and Information Technology, University of New South Wales, Canberra, ACT, Australia

Generalised Shepherding Notations

Notation	Meaning
Υ	The set of Shepherds
B	The set of Sheepdogs
(Π)	The set of Sheep
L (150u)	Length and width of environment
N (1–201)	Cardinality of Π
M (1–20)	Cardinality of B
$R_{\pi\beta}$ (65u)	π sensing range for β
$R_{\pi\pi}$ (2u)	π sensing range for π
$\Lambda^t_{\pi_i}$	Local centre of mass for π_i at time t
$\Lambda^t_{\beta_j}$	Local centre of mass for β_j at time t
$\Gamma^t_{\pi_i}$	Global centre of mass for π_i at time t
$\Gamma^t_{\beta_j}$	Global centre of mass for β_j at time t
$W_{\pi\pi}$ (2)	π repulsion strength from π
$W_{\pi\beta}$ (1)	π repulsion strength from β
$W_{\pi\Lambda}$ (1.05)	π attraction strength to Λ
W_{π_υ} (0.5)	Strength of π previous direction
$W_{e\pi_i}$ (0.3)	Strength of sheep π_j angular noise
$W_{e\beta_j}$ (0.3)	Strength of shepherd β_j angular noise
P^t_G	Position of goal at time t, classically remains unchanged
$P^t_{\beta_j}$	Position of agent β_j at time t
$P^t_{\pi_i}$	Position of agent π_i at time t
$P^t_{\beta_j\sigma_1}$	Position for driving by agent β_j at time t; $R_{\pi\pi}\sqrt{N}u$ behind the flock
$P^t_{\beta_j\sigma_2}$	Position for collection by agent β_j at time t; $R_{\pi\pi}u$ behind the furthest agent

(continued)

Notation	Meaning
$F^t_{\pi_i \beta_j}$	Force vector, repulsion of π_i agent away from β_j agents at time t
$F^t_{\pi_i \pi_{-i}}$	Force vector, repulsion of π_i agent away from other $\pi_{k \neq i}$ agent at time t
$F^t_{\pi_i \Lambda^t_{\pi_i}}$	Force vector, attraction to local centre of mass for the neighbours of a π_i agent at time t
$F^t_{\pi_i \epsilon}$	Force vector, jittering movements by the π_i agent at time t
$F^t_{\pi_i}$	Force vector, movement vector of the π_i agent at time t
$F^t_{\beta_j cd}$	Force vector, driving and collection vector of the β_j agent at time t
$F^t_{\beta_j \epsilon}$	Force vector, jittering movements by the β_j agent at time t
$F^t_{\beta_j}$	Force vector, movement vector of the β_j agent at time t
$\Omega_{\pi_i \pi}$	Set of π agents (neighbourhood) a π_i agent operates on
$\Omega_{\beta_j \pi}$	Set of π agents a β_j agent operates on
Ω^t	Set of π agents at time t that have not reached the goal; that is, the remaining agents to complete the mission
$\Omega^t_{\beta_j \pi}$	Set of π agents a β_j agent operates on at time t
$\|\Omega_{\pi_i \pi}\|$ (1–200)	Number of π agents (neighbourhood) a π_i agent operates on
$\|\Omega_{\beta_j \pi}\|$ (20)	Number of π agents a β_j agent operates on
$\|\Omega^t\|$	Number of π agents at time t that have not reached the goal; that is, the remaining agents to complete the mission
$\|\Omega^t_{\beta_j \pi}\|$	Number of π agents a β_j agent operates on at time t
S_π ($1\ u/t^2$)	Maximum speed of π per time step
S_β ($1.5\ u/t^2$)	Maximum speed of β per time step
$S^t_{\beta_j}$	Speed of β at time t
$\mathbb{D}$	The minimum distance required between any sheep and the goal to announce successful shepherding and task completion
η (0.05)	Probability of moving per time step while grazing
θ_β ($\pi/2$)	Blind angle for β

Chapter 1
Smart Shepherding: Towards Transparent Artificial Intelligence Enabled Human-Swarm Teams

Hussein A. Abbass and Robert A. Hunjet

Abstract The aim of this chapter is to uncover the beauty and complexity in the world of shepherding as we view it through the lens of Artificial Intelligence (AI) and Autonomous Systems (AS). In the pursuit of imitating human intelligence, AI researchers have made significant and vast contributions over decades. Yet even with such interest and activity from within industry and the academic community, general AI remains out of our reach. By comparison, this book aims for a less ambitious goal in trying to recreate the intelligence of a sheepdog. As our efforts display, even with this seemingly modest goal, there is a plethora of research opportunities where AI and AS still have a long way to go. Let us start this journey by asking the basic questions: what is shepherding and what makes shepherding an interesting problem? How does one design a smart shepherd for swarm guidance? What AI algorithms are required and how are they organised in a cognitive architecture to enable a smart shepherd? How does one design transparent AI for smart shepherding?

Keywords Explainable artificial intelligence · Interpretable artificial intelligence · Transparent artificial intelligence · Shepherding · Swarm control · Swarm guidance · Swarm ontology · Swarm tactics

The aim of this chapter is to uncover the beauty and complexity in the world of shepherding as we view it through the lens of Artificial Intelligence (AI) and Autonomous Systems (AS). In the pursuit of imitating human intelligence, AI researchers have made significant and vast contributions over decades. Yet even with such interest and activity from within industry and the academic community, general AI remains out of our reach. By comparison, this book aims for a less ambitious goal in trying to recreate the intelligence of a sheepdog. As our efforts display, even

---

H. A. Abbass (✉) · R. A. Hunjet
School of Engineering and Information Technology, University of New South Wales, Canberra, ACT, Australia
e-mail: h.abbass@unsw.edu.au; r.hunjet@adfa.edu.au

H. A. Abbass, R. A. Hunjet (eds.), *Shepherding UxVs for Human-Swarm Teaming*, Unmanned System Technologies,
https://doi.org/10.1007/978-3-030-60898-9_1

with this seemingly modest goal, there is a plethora of research opportunities where AI and AS still have a long way to go. Let us start this journey by asking the basic questions: what is shepherding and what makes shepherding an interesting problem? How does one design a smart shepherd for swarm guidance? What AI algorithms are required and how are they organised in a cognitive architecture to enable a smart shepherd? How does one design transparent AI for smart shepherding?

1.1 From Swarm Intelligence to Shepherding

The concept of swarm intelligence has been viewed from multiple perspectives in the literature. A biologically restricted definition [2] limits swarming to groups of social insects such as ants, termites, and many types of bees and wasps. The majority of the current literature sees swarm intelligence through a complex adaptive systems lens, and defines it as the interaction of simple agents that results in emergent self-organised behaviour, especially flocking. This latter definition usually relies on examples that include schools of fish and flocks of birds. We will assume this latter definition as a starting point in this chapter, then present a precise definition of swarming.

Consider a school of fish swimming in unison. Individual fish only have limited knowledge of their counterparts, i.e., they are unable to see the entire school, yet by adjusting their individual speed, alignment, and spacing from their nearby peers, they are able to swim cohesively.

Craig Reynolds [15] showed that a similar concept applies to the way birds flock in his seminal work on bird-like artificial objects that he named BOIDS (bird-oid object). His approach led to simulations of multi-agent systems which displayed flock-like behaviour only through local interaction with peer agents.

In classic robotics, every behaviour is fully engineered, resulting in more and more complex pieces of software for a single robot to display a reasonably complex behaviour. Swarm intelligence diverges from this concept on many fronts. First, the displayed complex behaviour on a group level can be generated from the interaction of simpler individuals. The complexity of behaviour is no longer encoded within each agent, but in the interaction space. Second, the simplicity of the rules allows an agent to respond and act fast, as opposed to executing complex software programs. Third, the computational burden of complex software programs requiring CPU and memory are eliminated, allowing agents with minimal processing power to display complex behaviours as a result of their interaction. Consequently, the energy and battery requirements are lower for these robots, allowing for extended operating time. Fourth, agents in the swarm normally rely on local information; thus, they have lesser demands on communication and complex sensors. This characteristic of the system has the ability to ameliorate congestion on the communication network.

The above discussion on swarm robotics, while encouraging, is missing two crucial requirements that are fundamental in technological solutions for most practical uses. One is related to the absence of a mission objective (how does the swarm know what it is meant to do?) and the second is related to the absence of a mechanism to instruct or command the swarm (how does one interact with the swarm?). Researchers in swarm intelligence have been innovative in addressing these gaps with approaches including external manipulation of a member within the swarm, which engenders a global effect due to the dynamic coupling amongst swarm members. Studying the impact of this form of benign or malignant control reveals that the swarm indeed could be influenced in that way [22].

Shepherding, we contend, provides a more disciplined approach to addressing both requirements. There are three basic agent types in shepherding: the human farmer, the sheepdog and the sheep. We assume a closed-world assumption, where these are the only three agent types. These agents are presented in descending order of cognitive complexity. The human farmer normally plays the role of the shepherd; the agent with the social responsibility and accountability towards the overall mission. The farmer possesses the cognitive capabilities and seeks to achieve the mission's objectives, which could range from herding the flock of sheep to patrolling them in an area away from danger.

However, the farmer does not have the physical capacity to run after the sheep, which can run as fast as 40 km/h. The fastest human speed recorded in history is close to 45 km/h[1]with the average speed in the Guinness World Record being 37.57 km/h. Clearly farmers are not expected to be top athletes. Even if they were, they do not have the endurance to sustain this speed for a long time. A sheepdog, however, can and is capable of running at speeds of up to 50 km/h.[2]The speed differential between the sheepdog, and the sheep is sufficient to offer an asymmetric physical advantage, in addition to cognitive advantage sheepdogs have over the sheep.

The above discussion creates an interesting partial order on these three agents. From cognitive ability perspective, the order is: Human > Sheepdog > Sheep. From a physical ability perspective, the order is: Sheepdog > Sheep > Human. These partial orders define the asymmetric relationship in the problem and explain the importance of a sheepdog to the human. One aspect that compounds this order is the environment, where the sheep (their location and mobility) combined with the environment (which may contain obstructions) can generate a level of complexity exceeding that, which can be addressed by the sheepdog's cognitive capabilities.

Within the shepherding task, the sheepdog acts as an actuator that augments the human physically, while the human could be perceived to augment the sheepdog cognitively. Together, this partnership is sufficient in providing cognitive and physical superiority over the sheep to complete the task successfully.

[1]https://www.guinnessworldrecords.com/products/books/superlatives/fastest.

[2]https://a-z-animals.com/.

1.2 Shepherding

The Cambridge Dictionary defines the verb to "shepherd" as "to make a group of people move to where you want them to go, especially in a kind, helpful, and careful way" [14] or "to move sheep from one place to another". [14]. We distil three characteristics of shepherding from these definitions:

1. Shepherding is concerned with the guidance of a group in some space. The concept of 'moving from one place to another' should not be limited to physical space; the mobility of an agent could occur in any space, such as land (see Chap. 10), air (see Chap. 11), sea (see Chap. 2), and information/cyber space.
2. Classical shepherding guides cognitive agents such as humans and animals, but the approach could also apply to artificial cognitive software or robotic agents. Guidance assumes that the agent to be guided is at least receptive to the guiding agent.
3. Guidance in shepherding is achieved through influencing an agent's position in a "responsible" way (see Chap. 9); that is, an agent's well-being is taken into account during the process, with guidance being "kind" and, we add, "ethical".

These three characteristics distinguish the research directions of shepherding from classic group guidance and formation control; they insinuate a level of sophistication and cognition in shepherding, requiring that actions are taken in-light of calculated risk. Moreover, it is important to have a model of the agents to be herded to modulate the influence in such a way as to be "kind" to them. In classic group guidance, the primary objective is simply to steer a group towards a stated target area. In shepherding, the welfare of the group (see Chap. 9) is explicitly stated as an objective.

As mentioned, in order to ensure influence, and therefore guidance, are carried out in a kind manner, shepherding requires the shepherd to have a cognitive model of the agents it intends to guide. This allows the shepherd to predict how the agents will respond to the action that it takes. Through feedback, it is able to adjust its action to the response of the agents it is guiding. The shepherd needs to have the cognitive capacity to infer from the state of the flock the activities and emotional (such as fear) states of the flock (see Chap. 7).

When implementing shepherding in a multi-agent framework (e.g., a robotic swarm) a level of anthropomorphism is required. Objectives mirroring those of caring and ethics must be displayed and are analogous to taking into account the dynamics of the robotic system to be herded and the associated internal states of the agents (for example, the chance of inducing collisions and the remaining energy/battery, memory and cognitive processing (CPU) levels). These objectives necessitate the design of a cognitive agent; i.e., one with a cognitive architecture which can model other agents, the environment, and itself.

In reactive agents (see Chaps. 2 and 3), the cognition of the agent is usually replaced with transfer functions that directly map information from sensors into actions for the agent to actuate on the environment, others, and self. To design a reactive sheepdog that is responsible for the welfare of the animals it shepherds, the designer needs to encode that welfare into the transfer function. For example, to avoid scaring a sheep, the sheepdog may need to stop before it violates the minimum separation it must remain from a sheep; a distance known as the stalling distance (see Chap. 4). Thus, the welfare objectives are encapsulated by reactive equations that can sense indicators about these objectives and adjust an agent's behaviour accordingly.

As discussed, in the human–animal world, farmers play the role of the shepherd. It is they who have the intent to complete the mission (shepherding tasks such as herding or patrolling) with all of its complexities and conflicting objectives (for example, move the flock as fast as possible, but minimise stress induced on the animals). In order to achieve their mission, they utilise their sheepdogs, but due to the cognitive disparity between these two agents, the shepherd transforms its intent into very precise and concrete tasks and issues commands that the sheepdog is trained to perform. If the mission is to herd the flock, shepherding offers an excellent practical application where a task is clearly decomposed into sub-tasks (e.g., collect and drive) that are used to teach the sheepdog how to shepherd.

Although they do not necessarily know of the overall mission objective, the sheepdogs have the intelligence to understand and action the commands issued by the shepherd. As cognitive agents, they can sense the sheep and determine an appropriate course of action that does not over stimulate the sheep. These task decomposition and teaching concepts offer a great deal of opportunities for robotic systems (see Chap. 10).

Despite that the concept of "shepherding" is well known and applies to human–human and human–animal interaction, it has not received significant attention from the modelling, simulation, and artificial intelligence communities proportional to its importance or potential practical impact. A recent review on the topic [11] reveals the sparse literature on shepherding, and identifies a few research directions on the topic.

Instead of repeating the review in [11], we aim to complement it by unfolding the complexity of shepherding and the opportunities such complexity offers to researchers in AI, autonomous systems, and swarm robotics. These opportunities will form a roadmap for researchers in these fields to design and develop smart artificial shepherds and sheepdogs, capable of efficiently performing cooperative shepherding tasks on a level of scale infeasible for their biological counterparts.

1.3 The Practical Significance of Shepherding

As fascinating as it is to see how nature has evolved the above dynamics complemented by intriguing mathematical and game theoretic characteristics, a discussion on the practical significance of shepherding requires us to delve deeper into the problem. The sheepdog–sheep relationship falls under the wider research area of predator–prey dynamics. Interestingly, a sheepdog is trained to look after the sheep, but the sheep react to it as a threat. What matters most to us from robotic and AI perspectives is the mathematical representation of this relationship.

A sheep is influenced by sheepdogs and other sheep. We will assume that the farmer does not influence sheep directly but uses the sheepdog for this purpose. Mathematically in this setup, a sheepdog can be abstracted as a state vector, where each element of the vector represents a parameter or attribute of the sheepdog. In its simplest form, the state vector is a position vector. Other state variables that could be added include, for example, the heading of the sheepdog relative to the sheep, body posture information, and eye contact. The state vector may include behavioural attributes such as how aggressive the sheepdog is and the sheepdog's energy level. Each sheep has a similar state vector.

This formulation allows us to model an influence on an agent as simply a change in that agent's state vector. The cause of this change in magnitude and direction of a state vector can be represented as a force vector. That is, the application of this force vector to an agent's state vector elicits a change in the agent's state. Given that multiple agents may simultaneously affect another agent, we will refer to an influence vector, which is formed by combining a number of force vectors. For example, the position information of one sheep may generate a repulsive force to avoid collision with another but also an attraction force to socialise, while simultaneously there is repulsion from the sheepdog also impacting the sheep's position.

All force vectors impacting a particular agent (a sheep or sheepdog) are summed to form the overall influence vector impacting that agent at each time step. Note that this force vector formulation allows the modelling of influence on any parameter within the state vectors of an agent, i.e., it is not limited to just position and speed, but can be used to model impact on emotional state such as fear level and fatigue. This vector representation offers significant advantages to robotic implementations in today's computing environment due to advances in Graphical Processing Units (GPUs), where orders of magnitude improvement in processing time could be achieved by cleverly structuring the problem space using tensor algebra.

The above English-based description will be transformed into mathematical equations in various chapters in this book. However, for the time being, the English description allows us to discuss the practical significance of shepherding. As a biologically inspired problem, shepherding may act as a source of inspiration to solving real-world problems; some of which are listed below:

- **Predator–Prey Dynamics:** Both real-world and simulation data could be used to understand the predator–prey dynamics between the sheepdog and sheep. Understanding these dynamics is important on both a fundamental science level and from an ecological perspective.
- **Ethics:** The original shepherding problem as defined above has two types of relationships: human–animal (shepherd–sheepdog) and animal–animal (sheepdog–sheep). When the biological sheepdog is replaced with a UxV such as a UAV (see Chaps. 9, 10, and 11), two additional relationships emerge: human–autonomy (shepherd–UxV) and autonomy–animal (UxV–sheep). Ethical

considerations are clearly required in the human–animal, human–autonomy, and autonomy–animal relationships. Shepherding therefore provides an application domain for researchers to ask and investigate ethical questions. It affords researchers the ability to run simulations and ethically approved, low-risk experiments in both synthetic and real environments [25].

- **Swarm Robotic Guidance and Control:** This is the most common use of shepherding in this book, where the sheepdog–sheep relationship offers a scalable model for swarm guidance. The sheep have been classically modelled as BOIDS, with one extra force representing a repulsive force from the sheepdog. This force vector could be modulated based on the sheepdog's proximity to the sheep [18]. The mere fact that the sheep respond to a sheepdog establishes dynamic coupling between the two agents, where one agent (the sheep) responds to the actions of the other agent (the sheepdog), putting the latter in a position to control the former. Moreover, shepherding offers an approach where a single sheepdog can manage a large number of sheep. The more disciplined the sheep are, the more of them a single sheepdog can manage. This makes shepherding a perfect model of inspiration to scale up swarm control, allowing few agents to control a larger number of agents.
- **Cyber Security:** The previous point made the implicit assumption that robots are physical entities such as unmanned aerial, ground, surface, underwater, or space vehicles. However, the mathematical models for shepherding are application agnostic and can be applied in abstract spaces modelling the information and cyber domains. The concept of mobility is conceptual and it can represent mobility of pieces of information in an abstract space, ideas in a community, people in a social network, or pieces of software in a computer network. Here the concept of shepherding becomes how one uses a few number of decision points to herd information through a network, ideas through a social network, or mobile intrusion detection systems (such as immune inspired intrusion detection systems) through a computer network.
- **Mission Cryptology:** In shepherding, the sheep do not know the intent of the sheepdog, neither do they know the intent of the farmer. As such, the goal is replaced with a timeseries of influence vectors that in their totality achieve the overall mission intent. Any of these influence vectors in isolation is insufficient to decode the intent of the mission. This form of encoding could be used to secure the plan and objectives of the mission, as agents do not need to explicitly communicate anything but a series of mathematical vectors. Moreover, the vector representation enables a further level of security via encryption, although in highly dynamic environments, this should be balanced against the associated increased processing time.
- **Human-Autonomy Teaming:** The human shepherd interacts with a manageable number of entities, which in turn exercise control over a large flock. The cognitive load on a human operator is clearly dramatically reduced from that required when explicitly controlling the individual members of the flock. The social interaction, cognitive load, trust, and the wider human factors and cognitive engineering considerations are worthy of further inquiry. The shepherding problem offers a

great deal of opportunities to explore some of the very fundamental problems in human–autonomy teaming, human–swarm interaction, and human–robot interaction (see Chaps. 12 and 13).

Each of the potential practical uses of shepherding discussed above raises the question of how to implement appropriate cognition within an artificial agent, such as a robot, to achieve the behaviours required in each problem. In the next sections, we will discuss reactive and cognitive designs of a shepherding agent.

1.4 Reactive vs Cognitive Shepherds and Sheepdogs

Different schools of thought exist on how to model an agent (see [24] for a detailed introduction on agents). Those prescribed to the reactive school transform humans' understanding of how the system should work into equations, rules, or models that directly associate a group of responses to a group of stimuli. One particular branch of research in reactive agents is Physicomimetics [20], whereby the world, be it social, biological, or cognitive, is modelled using physical principles and equations.

A model within a reactive agent could be seen as the set of shortcuts, sometimes represented as a set of event-condition-action statements, used to approximate stimuli-response or cause-effect relationships. The effort sits on the shoulders of the human designer to create the model and/or parameterise it. The human system designer is required to have sufficient knowledge of all conditions and inputs that could affect the agents' states and assign appropriate actions to these Event/Condition/Actions (ECA) tuples. These ECA tuples must appropriately capture all possible behaviours required by the agents. As tasks become more complicated, such approaches clearly do not scale well; further, interactions between the agents may lead to the emergence of unintended and detrimental system-level behaviour. Recent research in reactive behaviour uses AI to estimate the structure and/or parameters of the reactive model [17]. Such an approach might be considered similar to behavioural psychology with external observed stimuli and agent responses giving insight to agent behaviour.

Conversely there also exists the cognitive school of agent modelling, similar to cognitive psychology, whereby the agent is understood in terms of its cognitive components such as executive control functions and long term memory. The cognitive school attempts to equip agents with the processes to acquire knowledge, learn, plan, and adapt; as such cognitive approaches seek to instil a greater level of autonomy in the agents.

Cognitive agents tend to rely on more complex models, and therefore more computational resources, than their reactive counterparts. They are theoretically more robust to change in the environment and/or mission than reactive agents. Reactive agents are usually much faster and are more suitable for platforms with low computational resources. The implementation of a robust agent capable of operating in complex real-world settings would generally be a hybrid, with an appropriately chosen point of balance sitting between the two extremes of totally reactive and completely cognitive. A common design approach for such a control system utilises the required time-scales for responses (i.e., the decisions to be made) to determine the switching point between cognitive and reactive control. For example, in controlling the mobility of a robotic system, actions such as immediate collision avoidance could best be achieved using an ECA reactive approach to allow the agent to respond in a timely manner to an immediate collision. In contrast, determining an appropriate path to traverse to avoid known obstacles within an area would generally use a cognitive planning approach. Additional parameters, such as computational resources available to the agent, and complexity of the missions the agent is designed for, should also be considered when determining where this balance point between cognitive and reactive implementation lies.

In terms of academic critique within the literature, the reactive school is normally criticised due to the simplicity of its models, with most models sitting at a very high level of abstraction compared to classic control theoretical or cognitive AI models. This criticism is sometimes due to a fundamental misunderstanding driven by looking at reactive models as the complete solution to a problem. This criticism was discussed by Reynolds [16], the author of the infamous, reactive *Boids* concept, who proposed three modelling levels. The highest level is concerned with action selection, requiring strategies, goal setting, and mission-level planning. This highest level sends goals to the second level, which is concerned with path selection and planning. This second level sends the calculated path to the lowest level for locomotion to actuate. Here, we see that an early pioneer of reactive rules for swarm control understood where such rules should be utilised within a hybrid system. Such approaches essentially seek to ensure that the right reactive behaviour is selected at the appropriate time, a concept adhered to in many works [7, 19].

1.5 Swarm Ontology for Transparent Artificial Shepherding

While emergent behaviour of a large number of biological entities may be modelled and achieved through simple reactive local rules, these equations are clearly insufficient to capture the cognitive capabilities of biological agents. Consider, for example, a bird; although flocking can be represented and modelled by the equations which govern BOIDS, these equations are unable to sense and calculate the parameters required by the equations or map their outputs to actions through the bird's body.

The bird uses part of its brain to sense and perceive the environment, comprehend and project each instance it is situated in, decide what to do, and then execute this decision by actuating (flying). The brains of biological agents have evolved to cluster mental processes close to each other to reduce the energy required to perform these processes, forming a cognitive architecture of sorts. We contend that to appropriately embody swarming into agents, be they robotic or software, such a construct is required. This cognitive architecture organises the mental processing within artificial agents. In order for the swarm to be able to team with a human operator, this cognitive architecture, whether it be simple or complex, should be transparent, enabling the human to learn from it, use it, and to assure its trustworthiness.

While a discussion on transparency goes beyond the scope of this chapter, designing a swarm ontology for shepherding is an important enabler for transparency. An ontology has a few key advantages in this context:

- It can provide explanation on how an agent makes decisions using concepts that a human is familiar with, especially in a particular operational context.
- It allows mapping of an explanation from one form to another; thus, the same explanation could be interpreted differently based on the user's area of expertise. The ontology connects concepts between users; thus enabling interpretable decision making, whereby the decision is transformed from the representational language of one agent to a representation that can be understood by another agent.
- It can guide a machine learner (see Chap. 6) to constrain its search space, thereby gaining efficiency in learning time, and/or ensure that what an agent learns is interpretable by another agent.
- It can be used by a human to diagnose undesirable behaviours that an artificial agent may have expressed, and/or assure its performance and suitability in an operating environment.

We present our proposed shepherding ontology in Fig. 1.1. Such an ontology needs to capture a number of elements across the different layers the system is operating on.

First, it needs to capture the behavioural set of a single sheepdog and the skills that need to be acquired by a sheepdog to perform certain actions (see Table 1.1). For example, to drive a cluster of sheep to a home point, some of the basic skills required by a sheepdog include the ability to:

- locate itself relative to the goal
- approach the sheep
- collect indicators on its performance to identify appropriate corrective actions
- . . . and so forth

This behavioural set indicates to the shepherd the capabilities of an individual sheepdog. In the absence of knowing these concepts and their associated effects, a human is unable to know what to expect from the sheepdog when it responds to a command from the shepherd. Defining the capabilities of an agent in such a way also allows an observer to label and categorise the actions that the agent performs, and, as such, enables activity recognition (see Chap. 7).

Second, the ontology needs to capture the tactics each individual is capable of (see Table 1.2). A tactic is a series of organised behaviours displayed by an individual to achieve an intent. For example, singling out a specific sheep (removing a particular sheep from the flock) may require the sheepdog to first seek that sheep as it is standing, pursue it as it starts to run, then offset pursue it as it approaches the target area so that the sheepdog does not enter that area.

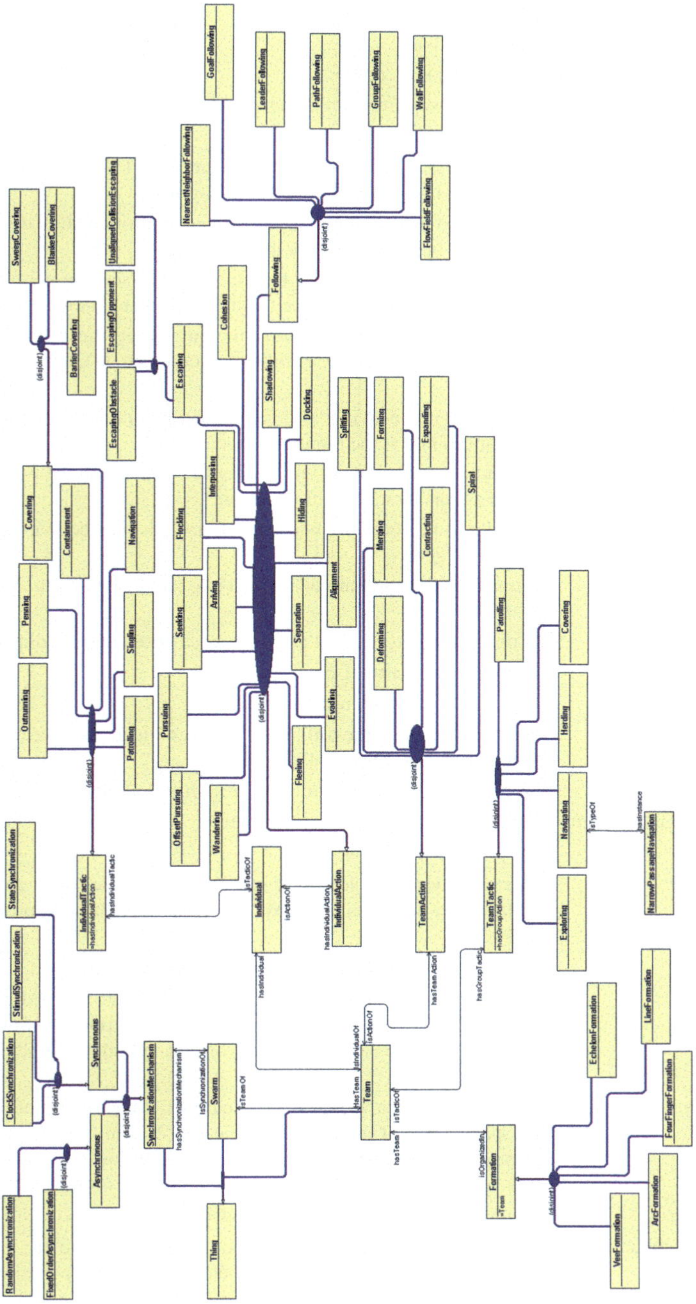

Fig. 1.1 A swarm ontology for shepherding

Table 1.1 Individual actions

Action	Definition
Wandering	A random walk.
Docking	Constrained orientation.
Evading	Steering away from a moving target.
Fleeing	Steering away from a static target.
Hiding	Set the target location to be behind an obstacle on the opposite side of an intruder and seek this location.
Seeking	Steering towards a static target, producing motion similar to that of a moth buzzing around a light bulb.
Pursuing	Steering towards a moving target. Similar to seeking with the static target position replaced with the predicted target position.
Offset pursuing	Steering near but directly into a moving target. Basically, the predator pursues the prey while maintaining a distance from the prey.
Arriving	Seeking with decaying speed as the agent approaches the target. The speed is 0 at the target location.
Interposing	Predict the centre of gravity among future positions of two or more agents then seek this position.
Escaping obstacle	Similar to flee but only when the obstacle is on a collision path with the agent.
Escaping opponent	Similar to flee but only when the opponent is on a collision path with the agent.
Shadowing	Approach the agent then align to match speed and heading.
Cohesion	Seeking the centre of gravity.
Separation	Steering away from nearby agents.
Alignment	Steering vector is the difference between the average velocity of neighbours and the agent's velocity.
Flocking	A combination of separation, cohesion and alignment.
Nearest neighbour following	Approaching the nearest neighbour
Goal following	Approaching the goal
Leader–follower	Arrival towards the leader. If in front of the leader, steer away sufficiently then resume arrival. Maintain separation to prevent crowding.
Wall following	Approach a wall then maintain certain distance.
Path following	Once a path is defined, margin is added to create a corridor. Path following is flexible alignment to the centreline of this corridor, which is a containment within a cylinder around the path's spine.
Flow field following	Alignment with a vector field. The flow field is a cloud of points in space. Each of these points is the tail of a unit vector, representing the direction an agent should follow when it reaches that point.
Group following	Approaching the majority of the group.
Unaligned collision escaping	Predict future locations of neighbours and accelerate or decelerate to get to the expected collision site before or after the intruder (thereby avoiding the collision).

Table 1.2 Individual tactics

Tactic	Definition
Outrunning	Moving in a pear-shaped trajectory with a wider arc as the sheepdog approaches the sheep until it reaches the point of balance; the latter is the point where the stressors on the sheep due to the sheepdog position places the sheep in a state of alert, while being at the edge of moving.
Penning	The flock is driven to a small enclosure. The door is closed when all sheep get inside the enclosure area.
Singling	A specific sheep is separated from the flock.
Patrolling	Preventing a group from exiting an area or keeping it at a fixed distance.
Covering	Navigation to multiple goals.
Containment	Motion of the agent is constrained within a region.
Navigation	Localisation in the environment and identification of appropriate routes.

With actions of a single sheepdog defined, let us turn our attention to what they may do as a team.

Definition 1.1 A team is a group of organised individuals joined together to execute team-level tactics and actions.

The definition above lists four concepts related to a team: organisation (such as a formation), team tactics, team actions, and the individuals making up the team.

Definition 1.2 A formation is a spatial organisation of a team of individuals.

Definition 1.3 A team action is a basic building block of what a team can do and is capable of generating an effect/outcome.

Definition 1.4 A team tactic is an organised set of team actions to achieve an intent or a higher-order effect.

Formations come in many forms. In this chapter, we limit these to some basic formations that could apply in land, air, and sea. The ontology in Fig. 1.1 lists five basic formations: vee (individuals are organised similar to the letter "V"), arc, line, echelon, and four-finger formations.

Team actions and tactics require the involvement of at least two individuals. Tables 1.3 and 1.4 define the team actions and tactics used in the ontology, respectively. Many of these definitions were synthesised from a variety of literature. Sometimes the concept is used in a manuscript without formal definition; we address this here by introducing such definitions. Occasionally different manuscripts use the same concept but refer to it by a different name; here we look to standardise the terminology. The team actions mostly derive from [12] and the individual actions mostly from [16]. The tactics are mostly from the rules of shepherding competitions and other sources including [4, 9]. We attempt to standardise the concepts, define them, and disambiguate their use to ensure consistency across the ontology.

Table 1.3 Team actions

Action	Definition
Splitting	The team needs to split into two or more clusters.
Merging	Two or more of the team clusters combine such that the radius of the combined team is less than a threshold.
Forming	The team moves to a specific formation.
Deforming	The team breaks formation into random organisation.
Contracting	The team reduces its radius.
Expanding	The team expands its radius.
Spiral	The team moves in three dimensions around an object as a spiral.

Table 1.4 Team tactics

Tactic	Definition
Exploring	Cover a region.
Foraging	Wander with resource seeking. Classically, the resource of interest in foraging is food.
Herding	Collecting and driving a group of agents from one location to a target area.
Covering	Team navigation to multiple goals.
Blanket covering	Statically arrange the team to maximise the detection rate of the opponents (predators or prey) in the covering area.
Barrier covering	Statically arrange the team into a barrier to minimise the probability of undetected opponent (predator or prey) penetration through the barrier.
Sweep covering	Move the team across an area to maximise the number of detections in the coverage area. For example, the shepherd sends the sheepdogs to the bush in a coordinated fashion to maximise the number of sheep detected in the coverage area.
Patrolling	Team prevention of another team from exiting an area or keeping it at a fixed distance.
Navigating	Team localisation in the environment and identification of appropriate routes.

Note that, the existence of a team does not guarantee coordinated behaviour. For example, in order for birds to flock, a team needs to swarm through appropriate synchronisation to create spatial and temporal alignment of the actions and tactics of the individuals.

Definition 1.5 A swarm is a team with actions of the individuals that are aligned spatially and/or temporally using a synchronisation strategy.

The above definition gives the entry point to defining a swarm with two main concepts: the concept of a team and the concept of synchronisation strategies, where what we refer to here is behavioural synchronisation, or the ability of the team members to produce actions at the same time.

Behavioural synchronisation normally falls into two categories: asynchronous and synchronous. The former requires individuals in the team to perform an action sequentially, or independently of what another agent is doing. Examples of asynchronous strategies include using a random order to perform actions. Another example may be taking input from one agent which triggers an action in another. Consider the case of a group of humans organising themselves in a line. Each individual may rely on the person standing to their right; when this person is stable, the next aligns with that person. This right-to-left ordering is based on the state of the person on the right and independent of any clock. It will trigger an action-chain from right to left. This is an asynchronous behaviour, in that the action of the individuals does not occur at the same time. We delineate here between behavioural and algorithmic synchronicity, where the latter seeks to ensure the order of operations at the algorithmic level; i.e., note that the asynchronous behaviour of humans forming a line is achieved here by using a synchronous algorithm where action in an individual is triggered by a state change of the person on the right.

In the synchronous approach, all team members perform the action simultaneously. This form of synchronisation requires team members to have a common point of reference to know when to perform the action. One popular approach is to synchronise the clock in each agent on a common clock and the actions are performed when the clock reaches a specific pre-agreed time. Another approach is to trigger the actions across multiple entities based on stimuli in the environment such as an event (e.g., farmer's whistling) that triggers all agents to act simultaneously. These are basic approaches to get a team to synchronise actions, but more sophisticated ones could be developed from these basic principles.

We do not seek to design shepherding based swarming systems for the sake of swarming, rather the swarming behaviours should be pertinent to achieving mission success. As such, the members of the swarm need to have mechanisms to represent and execute missions meaningfully with the behaviours elicited from the interaction of the individuals appropriate for the current environmental state. Smart swarming systems need to decide how to modulate their behaviours and control when to exhibit a particular swarming behaviour. To organise these mechanisms to execute a mission meaningfully, we need an architecture for shepherding to capture the different cognitive processing required by a sheepdog to receive and process information, make decisions, and actuate on the environment. We present such an AI architecture in the next section.

1.6 Artificial Intelligence Architecture for Shepherds and Sheepdogs

Cognition may be considered a system-of-systems. An artificial agent needs to be equipped with many skills to perform the simplest functions that a biological system performs. For example, for a sheepdog to detect that there is a sheep away from the flock, the sheepdog needs first to have sensors such as eyes, ears, and nose to see, hear, and smell where a sheep is, respectively. Each of these sensors offers the sheepdog different information, such as relative position of the sheep to the sheepdog, the body posture of the sheep and orientation of head, the state of fear as displayed by changes in body odour, and more. These cues are received by the sheepdog's brain in the form of signals that need to be transformed and encoded into actionable information.

The sheepdog takes the actionable information and matches it to its accumulated experience in its long term memory and natural instinct. The sheepdog then calls upon its *knowledge of self*, comprising its available actions, to select an appropriate course of action which it may actuate on the environment to achieve its intent.

What sounds to be natural instinct seemingly called upon and executed momentarily, when appropriately analysed, requires a series of skills that are either encoded in the sheepdog's innate knowledge due to evolutionary learning and natural selection, or acquired skills from the personal experience of the sheepdog in a particular habitat. Replicating such performance in an artificial agent is a tricky proposition. The designer is required to make a series of AI algorithms available within the artificial agent to enable the acquisition, adaptation, and adoption of the required skills. These could be algorithms to detect, classify, recognise, identify, verify, and track an object; algorithms to perceive, comprehend, and project events and situations; algorithms to localise the agent, map the environment, detect and avoid obstacles, plan the path and trajectory to navigate from one location to another; to control and decision making algorithms to move, direct sensors, communicate, negotiate, influence, shape, and actuate on self, other agents and the environment.

A designer might start with some of the modules, such as perception and decision making, and implement them with simple algorithms, then gradually add more sophisticated approaches to ensure a fully functional AI that is capable to adapt and handle unseen situations. One needs only to examine the growth in the literature of articles pertaining to AI to reach the conclusion that advances and contributions to the field are occurring at an astonishing rate. To appropriately endow the agent with the ability to cope with such complexity, the cognitive architecture of an artificial agent may need to host a variety of AI algorithms for different cognitive functions and different contexts.

We propose an architecture for shepherding and sheepdogs, and present it in two stages. The first stage will focus on the decision making architecture responsible for coming to and enacting a course of action. We then expand the architecture by incorporating perception, situation awareness, and situation assessment.

The architecture is applicable to both shepherds and sheepdogs; although we have stated that the work within this book seeks to create an artificial sheepdog, we present here also the concept of expanding this work to enable the command of the artificial sheepdog by an AI (albeit, scoped down to the purpose of conducting a shepherding mission) with the primary difference between these agent types being individual cognitive capacities, including memory and processing power, and consequently, the level of sophistication of the AI algorithms to be implemented. Although both are cognitive agents, the sheepdog has limited capability to comprehend complex situations when compared to a shepherd. The practical implication of this design choice is that the sheepdog will likely rely more on reactive models to allow it to respond fast, while being light weight. The artificial shepherd needs to be a cognitive agent with the capacity to represent and monitor the evolution of the context, track progress towards achieving the goal, plan ahead, and command the sheepdog to take corrective actions to steer the trajectory of events towards one which the shepherd believes will lead to a successful mission.

We do not underestimate the amount of time it will take to automate the cognitive functions required for real-time autonomous shepherding in all contexts, terrains, and biological agents (sheep, cattle, horses, or even rabbits). However, the cognitive architectures presented in the following subsections will enable the housing of the sophisticated algorithms we will continue to iteratively develop to achieve the technological goal.

1.6.1 Shepherds and Sheepdogs Autonomy Architecture

We expand Reynolds three-level model into a five-level AI architecture suitable for modelling both shepherds and sheepdogs in Fig. 1.2.

The top level of the architecture is concerned with the mission description language. Here, a mission is defined in terms of its objectives, the available resources and any mission-specific constraints such as the required mission completion time. This contextual information is sent to the goal planner to decompose the overall mission objective into sub-goals. The sub-goals are sent to the behaviour selection level to identify and extract from the behaviour database

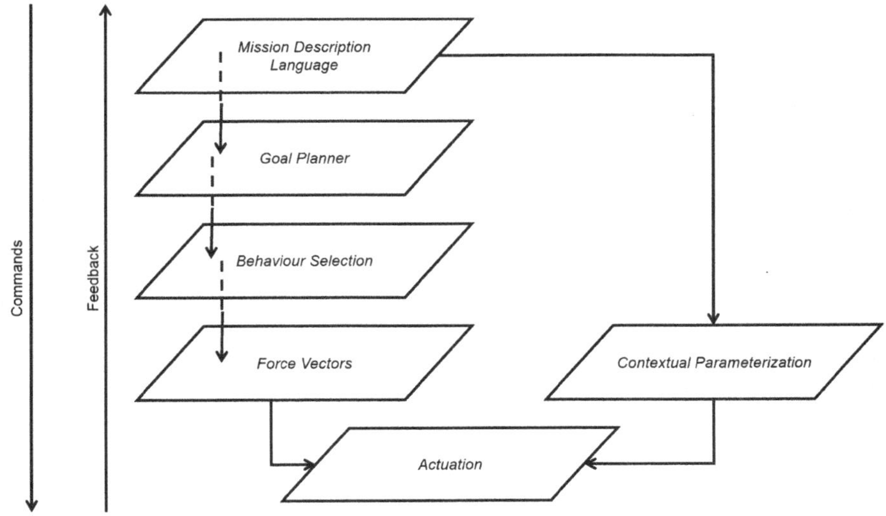

Fig. 1.2 An artificial intelligence architecture for the decision making functions for shepherding

the subset of behaviours required to achieve the set of sub-goals. The behaviours are transformed into force vectors and parameterised through the contextual parameterisation module. In this step, the vectors are combined to produce the aggregate trajectory of behaviours required to achieve the intended sub-goal. The information is sent to the low-level controllers to actuate on the environment.

We will illustrate the architecture using a smart artificial shepherd example. The mission could be stated as "Collect all sheep and bring them to the goal location within 30 minutes from now, while not stressing the sheep, using two sheepdogs of type A and three sheepdogs of type B". In the first part of this mission statement, we have the classic shepherding problem, whereby the objective is to collect and drive the sheep to a goal location. The second part sets a time constraint and an animal welfare objective. The third part defines the available resources for the mission, which in this particular case, we have five sheepdogs of two different types (note this may also be considered as a resource constraint).

The mission statement has all the information needed by the shepherd's goal planner to decompose the mission objective(s) into sub-goals that could feasibly be achieved given the constraints on the mission and available resources. The goal planner may output a sequence of sub-goals such as collect all sheep at location X within 20 min, drive the sheep to the goal within 7 min, then patrol the sheep to ensure they do not leave the goal for 3 min. These last 3 min could be allocated for contingencies if the first two sub-goals are delayed due to the uncertainty in the response of the sheep.

These three sub-goals are sent to the behaviour selection module. To collect the sheep, the module may select four behaviours: sense outlier sheep,sheepdog formation selection, sheepdog positioning behaviours, and specific pursuing behaviours for the sheepdogs to use to collect the sheep.

These behaviours are represented with attraction–repulsion equations that need to be parameterised and fused together to identify the aggregate force vector that a sheepdog needs to follow at each moment of time. To enable this, the individual force vectors associated with the behaviours are passed from the force vector module to the actuation module. Next, the contextual parametrisation module decides on the parameters and fusion weights based on contextual information such as mission resource constraints, the performance envelope of the sheepdogs and sheep in the environment, and the remaining energy for each sheepdog. This parametrisation information is also passed to the actuation module, which is responsible for using this information to appropriately weight and combine the individual force vectors to form the aggregate force vector. At this point the actuation module applies any actuation constraints and drives individual actuators commensurate with this aggregate force vector.

Figure 1.3 maps out the example used throughout this section on key components of the autonomy architecture.

The above example showcases an architecture that is purely focused on the fewest number of components required for a smart shepherd and sheepdog to make decisions. However, agents make decisions based on their perception of the environment. To transform this decision making focused architecture to operate on a

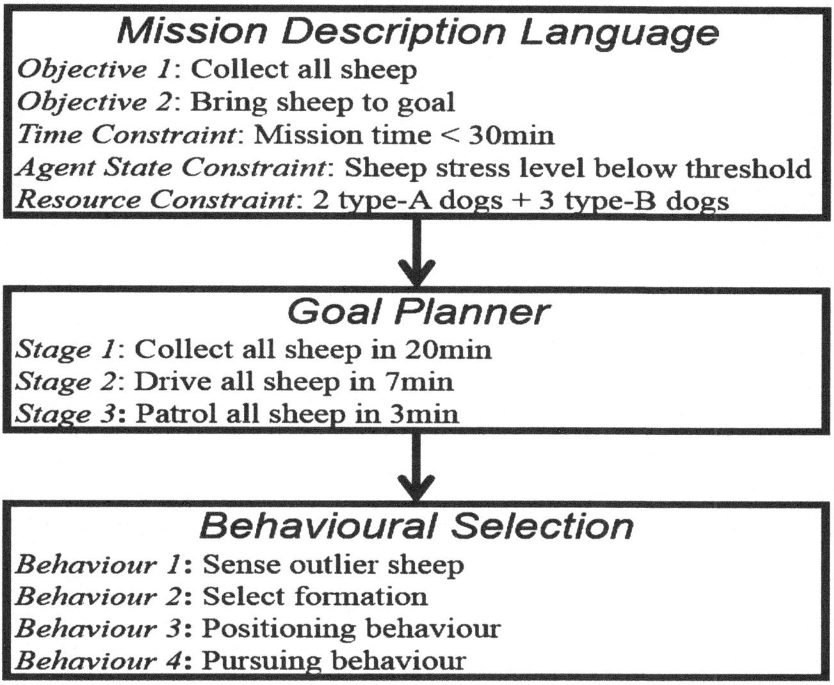

Fig. 1.3 A shepherding example using key components of the autonomy architecture

real robot, the shepherd and sheepdogs need to have the ability to sense and perceive the environment, including sensing one another. In addition, they need to be able to assess the risks associated with different parameterisations of the behaviours generated by the decision making architecture. Such assessment should consider the current situation, the overall mission context, and the sub-goals to be executed. The next sub-section will focus the discussion on the sensing and situation awareness architecture.

1.6.2 Shepherds and Sheepdogs Contextual Awareness Architecture

The contextual awareness architecture is presented in Fig. 1.4. It consists of three primary modules: situation awareness responsible for perception, comprehension, and projection; situation assessment focusing on activity recognition, course of action selection, and impact analysis; and context-driven parameter settings to parameterise the force vectors in the autonomy architecture based on the context, including the current situation.

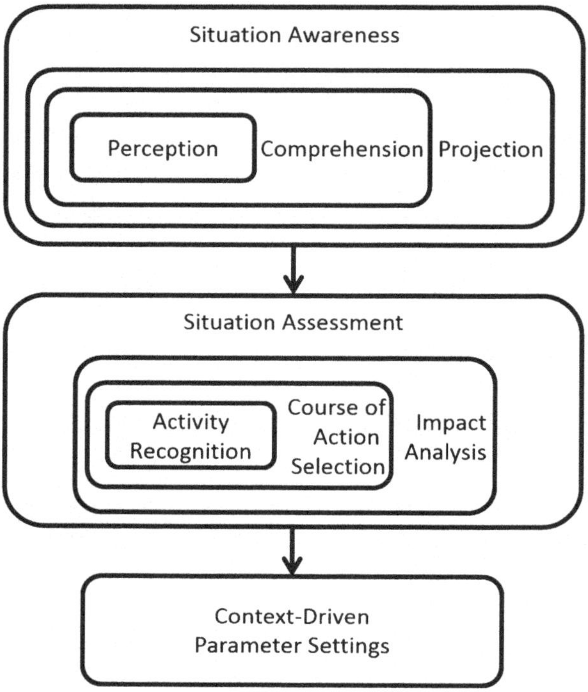

Fig. 1.4 A contextual awareness architecture spanning sensing, perception and context-driven parameter setting

Perception, on a functional level, transforms sensorial data streamed from the agent's sensors to features and indicators that the agent could act on. Comprehension transforms one or more features and indicators into summary statistics and flock- and context-level state information. Projection uses this information to anticipate and predict the evolution of these states into the future.

For example, a shepherd senses the location of sheep. The perception module transforms these raw data into features, such as the global centre of mass of the herd, the local centre of mass of the largest cluster in the herd, the relative direction between each of these two centre of masses and the goal, the location of the furthest sheep from the herd outside the largest cluster, and the relative direction between that sheep and the global centre of mass.

The comprehension module takes these features as inputs to calculate different flock-based state information such as "the herd is clustered" and "there is a stray sheep". The projection component is able to use this state information to estimate, for example, that the stray sheep will reach inaccessible bush area if it continues its current path for three minutes, or that the clustered herd will fragment into a large number of smaller clusters if left as is for ten minutes. Projection works on each state independently. The situation assessment module works on the interaction of these states.

The situation assessment module consists of three components: activity recognition, course of action selection, and impact analysis. Activity recognition is responsible for taking the features and state information from the perception and comprehension sub-modules and processing them to recognise the behaviour and/or functions and tasks an agent is undertaking. In essence, activity recognition transforms data of the individuals, and statistics of the group, to an understanding of the intent of both the individual and the coordinated behaviours of the group, as well as higher-order state information at a flock-level.

An activity recognition system applied to sheep data alone may indicate that the sheep are foraging, eating, or stressed. Other indicators could be produced by the activity recognition system such as confidence levels on each recognised activity or estimates of the parameters associated with an activity. For example, "the sheep are eating with a *high* stress level due to the sheepdogs standing in the vicinity of the sheep"; or "with a *low* stress level due to absence of a visible threat; the sheepdogs are standing far away and outside the field of view of the sheep". These examples highlight the importance of explainability in the activity recognition system, to indicate the reason an activity has been recognised and the confidence level associated with an activity.

The activity recognition system applied to the sheepdogs would produce information on the sheepdog level. For example, it could indicate that "sheepdogs are positioning themselves into an arc formation", or "the sheepdogs are running back to the water due to being thirsty". When the activity recognition system is applied to both the sheep and sheepdogs, the system attempts to recognise the reciprocal dynamics between the two agent types. For example, it could indicate that "the sheepdogs are driving the flock", or "one sheepdog is on its way to collect a stray sheep".

The information from the activity recognition system is key to inform the *course of action selection* module. A course of action in this module primarily represents the selection of strategies required to achieve an agent's intent. It is not an operational or tactical action. Note that, these courses of action are decided within the autonomy architecture. The contextual awareness architecture needs to operate ("think") on a higher level of abstraction sufficient to represent the strategic thinking of the agent, supporting it to evaluate the strategies required to achieve the mission objectives and goals. Increasing fidelity in the contextual awareness module should be done with utmost care due to the associated increase in computational resources and resulting reduction in response time.

The course of action selection may include course of action generation abilities. However, for simplicity, we will assume here that all courses of action are pre-designed and available for the agent to select from.

Continuing with the examples of the outputs of the activity recognition system discussed above, the course of action selection module may have a portfolio of high level strategies to select from such as: "fetching" or "singling".

The impact analysis module uses the current situation awareness picture formed by the agent and the activities that the agent has recognised the other agents are undertaking to assess the impact of the selected course of action on the mission objectives and goals. The impact analysis module may be tightly coupled with the course of action selection module and be part of the rationale for selecting a particular course of action. A computationally less intensive implementation would keep the two modules decoupled; in this simplified form, course of action selection may be driven by a simple decision tree, with the impact analysis module evaluating the resultant course of action without necessarily changing it, as required changes could be made at the next iteration through the contextual awareness loop.

In the case that the selected course of action is, for example, "singling", examples of the outputs from the impact assessment module could be the singled sheep is a leader amongst the flock and removing it will increase the complexity of the driving task, or that the singled sheep is an unsettled leader that will decrease the complexity of the driving task. In essence, the impact assessment module offers the agent with the ability to assess consequences and risks.

Figure 1.5 maps out the example used throughout this section on key components of the contextual awareness architecture.

The two architectures discussed so far do not touch on the communication and commands exchanged between the shepherd and sheepdog. In essence, a command that is issued from the shepherd to a sheepdog is an order to achieve a sub-goal that the shepherd decided should be executed by the sheepdog. The shepherd and the sheepdog, due to the difference in their cognitive capacity, operate at differing levels of mission complexity and over different time-scales. The command issued by the shepherd to the sheepdog (which is a sub-goal of the shepherd's mission) appears as the sheepdog's mission in its instantiation of the architecture. These commands do not require any changes in the architecture. The issuing of a command is a sensorial input to the receiving agent and action actuated by the commanding agent. In the next section, we bring these two architectures together to form the cognitive architecture for both the shepherd and sheepdogs.

1.6.3 Smart Shepherds and Sheepdogs Overall Architecture

To implement a smart shepherd and/or sheepdog, the designer should take a system-of-systems approach to structure the design of cognition in a nested and modular fashion. The architecture presented in this chapter uses modularisation to ensure a neat, efficient, and easy-to-maintain design. Nevertheless, there will always be bridges between modules in the system to pass essential information. These bridges should be minimised to reduce system complexity, but cannot be eliminated without negative consequences on the performance of the system. They provide the feedback loops which enable modules to make informed decisions.

Figure 1.6 highlights the links between the autonomy and contextual awareness architectures, and introduces the sensors and actuators. Seven links in total have been added in the figure, two of them are links to sensors and actuators. The interdependency between a module in the autonomy architecture and a module in the contextual awareness architecture is necessary for the agent to function as a smart agent, while maintaining a well-designed modular cognitive architecture.

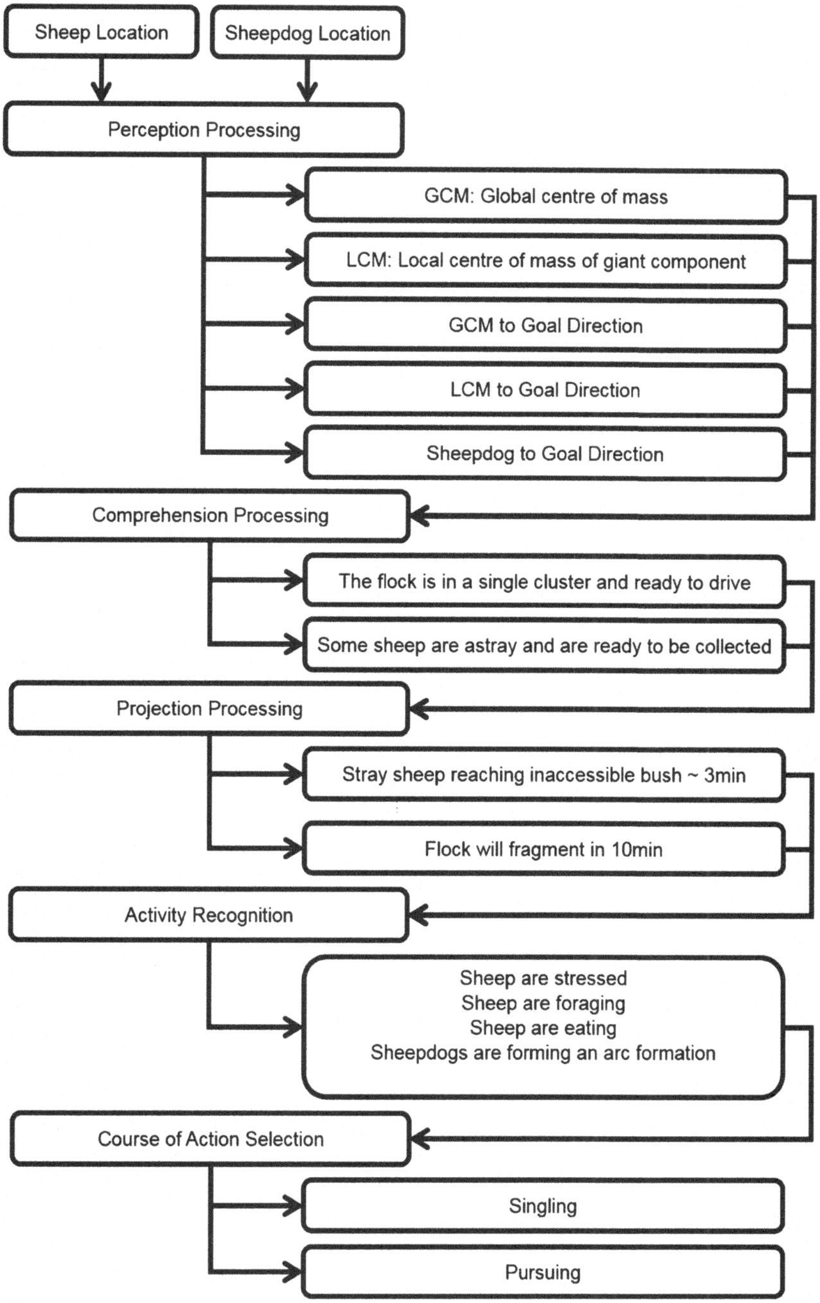

Fig. 1.5 A shepherding example using key components of the contextual awareness architecture

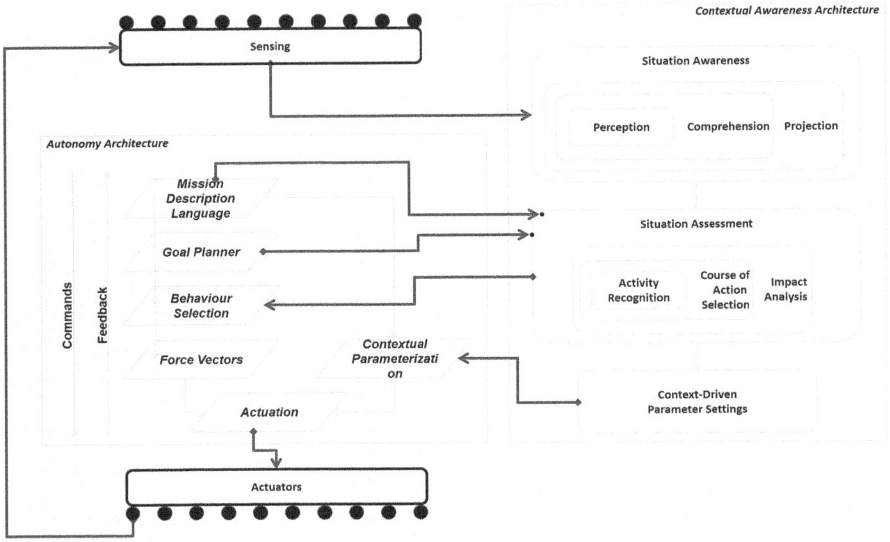

Fig. 1.6 An architecture for smart shepherds and sheepdogs, linking together the previous two sub-architectures. It empowers mission planning and decision making with the ability to perceive the environment and reason about it as well as the agent's actions. The lines for the two sub-architectures responsible for autonomy and contextual awareness are faded out to emphasise the additional links that connect the two sub-architectures

The links between some of the actuators and sensors allow for direct interaction with sensors. For example, the situation assessment may result in course of action to actively direct the sensors towards specific regions in the environment to collect missing information. Active sensing allows the agent to independently control its own orientation in the environment and the relative orientation of its sensors to its body, such as orienting a gimbal mounted camera on a robot sheepdog towards the furthest sheep location without changing the orientation of the robot itself. It also allows for dynamic configuration of its sensors, such as in the case of adjusting camera zoom in and out.

It is assumed that agent-to-agent communication is possible in order to form common contextual awareness across multiple shepherding agents. This is enabled through the architecture in Fig. 1.6 under the assumption that radio receivers may be represented as sensors and radio transmitters may be represented as actuators. The decision to communicate requires actions and actuation. Therefore, communication should not be seen in this architecture as a mere exchange of information. It is an intentional act to either improve another agent's situation awareness or to delegate the execution of a sub-goal to another agent. An agent may communicate to request information to improve its own situation awareness. In this case, communication is still an action of that agent and is adequately captured by this architecture.

In the contextual awareness architecture, the situation assessment module requires information about the mission and the sub-goal(s) the agent intends to execute to inform the three situation assessment sub-modules. The resultant situation assessment informs the module responsible for parameters' settings, which are sent to the contextual parametrisation modules in the autonomy architecture (ultimately enabling appropriately weighted and parameterised force vector aggregation). Actions are generated in the situation assessment module. These actions represent the tactics of an individual. The behavioural selection module selects the individual behaviours to achieve these tactics. The separation between tactics and individual skills in the two architectures is intentional. It allows the agent to change its skill sets available to achieve the same tactics. It also allows the agent to change its tactics by reusing its skills. In essence, tactics exist to achieve the sub-goals selected by the autonomy architecture, while behaviours are the sequence of actions to achieve a tactic.

AI methods and algorithms sit at the core of each module. Even though we propose this architecture to enable cognitive agents as a step towards realising true autonomy, a spiral design approach could see the overall architecture implemented with very simple rules in each module. For example, the rule-based system used by Bayazit et al. [1, 3], or the repulsion–attraction equations approach followed by many researchers such as Lien et al. [8, 10] Miki and Nakamura [13], Lee and Kim [9], Tsunoda et al. [23], Strömbom et al. [21], and Hoshi et al. [5, 6] could all be used to design a functional, artificial, and reactive sheepdog. In this case many of the modules in the architecture could arguably be empty, or trivial, given that these models assume flock state information is readily available and/or that the agent has global information. Such assumptions preclude the need for active sensing; the situation awareness module would be bypassed, as all required information is readily available in the agent. Similarly, the situation assessment module would be empty due to the pure reactive nature of the agent. Our architecture allows for the expansion of such approaches for implementation where overly favourable assumptions surrounding availability of knowledge are relaxed.

1.7 Conclusion

With the concept of shepherding introduced and its applicability to the many real-world problems explained, we have argued the need for cognition in order to enable effective swarm control through the use of artificial shepherds and sheepdogs. In order to improve transparency in decision making of such entities and to ease the learning burden upon AI systems in this pursuit, we have provided an ontology for shepherding. To enable implementation of smart shepherding agents we have introduced a modular architecture combining autonomy in decision making, and contextual awareness applicable to both artificial shepherds and sheepdogs. We contend that the complexity associated with implementing a smart shepherding agent lies in providing the agent with the right information at the right time. An artificial sheepdog does not need to be an exact replica of a biological one, it needs only to achieve the same objective. As such, it may not need to act on the same information sensed and used by a real sheepdog, however, the information it does obtain needs to be acquired and processed in a timely fashion given the dynamic nature of its task.

To emulate the functionality of biological sheepdog, the artificial incarnation requires AI algorithms ranging from those analysing data such as clustering, classification, and point prediction to those making plans and decisions such as Bayesian belief networks and reinforcement learning. For example, classification algorithms might be needed in the activity recognition module, clustering in the comprehension module, point prediction in the projection module, reinforcement learning in the goal planning module, and more. The course of action selection and impact analysis modules primarily rely on recommender systems with appropriate optimisation, simulation, and search algorithms.

While the behaviours of an agent here are expressed in the form of force vectors similar to those presented in Miki and Nakamura [13], the information needed as inputs to these force vectors and the parameterisation of these equations, including the selection of which behaviours to exhibit in which situation, all require AI algorithms.

We conclude that the advent of the artificial sheepdog should further enhance shepherding capability. Distributed AI algorithms in the areas of task allocation, decision making, and planning are possible upon radio equipped artificial shepherding agents and would enable optimised coordination upon the sheepdog pack; a capability alien to their biological counterparts.

The remaining chapters in this book are of various granularity. Some zoom out and discuss the complexity of a topic, such as activity recognition or human-swarm teaming, offering conceptual designs without concrete functional implementation. Others zoom in on a particular algorithm to solve a problem, such as those looking at deep learning to design controllers for shepherding or swarm decision making algorithms to disambiguate a global system-level state from local information collected by many agents. The variety of the contributions in the following chapters showcase the richness of the problem space, which is one of the reasons we are drawn to it. Clearly, implementing a fully autonomous smart sheepdog is a non-trivial system-of-systems problem, with potential application in many domains. It begs for the consolidation of different AI algorithms, and engineering modules; we hope that our architecture can help enable this and invite you on our journey.

Acknowledgement The work in this chapter is funded by the Office of Naval Research Global.

References

1. Bayazit, O.B., Lien, J.M., Amato, N.M.: Better Group Behaviors Using Rule-Based Roadmaps, pp. 95–111. Springer, Berlin (2002). https://doi.org/10.1007/978-3-540-45058-0-7
2. Bonabeau, E., Dorigo, M., Marco, D.d.R.D.F., Theraulaz, G., Théraulaz, G., et al.: Swarm Intelligence: From Natural to Artificial Systems. Oxford University Press, Oxford (1999)

3. Burchan Bayazit, O., Jyh-Ming Lien, Amato, N.M.: Roadmap-based flocking for complex environments. In: 10th Pacific Conference on Computer Graphics and Applications, 2002. Proceedings., pp. 104–113 (2002). https://doi.org/10.1109/PCCGA.2002.1167844
4. Gage, D.W.: Command control for many-robot systems. Technical Report, Naval Command Control and Ocean Surveillance Center RDT And E Div., San Diego, CA (1992)
5. Hoshi, H., Iimura, I., Nakayama, S., Moriyama, Y., Ishibashi, K.: Computer simulation based robustness comparison regarding agents' moving-speeds in two- and three-dimensional herding algorithms. In: Joint 10th International Conference on Soft Computing and Intelligent Systems (SCIS) and 19th International Symposium on Advanced Intelligent Systems (ISIS), pp. 1307–1314 (2018). https://doi.org/10.1109/SCIS-ISIS.2018.00205
6. Hoshi, H., Iimura, I., Nakayama, S., Moriyama, Y., Ishibashi, K.: Robustness of herding algorithm with a single shepherd regarding agents' moving speeds. J. Signal Process. **22**(6), 327–335 (2018). https://doi.org/10.2299/jsp.22.327
7. Hunjet, R., Stevens, T., Elliot, M., Fraser, B., George, P.: Survivable communications and autonomous delivery service a generic swarming framework enabling communications in contested environments. In: MILCOM 2017–2017 IEEE Military Communications Conference (MILCOM), pp. 788–793. IEEE, Piscataway (2017)
8. Lien, J.-M., Rodriguez, S., Malric, J., Amato, N.M.: Shepherding behaviors with multiple shepherds. In: Proceedings of the 2005 IEEE International Conference on Robotics and Automation, pp. 3402–3407 (2005). https://doi.org/10.1109/ROBOT.2005.1570636
9. Lee, W., Kim, D.: Autonomous shepherding behaviors of multiple target steering robots. Sensors **17**(12), 2729 (2017)
10. Lien, J.M., Bayazit, O.B., Sowell, R.T., Rodriguez, S., Amato, N.M.: Shepherding behaviors. In: IEEE International Conference on Robotics and Automation, vol. 4, pp. 4159–4164. Citeseer (2004)
11. Long, N.K., Sammut, K., Sgarioto, D., Garratt, M., Abbass, H.A.: A comprehensive review of shepherding as a bio-inspired swarm-robotics guidance approach. IEEE Trans. Emer. Topics Comput. Intell. **4**, 523–537 (2020)
12. Masehian, E., Royan, M.: Cooperative control of a multi robot flocking system for simultaneous object collection and shepherding. In: Computational Intelligence, pp. 97–114. Springer, Berlin (2015)
13. Miki, T., Nakamura, T.: An effective rule based shepherding algorithm by using reactive forces between individuals. Int. J. InnovativeComput. Inf. Control **3**(4), 813–823 (2007)
14. Org, I.: Cambridge dictionary. https://dictionary.cambridge.org/dictionary/english/
15. Reynolds, C.: Flocks, herds and schools: A distributed behavioral model. In: Proceedings of the 14th Annual Conference on Computer Graphics and Interactive Techniques, SIGGRAPH '87, pp. 25–34. ACM, New York (1987)
16. Reynolds, C.W.: Steering behaviors for autonomous characters. In: Game Developers Conference, vol. 1999, pp. 763–782. Citeseer (1999)
17. Schultz, A., Grefenstette, J.J., Adams, W.: Roboshepherd: learning a complex behavior. Rob. Manuf. Recent Trends Res. Appl. **6**, 763–768 (1996)
18. Singh, H., Campbell, B., Elsayed, S., Perry, A., Hunjet, R., Abbass, H.: Modulation of force vectors for effective shepherding of a swarm: A bi-objective approach. In: 2019 IEEE Congress on Evolutionary Computation (CEC), pp. 2941–2948. IEEE, Piscataway (2019). https://doi.org/10.1109/CEC.2019.8790228
19. Smith, P., Hunjet, R., Khan, A.: Swarm learning in restricted environments: An examination of semi-stochastic action selection. In: 2018 15th International Conference on Control, Automation, Robotics and Vision (ICARCV), pp. 848–855. IEEE, Piscataway (2018)
20. Spears, W.M., Spears, D.F.: Physicomimetics: Physics-Based Swarm Intelligence. Springer Science & Business Media, Berlin (2012)
21. Strömbom, D., Mann, R.P., Wilson, A.M., Hailes, S., Morton, A.J., Sumpter, D.J.T., King, A.J.: Solving the shepherding problem: heuristics for herding autonomous, interacting agents. J. R. Soc. Interf. **11**(100) (2014). https://browzine.com/articles/52614503

22. Tang, J., Leu, G., Abbass, H.A.: Networking the boids is more robust against adversarial learning. IEEE Trans. Netw. Sci. Eng. **5**(2), 141–155 (2017)
23. Tsunoda, Y., Sueoka, Y., Osuka, K.: On statistical analysis for shepherd guidance system. In: 2017 IEEE International Conference on Robotics and Biomimetics (ROBIO), pp. 1246–1251 (2017). https://doi.org/10.1109/ROBIO.2017.8324588
24. Wooldridge, M.: An Introduction to MultiAgent Systems. Wiley, New York (2009). https://books.google.com.au/books?id=X3ZQ7yeDn2IC
25. Yaxley, K.J., Joiner, K.F., Bogais, J., Abbass, H.A.: Life-learning of smart autonomous systems for meaningful human-autonomy teaming. In: Handley, H., Tolk, A. (eds.) A Framework for Human System Engineering Applications and Case Studies.

Part I
Shepherding Simulation

Chapter 2
Shepherding Autonomous Goal-Focused Swarms in Unknown Environments Using Hilbert Space-Filling Paths

Nathan K. Long, Matthew Garratt, Karl Sammut, Daniel Sgarioto, and Hussein A. Abbass

Abstract A novel technique has been developed for autonomous swarm-based unknown environment scouting. A control method known as swarm shepherding was employed, which replicates the behaviour seen when a sheepdog guides a herd of sheep to an objective location. The guidance of the swarm agents was implemented using low computation cost, force-based behaviours. The exploration task was augmented by introducing swarm member role assignments, including a role which imposes a localised covering area for agents which stray too far from the swarm global centre of mass. The agents then proceeded to follow a Hilbert space-filling curve (HSFC) path within their localised region. The simulation results demonstrated that the inclusion of the HSFC paths improved the efficiency of goal-based exploration of the environment, which became more prominent with an increase in the density of the number of goals in the environment.

Electronic Supplementary Material: The online version of this chapter (https://doi.org/10.1007/978-3-030-60898-9_2) contains supplementary material, which is available to authorized users.

N. K. Long (✉)
School of Engineering and Information Technology, University of New South Wales, Canberra, ACT, Australia
e-mail: nathan.long1@student.adfa.edu.au

M. Garratt · H. A. Abbass
School of Engineering and Information Technology, University of New South Wales, Canberra, ACT, Australia
e-mail: m.garratt@adfa.edu.au; h.abbass@adfa.edu.au

K. Sammut
College of Science and Engineering at Flinders University, Bedford Park, SA, Australia
e-mail: karl.sammut@flinders.edu.au

D. Sgarioto
Maritime Division, Defence Science and Technology Group, Port Melbourne, VIC, Australia
e-mail: Daniel.Sgarioto@dst.defence.gov.au

H. A. Abbass, R. A. Hunjet (eds.), *Shepherding UxVs for Human-Swarm Teaming*, Unmanned System Technologies,
https://doi.org/10.1007/978-3-030-60898-9_2

Keywords Shepherding · Swarm robotics · Hilbert space-filling curves · Herding · Robotic exploration

A novel technique has been developed for autonomous swarm-based unknown environment scouting. A control method known as swarm shepherding was employed, which replicates the behaviour seen when a sheepdog guides a herd of sheep to an objective location. The guidance of the swarm agents was implemented using low computation cost, force-based behaviours. The exploration task was augmented by introducing swarm member role assignments, including a role which imposes a localised covering area for agents which stray too far from the swarm global centre of mass. The agents then proceeded to follow a Hilbert space-filling curve (HSFC) path within their localised region. The simulation results demonstrated that the inclusion of the HSFC paths improved the efficiency of goal-based exploration of the environment, which became more prominent with an increase in the density of the number of goals in the environment.

2.1 Introduction

The emergence of intelligent autonomous robots is opening up exciting new opportunities for humanity. Arduous, monotonous, and dangerous tasks, normally carried out by humans, can be delegated to robotic systems, which have the potential to increase efficiency and reduce human casualties when operating in adverse environments.

The exploration of unknown environments, in particular, often presents unforeseen challenges and inherent risks due to the uncertainties involved [18]. Examples of dangerous human exploratory missions to uncharted territories include Antarctic expeditions [13], the navigation of cave systems [14], and the continual exploration of the world's oceans [15].

While single uninhabited systems are capable of completing complex missions, the introduction of multi-robotic teams can permit an increased level of efficiency to a given task, especially when these tasks are tedious or time-consuming [1]. Moreover, the use of swarm robotic systems allows for a higher degree of flexibility and redundancy when analysing unfamiliar surroundings [20].

This study is a step towards solving a problem that has previously been posed by the authors [9], which aimed to distribute a network of uninhabited surface vessels, with each acting as a sea state estimation sensor, in order to determine the wave properties of complex wave systems. Unknown sea states can pose major risks to the maritime industry as wave-induced loads are capable of damaging marine structures and destabilising ocean platforms [6].

Described in this chapter is a novel method for the exploration of unknown environments (with a focus on aquatic environments with wave systems present) which uses a swarm of agents guided by a shepherd to a number of goal locations. The use of shepherding for environment exploration has seldom been explored before (although shepherding for covering was briefly discussed by Lien et al. [8]). The swarm is tasked with finding the goals within an environment, where no information is assumed about the environment prior to the start of mission. Each member in the swarm only has a localised view of the environment, only possessing memory to respond to their assigned role.

The shepherding guidance technique, then, replicates the behaviour observed by sheepdogs when herding sheep to objective destinations [16]. Additionally, certain agents which become separated from the swarm are forced to follow a Hilbert space-filling curve (HSFC) path within localised regions of the mission environment. This allows the swarm agents to explore the environment for goal locations without the aid of the shepherd, where the use of multiple agents in the classic shepherding case could leave the system overly redundant (few agents could be capable of reaching all goals).

The goal-based exploration model allows the swarm of agents to analyse specific locations of potential interest within an unknown environment. Concurrently, it is hypothesised that the secondary objective of the envisioned sea state estimation application, to gather additional information along the path between the goals, could also be achieved. This would lead to an enhanced understanding of the mission environment, allowing the identification of regions that require further examination, without the need for an agent to cover every possible point within the environment.

Practically, the goal locations would represent areas which are predetermined to be of interest, thus requiring an agent to further investigate that region. The agents would, moreover, gather information along the trajectories between the goals to establish if additional areas of interest exist within the environment. It is hypothesised that the development of a modified autonomous shepherding method, using HSFC paths, will improve the speed at which a set of goal locations can be reached within a fixed space. Furthermore, it is hypothesised that the introduction of the HSFC paths will permit greater coverage of the environment within a reduced time period. A comparison with the classic sheepdog shepherding technique is, therefore, conducted as a baseline for the performance of the HSFC-based method.

The main contribution of this work is an enhanced temporal efficiency for environment exploration, which is permitted when the HSFC for local environment covering is introduced to a herding task. However, the chapter also presents a unique force modulation technique to better represent the interactions between sheep and sheepdogs, as well as a circular path planning method which helps to reduce unwanted interactions between the shepherd and swarm agents.

It should be noted that the traditional shepherding task relates more specifically to a searching exercise than an exploration exercise, while the newly created technique for this study extrapolates the original task to one closer to environment exploration.

The technique developed, however, is generic and could be applied to a plethora of exploration applications, including search and rescue operations, mineral prospecting, foraging, and the scientific investigation of hostile environments.

2.2 Background Research

Swarm intelligence originates from the local interactions between a group of relatively simple agents which lead to complex behaviour of the system as a whole. The intelligent behaviours arise without any external control or regulation, defined as being an emergent property of the interactions [20]. Swarm robotic systems introduce numerous advantages, including relative simplicity in construction and programming, high robustness, and ease of scalability.

One of the first swarm control techniques was introduced by Reynolds [11] in 1987. Reynolds' control framework was based on the flocking behaviour of birds, and was originally developed to graphically simulate such behaviour. Three main behaviours were implemented to replicate the flock motion: velocity alignment, flock centring, and collision avoidance. A series of force vectors were used to enforce the three behaviours. The results closely replicated the flocking motion of real birds. Reynolds' major contribution from the research was the concept of a *boid* agent; an object which is able to sense both location and orientation information.

An ocean-based exploration study was conducted by Varughese et al. [17], who created a model which employs swarms of robotic agents to explore an underwater environment. Two different types of robotic swarms were used to conduct the exploration task: one which identifies areas of interest (the aFish), and another which stores information about the environment and identifies new regions to explore (the aMussels). The aFish mimic the scouting behaviour of bees, moving about at random until they find a source of pollen. During their simulations, if an aFish found a source of interest along the sea floor, they would scout for an aMussel, transmit the coordinates of that location to it, then the aMussel would transmit back to the aFish whether it should return to the source for further investigation, or continue scouting in a different region. The results of the authors' simulations showed that their model was effective at exploring regions of up to a hectare in size over a time period of 10–14 h. Varughese et al. noted, however, that there exists little previous research for comparison with their study.

A further bio-inspired control technique based on the herding of sheep by sheepdogs to an objective location, swarm shepherding, has also been applied to swarm agents. Strömbom et al. [16] developed a heuristic for replicating the herding behaviour of both sheep and sheepdogs which implemented the boid model and force vector-induced behaviour of Reynolds [11]. The sheep were controlled using three behaviours: sheep-to-sheep repulsion, sheep-to-sheepdog repulsion, and sheep attraction to their local centres of mass. The sheepdog enforced these behaviours on the sheep by taking one of two actions. If the sheep herd was dense enough, then the sheepdog would move into a driving position behind the herd relative to their objective destination, repelling the herd in that direction. Otherwise, the sheepdog would collect the furthest outlying sheep by moving to a position behind it relative to the herd's global centre of mass (GCM). Strömbom et al.'s simulated results showed sheep and sheepdog movements which were similar to those observed during real herding exercises.

Shepherding has now been generalised to be defined as the external guidance of a swarm, or group of agents, in order to achieve a shepherd's objective [4]. As such, shepherding has been used to control a variety of swarm systems. Of particular interest is a study by Chaimowicz and Kumar [2] who conducted a number of experiments using shepherding to command a swarm of uninhabited ground vehicles (UGVs) using multiple aerial shepherds (uninhabited aerial vehicles). The objective of the authors' research was the exploration of unknown environments, particularly for defence applications. In order to manoeuvre around obstacles and through suburban terrains, the swarm exhibited two behaviours: merging and splitting. The UGVs would split into smaller groups to fit through narrow passages and building cavities, then merge again once their surrounding environment expanded. A near-three-dimensional awareness of the environment was permitted by combining video footage from both the aerial shepherds and UGVs. While Chaimowicz and Kumar's experiments provided support for the use of shepherding as a control mechanism for swarm exploration, their methodologies and discussion lacked the depth required for replication and validation.

Furthermore, Clayton and Abbass [3] argued that all previously developed methodologies used to replicate shepherding behaviours oversimplify the complex relationship between a sheepdog and herd of sheep. As such, no standard shepherding control framework exists, and each application requires a unique shepherding control design. This research attempts to create a generalised technique for shepherd swarm guidance for the purpose of exploration.

2.3 Methodology

This section describes the simulation setup which involves a goal-based exploration task. A shepherd guides a swarm of agents to a series of goal locations within the environment. The task is repeated using a swarm act based on classical shepherding behaviours (as defined by Strömbom et al. [16]), as well as a modified approach (outlined in this section), which incorporates a localised HSFC covering behaviour for varying local environment sizes. A modulated force is introduced to better represent a sheep's response to an approaching sheepdog, as well as a simple path planning method to reduce unwanted interactions between the shepherd and swarm agents.

In each simulation, the objective of the shepherd (β) was to guide the swarm agents (π_i) to a series of goals (G_h) generated with a fixed sequential order at random within an environment. Once one of the agents reached a distance of R_G (or less) from any goal, then the goal vanishes. The shepherd achieved its objective once all goal locations disappeared. The objective of the swarming exercise, however, was only to reach every goal location within the environment.

The swarm agents were depicted using a variation of Reynolds' boids [11] and Strömbom et al.'s particle model [16]. Each swarm member was represented by an object containing an index (i), position coordinates ($P^t_{\pi_i}$), and an agent role as attributes. The shepherd agent was only characterised by its position (P^t_β). A total of N_π agents comprised the swarm, with one shepherd guiding them.

The relative locations and statuses of the swarm agents dictate the behaviour exhibited by the swarm agents themselves, as well as the behaviour of the shepherd, at each time step (t). The shepherd was able to sense the location of all of agents at time t, as well as being able to detect and dictate their roles. The swarm agents were also able to sense the location of other agents and their roles; however, they had no information regarding the objective of their motion and were not capable of decision-making.

The three main behaviours applied to the sheep in the simulations of Strömbom et al. [16] were adopted for the swarm agents in this study. Thereby, each swarm agent was repelled from each other swarm member ($\pi_{k\neq i}$) by force vector $F^t_{\pi_i\pi_k}$ if they come within a distance of $R_{\pi\pi}$ from one another (note that $R_{\pi\pi}$ represents the radius of influence of the swarm agents, not their sensing range). The swarm members were, likewise, repulsed by the shepherd when within a distance of $R_{\pi\beta}$ from them by force vector $F^t_{\pi_i\beta}$. Further, the swarm agents were attracted to their global centre of mass ($\Gamma^t_{\pi_i}$), when within sensing range ($R_{\pi\Gamma}$) of other agents, by force vector $F^t_{\pi_i\Gamma^t_{\pi_i}}$ (thus, slightly modified from [16]). The weights of the forces, $W_{\pi\pi}$, $W_{\pi\beta}$, and $W_{\pi\Gamma}$ respectively, where $W = |F|$, were calculated based on their distance from the source of the force (see Sect. 2.3.2). The overall force acting on the swarm agents when being herded was then calculated using Eq. (2.1), where $F^t_{\pi_i e}$ is a random jittering noise force which replicates the non-precise movement of sheep or robotic systems.

$$F^t_{\pi_i} = \sum_{\pi=1,k\neq i}^{\pi=N} F^t_{\pi_i\pi_k} + F^t_{\pi_i\beta} + F^t_{\pi_i\Gamma^t_{\pi_i}} + F^t_{\pi_i e} \tag{2.1}$$

Similarly, the two main behaviours of the sheepdog in Strömbom et al.'s heuristic, driving and collecting, governed the shepherd during this study. If one (or more) of the swarm agents moved outside of swarm density radius $R_{\pi\rho}$, with $\Gamma^t_{\pi_i}$ at its origin, then the shepherd moved to the target position $P^t_{T_C}$, located at distance $\mathbb{D}_Q$ behind the furthest outlying agent (π_O) with respect to $\Gamma^t_{\pi_i}$. If all agents lay within $R_{\pi\rho}$, then the shepherd moved to the target position $P^t_{T_D}$ at distance $\mathbb{D}_Q$ behind $\Gamma^t_{\pi_i}$ with respect to the current goal location. $R_{\pi\rho}$ was calculated as a function of N_π using Eq. (2.2), the same as used in [16].

$$R_{\pi\rho} = R_{\pi\pi} N^{2/3} \tag{2.2}$$

The novel aspect of this study, departing from that of Strömbom et al. [16], was to assign each swarm agent one of four roles: *herd member*, *immobilised*, *torpedo*, or *explorer*. During an exploration exercise, the herd members navigated their environment according to the aforementioned three classical shepherding behaviours, being influenced by the repulsion and attraction forces of the swarm and shepherd (the only role present in [16]). However, if an agent reached a goal location (within distance R_G), which was not required to be the current goal sought by the shepherd, then the agent's role would switch to immobilised, and it would become stuck at the goal location for a count of T_f time steps. Once the T_f steps are complete, the agent's role switches to torpedo. The agent is then pushed by force vector $F^t_{\pi_i B}$, of constant weight $W_{\pi B}$, in a randomly generated direction for a count of T_b time steps, as a means for further environment coverage.

Following the T_b steps, the agent's role will either be converted back to a herd member, or switches to an explorer (depending on its distance from $\Gamma^t_{\pi_i}$). The final role, explorer, is initialised if an agent reaches a radial distance of R_H from $\Gamma^t_{\pi_i}$. At this point, it is assumed that the agent is outside of sensing range of the other swarm agents, and the shepherd recognises it as separate from the rest of the swarm, which it is collecting and driving to the next goal location. Therefore, any swarm agents outside of R_H are not included in the calculation of $\Gamma^t_{\pi_i}$. The final role, explorer, is initialised if an agent reaches a radial distance of R_H from $\Gamma^t_{\pi_i}$. Once assigned the explorer role, the agent restricts itself to a local environment of size L_l and begins following a Hilbert space-filling curve (HSFC), of order H_O (more information on HSFCs can be found in Sect. 2.3.4), acting as an area covering path. Once the path is complete, the agent's role will switch to a torpedo. The swarm role assignment methodology is captured in Algorithm 1, where roles are assigned as herd member (1), immobilised (2), torpedo (3), and explorer (4), updated at each time step.

Figure 2.1 gives an example of a simulation, where the shepherd and three swarm agents are shown with their corresponding repulsion force radii (and their influential spheres shown surrounding them). The outlying agent (π_O) is outside of R_H, therefore, its role would switch to explorer, while the rest of the agents are herd members, lie within $R_{\pi\rho}$, and are, therefore, dense enough to be driven towards the goal.

The swarm agents were randomly initialised in the north-east quadrant of the environment at positions $P^{t=1}_{\pi_i}$, and the shepherd was initialised in the south-west quadrant at position $P^{t=1}_{\beta}$. All swarm members began with the role of a herd member.

2.3.1 Simulation Setup

Two types of shepherding were compared during the simulations. The first was a general sheepdog herding sheep to multiple goals shepherding exercise, where only a single sheep was required to reach any single goal location, rather than the entire herd. The second introduced a new method for exploring an environment using goal locations and HSFC paths.

The second set of simulations, which introduce the Hilbert exploration technique, have the same objective as the first, to herd the swarm agents to each goal location.

Algorithm 1 Swarm agent role assignment

for all i **do**
  for all h **do**
    if $\| P^t_{\pi_i} - P_{G_h} \| \leq R_G$ **then**
      $\pi_i(role) == 2$
      $\pi_i(immobilise\_count) = 0$
    end if
  end for
  if $\pi_i(role) == 1$ **then**
    if $\| P^t_{\pi_i} - \Gamma^t_{\pi_i} \| \geq R_H$ **then**
      $\pi_i(role) == 4$
      $\pi_i(HSFC\_position) = 0$
    end if
  else if $\pi_i(role) == 2$ **then**
    if $\pi_i(immobilise\_count) == T_f$ **then**
      $\pi_i(role) == 3$
      $\pi_i(torpedo\_count) = 0$
    else
      $\pi_i(immobilise\_count) = \pi_i(immobilise\_count) + 1$
    end if
  else if $\pi_i(role) == 3$ **then**
    if $\pi_i(torpedo\_count) == T_b$ **then**
      if $\| P^t_{\pi_i} - \Gamma^t_{\pi_i} \| \geq R_H$ **then**
        $\pi_i(role) == 4$
        $\pi_i(HSFC\_position) = 0$
      else
        $\pi_i(role) == 1$
      end if
    else
      $\pi_i(torpedo\_count) = \pi_i(torpedo\_count) + 1$
    end if
  else if $\pi_i(role) == 4$ **then**
    if $\pi_i(HSFC\_position) == (length(HSFC\_path) - 1)$ **then**
      $\pi_i(role) = 3$
      $\pi_i(torpedo\_count) = 0$
    else
      $\pi_i(HSFC\_position) = \pi_i(HSFC\_position) + 1$
    end if
  end if
end for

However, in this setup, if an agent strays a distance of $R_{\pi H}$ from $\Gamma^t_{\pi_i}$, then it re-initialises itself within a local environment and starts following an HSFC path, hoping to reach one of the goal locations along its trajectory.

If the swarm agent reaches one of the goals, then it becomes bound to the goal (as the straight herding setup), before being torpedoed away. Similarly, if an agent arrives at the end of a HSFC path without encountering any goals, then the agent gets torpedoed in a random direction and either rejoins the herd or begins following a new HSFC path.

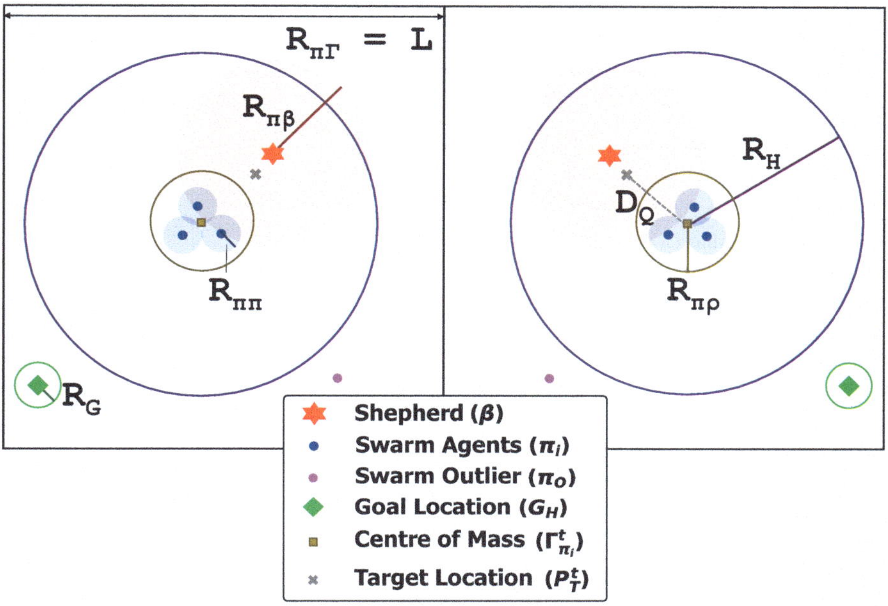

Fig. 2.1 Split mirror illustrations of experimental setup including a shepherd, swarm agents, and a goal. The left image, additionally, shows the shepherd repulsion radius, sheep repulsion radius, goal radius, and environment length and global centre of mass attraction radius. The right image includes the Hilbert radius, swarm density radius, and shepherd target distance. Note that the parameters are not correctly scaled

Table 2.1 outlines each of the simulation parameters and their corresponding values used. As this study is the first of its kind, the environment size, swarm repulsion radius, shepherd repulsion radius, swarm density radius, and shepherd and sheep step size were all given the same values as used by Strömbom et al. [16] as a foundation, while the other values were chosen intuitively to complement Strömbom et al.'s original study.

The first experiment undertaken was to test the effect of the size of the localised HSFC environment (L_l) on the time it took to reach 10 goals for varying maximum influential force weight (W_{max}). Three L_l values were simulated: $L_l = \frac{L}{N}$, $L_l = 0.50 * \frac{L}{N}$, and $L_l = 1.50 * \frac{L}{N}$. Force weightings between $10N$ and $200N$ were tested at intervals of $10N$ separately for $W_{\pi\pi}$, $W_{\pi\beta}$, and $W_{\pi\Gamma}$. While one maximum weighting was modified, the other two were fixed at $100N$.

The second experiment was completed to examine how the number of goals (N_G) within a fixed environment, and thus goal density, affected the performance of the developed HSFC exploration technique. N_G values ranging from 5 to 50 were simulated at intervals of 5 for each of the local environment lengths described above.

While the goals were generated with the perimeter of the environment, the swarm agents and shepherd were not restricted to boundaries. Each simulation setup was

Table 2.1 Parameters for shepherding experiments

Parameter	Value
Maximum number of time steps (t_{max})	5000
Environment length (L)	150m
Number of swarm agents (N_π)	10
Swarm repulsion radii ($R_{\pi\pi}$)	2m
Swarm speed (V_π)	$1\frac{m}{t}$
Swarm density radius ($R_{\pi\rho}$)	9.28m
Number of shepherds (N_β)	1
Shepherd repulsion radius ($R_{\pi\beta}$)	65m
Shepherd speed (V_β)	$1.5\frac{m}{t}$
Shepherd target distance ($\mathbb{D}_Q$)	40m
Shepherd alignment distance ($\mathbb{D}_A$)	25m
GCM attraction radii ($R_{\pi\Gamma}$)	150m
Noise force weight ($W_{\pi_i n}$)	0.3N
Number of goals (N_G)	10
Goal radius (R_G)	5m
Circular path radius (R_{CO})	65m
Circular path angular speed ($\Delta\theta_\beta$)	$0.025rad/t$
Fixed count (T_f)	100t
Torpedo force weight ($W_{\pi B}$)	100N
Torpedo force count (T_b)	20N
HSFC order (H_O)	3
Hilbert radius (R_H)	39.28m

repeated 30 times with different seeds, then the results for that setup were averaged. The simulations were run until all goals had been reached, or the maximum of 5000 time steps was completed. Both experiments were compared to the performance of the pure herding shepherding to reach all goals. Both experiments compare the HSFC-based shepherding to the classic shepherding (derived from [16]).

2.3.2 Force Modulation

Strömbom et al. [16] applied a discrete force when a sheep fell within the ranges of influence of other agents. In this study, it was assumed that the influence force weightings acting on a swarm agent were applied with a natural exponential trend. The repulsion force weakens, and attraction force strengthens, as the radial distance from their source increased. Equation (2.3) defines the repulsion force as a function of radial distance ($r_{\pi R}$) from the repulsion force origin, while Eq. (2.4) describes the attraction force as a function of radial distance ($r_{\pi \Lambda}$) from the attraction source. $W_{\pi R}$ is the modulated weight of the repulsion force and W_{Rmax} is the maximum

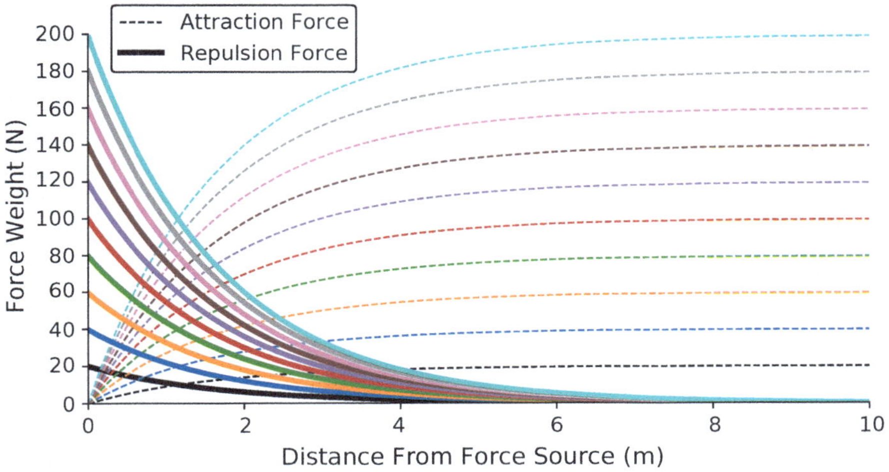

Fig. 2.2 Influential force modulation for a series of maximum force weights. Each colour represents the matching repulsive and attractive force weights applied at a given radius

force weight (replaced by $W_{\pi\pi}$ or $W_{\pi\beta}$). Likewise, $W_{\pi A}$ is the modulated weight of the attraction force and $W_{\pi\Gamma}$ is the maximum possible attraction force strength. $R_{\pi R}$ represents the radii $R_{\pi\pi}$ and $R_{\pi\beta}$. Figure 2.2 depicts the natural exponential relationship between force strength and radial distance from the source (here, at the origin [0,0]), for an influential radius of $6m$. Beyond $6m$, the force tends to $0N$, so has no significant effect on the behaviour of the swarm agents.

$$W_{\pi R} = W_{Rmax} * e^{\left(\frac{-r_{\pi R}}{R_{\pi R}} * 6\right)} \tag{2.3}$$

$$W_{\pi A} = W_{\pi\Gamma} * \left(1 - e^{\left(\frac{-r_{\pi\Gamma}}{R_{\pi\Gamma}} * 6\right)}\right) \tag{2.4}$$

2.3.3 Path Planning

As discussed, when herding sheep, a sheepdog exhibits two main behaviours: driving and collecting [16]. When the shepherd moves to a target location P_T^t (representing $P_{T_C}^t$ or $P_{T_D}^t$), it must position itself behind the swarm with respect to the direction of G_h, or behind π_O with respect to $\Gamma_{\pi_i}^t$. As such, in order to achieve this without driving the swarm agents towards an undesired direction, the shepherd must plan its path to P_T^t.

A simple method of achieving this was to create a circular path with its origin (P_{CO}^t) centred on either the $\Gamma_{\pi_i}^t$ or π_O, with a radius large enough to render the influential force of the shepherd ineffective (in this case, $R_{\pi\beta}$). With each step, the shepherd calculates its distance $\mathbb{D}_T$ from P^t_T, and distance $\mathbb{D}_{CO}$ from P^t_{CO}. The shepherd moves towards P^t_T until it reaches a point along the circular path, within a tolerance of ϵ, such that $P^t_{CO} = R_{\pi\beta} \pm \epsilon$. The shepherd then follows the path until positioned behind its desired P^t_T, finally moving towards it once within an alignment distance of $\mathbb{D}_A$ from it, where $\mathbb{D}_A = R_{\pi\beta} - \mathbb{D}_Q$. If the shepherd already lies within the circle, then it simply moves in a straight path towards P^t_T. Algorithm 2 gives a breakdown of the path planning algorithm used.

Algorithm 2 Circular-path planning algorithm

while $t < 5000$ **do**
  Calculate $\mathbb{D}_T$
  Calculate $\mathbb{D}_{CO}$
  if $\mathbb{D}_T < \mathbb{D}_A$ **then**
    $P^{t+1}_\beta(x, y) = P^t_\beta(x, y) + V_\beta * \frac{P^t_T(x,y) - P^t_\beta(x,y)}{\| P^t_T(x,y) - P^t_\beta(x,y) \|}$
  else if $\mathbb{D}_T > (R_{\pi\beta} + \epsilon)$ **then**
    $P^{t+1}_\beta(x, y) = P^t_\beta(x, y) + V_\beta * \frac{P^t_T(x,y) - P^t_\beta(x,y)}{\| P^t_T(x,y) - P^t_\beta(x,y) \|}$
  else
    $P^{t+1}_\beta(x) = P^t_{CO} + R_{\pi\beta} * \cos \Delta\theta_\beta$
    $P^{t+1}_\beta(y) = P^t_{CO} + R_{\pi\beta} * \sin \Delta\theta_\beta$
  end if
end while

Figure 2.3 gives an example simulation of such a path. The simulation was such that the swarm, of size $N_\pi = 5$, was forced to remain stationary to prove the path planning concept. Once initialised (where $P^{t=1}_\beta$ is represented by a larger marker), the shepherd makes a decision on whether to drive the herd towards the single goal at $P_G = [10, 10]$, or collect an outlier agent. In this case, the swarm is dense enough to drive, so the shepherd approaches $P^t_{T_D}$ until it reaches the radial distance $R_{\pi\beta}$ from $\Gamma^t_{\pi_i}$, then tracks along the circular path (represented by the thin black line) until it arrives at a location approximately behind $P^t_{T_D}$ and moves towards it. $R_{\pi\beta}$ was set to $30m$ and $\mathbb{D}_A$ was set to $10m$. R_G is displayed as the green circle surrounding the goal.

2.3.4 *Hilbert Space-Filling Curves*

A space-filling curve, within a square boundary, is a continuous path which moves through each point within the square. Concepts for space-filling curves were originally invented by Peano [10], Hilbert [7], and Sierpinski [12].

Hilbert curves are one of the simplest space-filling, or fractal, curves as there are only four directions that the path can move in at any given step [5]. These directions can be represented as the four cardinal directions: north (Nth), south (Sth), east (Est), and west (Wst). The shapes constructed during the Hilbert space-filling walk

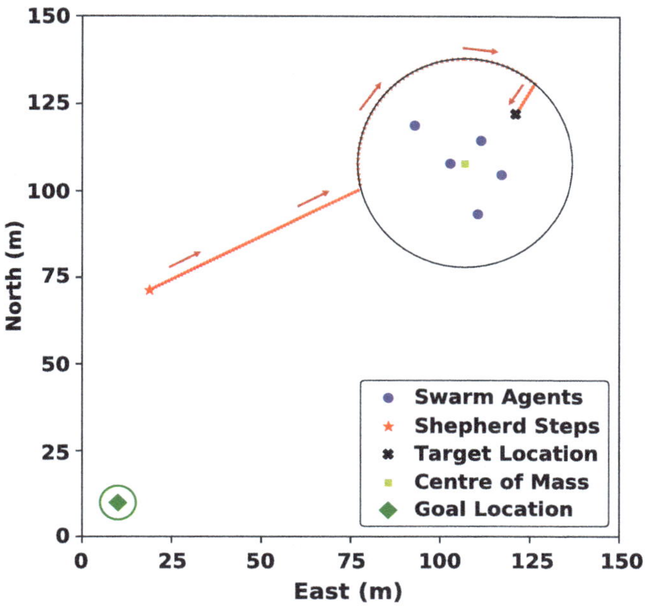

Fig. 2.3 Shepherd circular path to driving target location

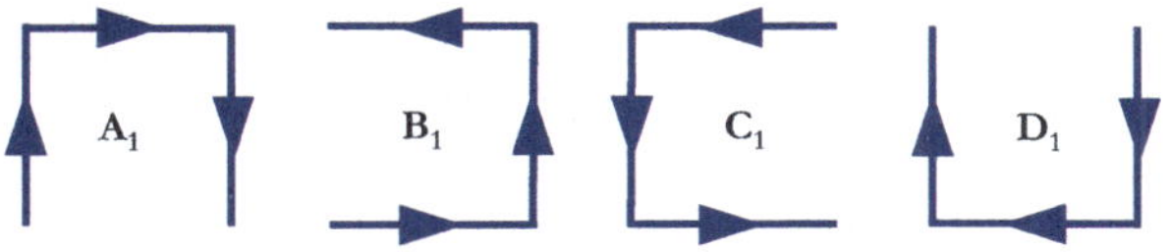

Fig. 2.4 First order Hilbert space-filling curve path directions

are repeated square patterns where one edge is removed. The path is followed in a clockwise direction if the square has only one horizontal edge, and in an anti-clockwise direction if it has only one vertical edge. The four open-square patterns which are formed along the walk, collectively defined as the first order HSFC (H_1), and their corresponding directions of motion, are given in Fig. 2.4.

These units are represented by the identifiers: A_1, B_1, C_1, and D_1. For a second order Hilbert curve (H_2), two of the H_1 units are linked together, and joined to other H_1 units which are traversed in an opposite sense. Wirth [19], then, defined the set of Eqs. (2.5)–(2.8) to find an mth order HSFC, where each sequential order is comprised of sets of the previous order with alternating senses. The first to fourth order HSFCs are shown in Fig. 2.5.

$$A_{m+1} = [B_m, Nth, A_m, Est, A_m, Sth, C_m] \tag{2.5}$$

$$B_{m+1} = [A_m, Est, B_m, Nth, B_m, Wst, D_m] \tag{2.6}$$

$$C_{m+1} = [D_m, Wst, C_m, Sth, C_m, Est, A_m] \tag{2.7}$$

$$D_{m+1} = [C_m, Sth, D_m, Wst, D_m, Nth, B_m] \tag{2.8}$$

The HSFC was selected as the covering method for this study due to its simplicity, which matches the advantages of using a swarm-based system for exploration, where each swarm agent should remain rudimentary in its own capacity to maintain the robustness and flexibility of swarm system as a whole. Further, the third order HSFC was used to match the L_l, L, and R_G used, with lesser orders not covering enough points within the environment, and larger orders covering too many.

2.4 Results and Discussion

A preliminary simulation was run for both the classic herding (Fig. 2.6) and Hilbert shepherding (Fig. 2.7) setups in order to give an example comparison of the methods. The same goals were generated, which followed the same sequence

Fig. 2.5 Hilbert space-filling curves of orders 1–4

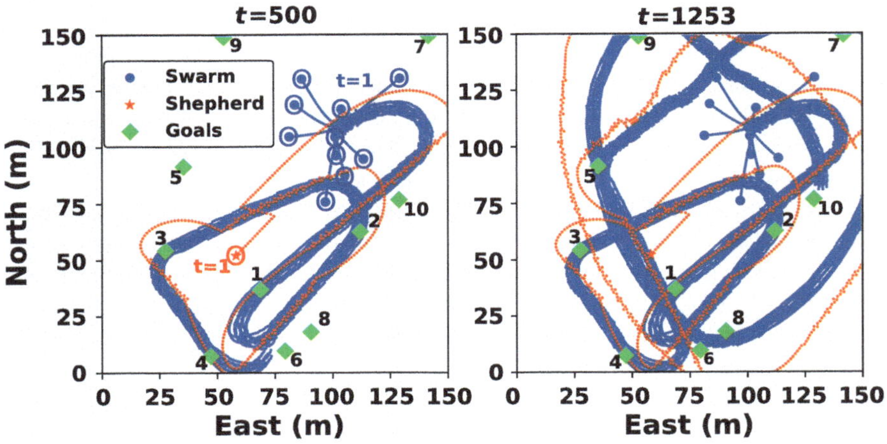

Fig. 2.6 Herding swarm shepherding exploration paths for $t = 500$ and $t = 1253$ time steps, and with goals G_1 to G_{10}

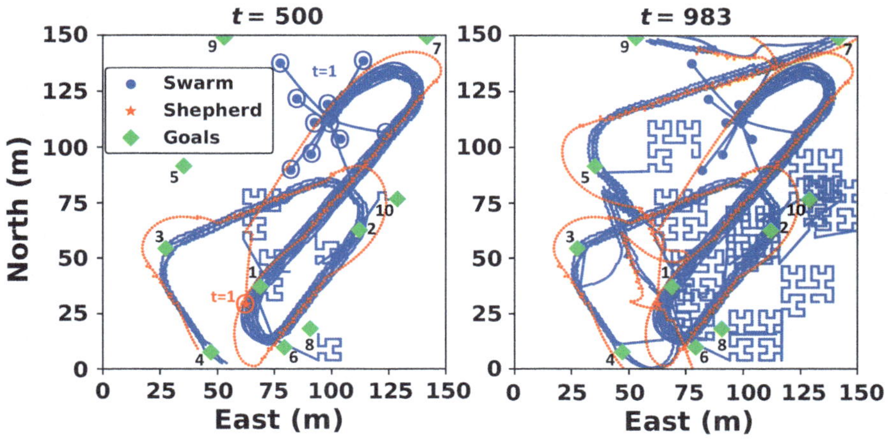

Fig. 2.7 HSFC-based swarm shepherding exploration paths for $t = 500$ and $t = 983$ time steps, and with goals G_1 to G_{10}

Table 2.2 Goal coordinates and order for example simulations

Goal (G_H)	Eastern coordinate (m)	Northern coordinate (m)
1	68.96	37.28
2	112.22	62.81
3	27.69	54.42
4	47.31	7.75
5	35.43	91.74
6	79.62	9.94
7	141.84	149.57
8	90.80	18.37
9	53.10	149.23
10	128.99	76.92

(given in Table 2.2), and the total paths of the shepherd and swarm agents were plotted. Finally, the maximum influential force weights were equated as $W_{\pi\pi} = W_{\pi\beta} = W_{\pi\Gamma} = 100N$.

The pure herding case, given in Fig. 2.6, is split into two sections. The first shows the overall paths for the shepherd and swarm agents at $t = 500$, and the second displays the completion of the exercise, at $t = 1253$. $P_{\pi_i}^{t=1}$ and $P_{\beta}^{t=1}$ are represented by a larger marker than was used for their corresponding paths, with a surrounding circle included on the left image, while each of the goals is numbered (as given in Table 2.2). The motion of the shepherd was found to be quite similar to that of Strömbom et al. [16] when driving and collecting the agents, with a small-scale zigzag type manoeuvre observed keeping the herd centred about $\Gamma_{\pi_i}^t$. Upon initialisation, the swarm agents were quickly attracted to $\Gamma_{\pi_i}^t$, forming a herd. The shepherd moved towards a target position at first (likely $P_{T_C}^t$), before briefly following the circumference of the circular path, then proceeded to move towards $P^t_{T_D}$ as the herd condensed and it fell within $R_{\pi\beta}$. The agents were then herded to G_1 to G_4 in the prescribed order, before reaching G_6 at $\sim t = 500$. In the final graph, it can be seen that the shepherd herds the swarm to the goals in the correct sequential order (once G_6 was disregarded as it had already been reached), passing outside of the environment bounds briefly between G_7 and G_8, then again between G_9 and G_{10} (due to the goals being generated near the borders). The shepherd spent very little time following the circumference of the circle defined by the path planning algorithm, though, as once it reached a P^t_T, it would generally remain within a distance of $R_{\pi\beta}$ from the circular path origin, instead continually moving along a straight line towards P^t_T. The reduction of $R_{\pi\beta}$ could solve this problem and further mitigate unwanted repulsion by the shepherd.

An equivalent simulation for the HSFC-based exploration technique is given in Fig. 2.7 for $L_l = 22.5$ m at $t = 500$ and $t = 983$ (the end of the simulation). At $t = 500$ it can be seen that the path of the herding agents begins quite similar to that of the pure herding case, with the herd of agents reaching G_1 to G_4. However, the agent which became immobilised at G_2 had torpedoed north-east, switched to an explorer, then quickly reached G_{10}. Similarly, an agent which became immobilised at G_6 was torpedoed south-east before following an HSFC path until it reached G_8. A second agent, after becoming immobilised at G_1 for time T_f, had begun following a Hilbert path, while two other agents had become immobilised at G_1. At the completion of the environment exploration, it can be seen that the agents which reached G_6 and G_{10} remained outside of R_H stayed within the same regions and continued following HSFC paths. Likewise, the agents which were torpedoed from G_4 and G_5 started exploring without rejoining the herd, while the agent immobilised at G_3 was collected before G_7 and G_9 were finally covered. As with the herding case, there was only a very brief moment, just after the shepherd was initialised, when it followed the circumference of the circular path.

It can be seen that the Hilbert exploration setup was able to reach the 10 goals faster than the herding exploration setup, taking 78% of the number of time steps. It also appears that the HSFC-based method covered more of the environment over the shorter time period. While no reliable conclusions can be made from one simulation, the example gives a good indication of the pattern which emerged from the extensive experimental campaign.

2.4.1 *Force Weights*

Figure 2.8 displays the amount of time it took to reach 10 goals for the pure herding and Hilbert shepherding exploration experiments, averaged over the 30 simulations run for each of the HSFC local environment sizes and classic herding cases. This was done for the three local environment sizes, over each of the force weight ranges tested for $W_{\pi\pi}$, $W_{\pi\beta}$, and $W_{\pi\Gamma}$. The size of L_l, however, had little affect on the time taken to cover all of the goals for any of the maximum force weights.

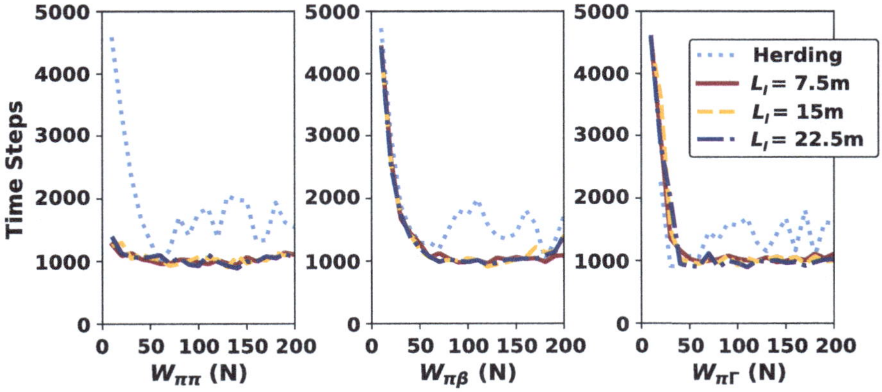

Fig. 2.8 Maximum force weighting versus time

When varying $W_{\pi\pi}$, the HSFC-based exploration technique outperformed the pure herding method each for all weights, except at $W_{\pi\pi} = 60$, where the herding method was able to reach all 10 goals slightly quicker than the Hilbert method for $L_l = 22.5m$. The Hilbert technique was able to cover all goals much faster than the traditional herding method for low sheep-to-sheep repulsion values, likely due to the herd forming a much tighter group, thus, requiring more precise guidance to attain a goal. The performance of each of the methods remained the same for low $W_{\pi\beta}$ values, as the herd would have been less likely to break apart due to the presence of the shepherd. As the weight increased, the Hilbert method was shown to be more efficient. When the maximum weight for attraction to $\Gamma^t_{\pi_i}$ was $10N$, the shepherd was unable to guide the swarm to all 10 goals. This showcases the difficulty confronting the shepherd if the swarm agents are not attracted to their herd, where collection would likely become more time consuming and take up the majority of the shepherd's energy. In general, the maximum repulsion and attraction force weightings were found to not significantly affect the temporal efficiency of the HSFC-based shepherding task beyond a value of approximately $50N$; however, the classic shepherding case was found to fluctuate significantly, likely due to the greater percentage of swarm agents acting as herd members.

2.4.2 Number of Goals

The effectiveness of the developed HSFC exploration method increased as the number of goals in the environment increased. This can be explained by the fact that the HSFC-based technique results in more coverage of an environment over a shorter time period (as can be seen in Figs. 2.7 and 2.8). Therefore, with a higher density, the Hilbert paths became more advantageous. The traditional herding exploration method was unable to cover 50 goals within the 5000 time steps, and barely able

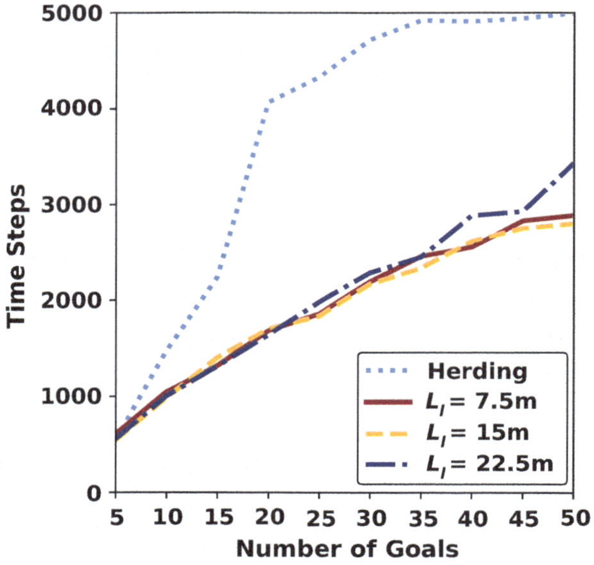

Fig. 2.9 Number of goals versus time

to reach more than 35 goals. The performance of each of the L_l values remained fairly constant until N_G increased above 35, at which point the larger Hilbert covering area ($L_l = 22.5m$) reduced the efficiency of the task. If there are only a few goal destinations required to be visited within an area (for $N_G = 5$), then the herding method performed just as well as the Hilbert method. The quantity of areas of interest within a given environment should, therefore, dictate the type of exploration technique used. If there are a few very specific areas of significance requiring exploration within an environment, then classic shepherd herding control should be adopted, however, for regions with more uncertainty or a vast quantity of goal destinations, then the Hilbert method was shown to be more appropriate. The local environment size for the HSFCs must then be considered, where one too large may not perform as well given a very high goal density (Fig. 2.9).

2.5 Conclusion

The use of shepherding for environment exploration using a goal-based system has been investigated. The autonomous swarm shepherding method developed using HSFC paths appears to be more effective at exploring an environment than traditional herding using goal-based waypoints. The time taken to cover all goals within the mission environment was significantly lower for the majority of maximum force weights tested. The improvements upon the herding technique became more apparent with the greater number of goals within the environment. Further, it seemed as though the Hilbert method enabled greater coverage of the environment than the traditional herding technique.

The simple circular path planning algorithm created for the exploration task to enable the shepherd to manoeuvre around the swarm agents without imparting unwanted force on them appeared to have been used primarily at the beginning of the exercise, then seldom again. More research must be conducted to investigate the effect of the circular path radius as the shepherd often lay within the circular path during the simulations, signifying that it rarely tracked along its circumference.

Future work will test the effects of the HSFC order on the performance of the goal-based exploration task, as well as the variation of the number of swarm agents involved. Moving beyond the described setup, additional research should address dynamic goals, uncertainty in their locations and seek to compare single and multi-agent approaches for exploration to single and multi-shepherding agent approaches.

The initial results, however, show promise for the use of shepherding guidance of swarms as a solution for autonomous environment exploration. Furthermore, the presented simulations have given an example of the feasibility of the sea state determination problem.

References

1. Burgard, W., Moors, M., Fox, D., Simmons, R., Thrun, S.: Collaborative multi-robot exploration. In: Proceedings 2000 ICRA. Millennium Conference. IEEE International Conference on Robotics and Automation. Symposia Proceedings, pp. 476–481 (2000)
2. Chaimowicz, L., Kumar, V.: Aerial shepherds: Coordination among UAVS and swarms of robots. In: Alami, R., Chatila, R., Asama, H. (eds.) Distributed Autonomous Robotic Systems, vol. 6, pp. 243–252. Springer Japan, Tokyo (2007)
3. Clayton, N.R., Abbass, H.: Machine teaching in hierarchical genetic reinforcement learning: Curriculum design of reward functions for swarm shepherding. In: Proceedings of the IEEE Congress on Evolutionary Computation (2019)
4. Gee, A., Abbass, H.: Transparent machine education of neural networks for swarm shepherding using curriculum design. In: Proceedings of the International Joint Conference on Neural Networks (2019)
5. Griffiths, J.: Table-driven algorithms for generating space-filling curves. Comput.-Aided Design **17**(1), 37–41 (1985)
6. Gu, X., Moan, T.: Long-term fatigue damage of ship structures under nonlinear wave loads. Marine Technol. **39**(2), 95–104 (2002)
7. Hilbert, D.: Über die stetige Abbildung einer Linie auf ein Flächenstück. In: Dritter Band: Analysis· Grundlagen der Mathematik· Physik Verschiedenes, pp. 1–2. Springer, Berlin (1935)
8. Lien, J.M., Bayazit, O.B., Sowell, R.T., Rodriguez, S., Amato, N.M.: Shepherding behaviors. In: IEEE International Conference on Robotics and Automation, vol. 4, pp. 4159–4164. Citeseer (2004)
9. Long, N.K., Sammut, K., Sgarioto, D., Garratt, M., Abbass, H.A.: A comprehensive review of shepherding as a bio-inspired swarm-robotics guidance approach. IEEE Trans. Emer. Topics Comput. Intell. **4**, 523–537 (2020)
10. Peano, G.: Sur une courbe, qui remplit toute une aire plane. Math. Ann. **36**(1), 157–160 (1890)

11. Reynolds, C.: Flocks, herds and schools: A distributed behavioral model. In: Proceedings of the 14th Annual Conference on Computer Graphics and Interactive Techniques, SIGGRAPH '87, pp. 25–34. ACM, New Yrok (1987)
12. Sierpínski, W.: Sur une nouvelle courbe continue qui remplit toute une aire plane. Bull. Acad. Sci. Cracovie (Sci. Math. et Nat. Serie A) 462–478 (1912)
13. Smith, M.: I Am Just Going Outside: Captain Oates-Antarctic Tragedy. Gill & Macmillan, Dublin (2002)
14. Stella, A.C., Vakkalanka, J.P., Holstege, C.P., Charlton, N.P.: The epidemiology of caving fatalities in the United States. Wild. Environ. Med. **26**(3), 436–437 (2015)
15. Stevens, S.C., Parsons, M.G.: Effects of motion at sea on crew performance: a survey. Marine Technol. **39**(1), 29–47 (2002)
16. Strömbom, D., Mann, R.P., Wilson, A.M., Hailes, S., Morton, A.J., Sumpter, D.J.T., King, A.J.: Solving the shepherding problem: heuristics for herding autonomous, interacting agents. J. R. Soc. Interf. **11**(100) (2014). https://browzine.com/articles/52614503
17. Varughese, J.C., Thenius, R., Leitgeb, P., Wotawa, F., Schmickl, T.: A model for bio-inspired underwater swarm robotic exploration. IFAC-PapersOnLine **51**(2), 385–390 (2018)
18. Whaite, P., Ferrie, F.P.: Autonomous exploration: driven by uncertainty. IEEE Trans. Pattern Anal. Mach. Intell. **19**(3), 193–205 (1997). https://doi.org/10.1109/34.584097
19. Wirth, N.: Algorithms+ data structures= programs. Technical Report (1976)
20. Zoghby, N.E., Loscri, V., Natalizio, E., Cherfaoui, V.: Chapter 8: Robot Cooperation and Swarm Intelligence, Chap. 8, pp. 163–201 (2013). https://doi.org/10.1142/9789814551342_0008

Chapter 3
Simulating Single and Multiple Sheepdogs Guidance of a Sheep Swarm

Daniel Baxter, Matthew Garratt, and Hussein A. Abbass

Abstract Shepherding is a specific class of flocking behaviour where external agents (the shepherd) influence the movements of a group of agents (the flock). In nature, a powerful example is herding a flock of sheep by an influential sheepdog. When the sheepdog encroaches on the sheep's influence zone, the sheep is essentially "repelled". Optimising this phenomenon has many engineering applications, such as environmental protection, security, crowd, and agricultural control. In this chapter, we build on Strömbom et al's adaptive switching algorithm to incorporate multiple sheepdog agents programmed with basic swarm intelligence rules. Our simulation results show that synergising shepherds with swarming behaviours improves the effectiveness of the shepherds as measured by the speed of collecting and driving the sheep towards a target destination while maintaining a more compact and cohesive flock.

Keywords Swarming · Shepherding · Multi-agent shepherding

Shepherding is a specific class of flocking behaviour where external agents (the shepherd) influence the movements of a group of agents (the flock). In nature, a powerful example is herding a flock of sheep by an influential sheepdog. When the sheepdog encroaches on the sheep's influence zone, the sheep is essentially "repelled". Optimising this phenomenon has many engineering applications, such as environmental protection, security, crowd, and agricultural control. In this chapter, we build on Strömbom et al's adaptive switching algorithm to incorporate multiple

Electronic Supplementary Material: The online version of this chapter (https://doi.org/10.1007/978-3-030-60898-9_3) contains supplementary material, which is available to authorised users.

D. Baxter (✉) · M. Garratt · H. A. Abbass
School of Engineering and Information Technology, University of New South Wales, Canberra, ACT, Australia
e-mail: daniel.baxter@student.adfa.edu.au; m.garratt@adfa.edu.au; h.abbass@adfa.edu.au

H. A. Abbass, R. A. Hunjet (eds.), *Shepherding UxVs for Human-Swarm Teaming*, Unmanned System Technologies,
https://doi.org/10.1007/978-3-030-60898-9_3

sheepdog agents programmed with basic swarm intelligence rules. Our simulation results show that synergising shepherds with swarming behaviours improves the effectiveness of the shepherds as measured by the speed of collecting and driving the sheep towards a target destination while maintaining a more compact and cohesive flock.

3.1 Introduction

We consider the shepherding problem of non-cooperative herding, analogous to herding performed by farmers (shepherds agents) and dogs (sheepdogs agents) to drive a flock (sheep agents) to a target destination. In real world systems, sheep are repelled by the sheepdogs, with the repulsion dependent on the sheepdogs' position and locomotion. Sheepdogs in this chapter will also play the role of shepherds due to the autonomous role they play without the interference of a human mediator.

The proposed model reflects a self-propelled particle model with basic local attraction-repulsion swarm behaviour to model the herding of a group of interacting agents by multiple shepherds to a predetermined target destination. The shepherds have been programmed using Strömbom et al's biologically sound dynamic switching algorithm [15] with the inclusion of the same basic swarm rules used for flocking agents. The inclusion of basic swarm behaviour was a means to promote collective intelligence [10].

We study a collective intelligence approach for the shepherds as a means to coordinate their locomotion collectively to herd the flock. The system is a non-cooperative, multi-robot shepherding system as the sheep are inclined to randomly graze until they get influenced to move by the presence of the shepherd—whose goal is to forcefully relocate the sheep to a target destination; a destination that is not known to the sheep. If the shepherd is not within the influence range appropriate to affect any sheep's movements, the sheep continue to happily graze the paddock.

The proposed multi-shepherd swarm guidance algorithm has a number of applications in a variety of fields. For example, wildlife control at airports and national parks [6], large hazardous chemical spills [5], crowd control [8, 9], and patrolling unsafe or restricted areas [3, 12]. The main aim, however, is to reduce the $\sim$12 lives lost per annum on Australian farms involving quad-bikes, motorbikes, and helicopters used for herding purposes [13, 14] by implementing the swarm control algorithm on multiple UxVs to fully automate the herding process.

Commercial and private uses of UxVs have proven to be a safe practice, with no recorded deaths or serious injuries in Australia. Over the period between 2012 and 2016 (inclusive), there were 180 recorded incidents involving UxVs—of which, only $\sim$8 per annum involved accidental crashes into terrain resulting in minor injuries. These were predominately in highly populated areas and rarely occurred to the remote pilot [1].

In this chapter, our aims are to identify how multiple shepherds can work cooperatively using only local knowledge from the environment to effectively guide a group of non-cooperative sheep—without shepherd-to-shepherd communication or a centralised control mechanism. The hypothesis is that synergising swarm intelligence (*SI*) and shepherding behaviours in multi-shepherd systems will result in decreased task completion times while maintaining a more compact, cohesive flock when compared with a single shepherd working to the same set of herding and shepherding rules.

Additionally, we hypothesise that these results may be further enhanced by developing a method that encourages the shepherds to position themselves more intelligently during shepherding tasks. In order to investigate the above, the study begins by designing a series of simulation tests using a single shepherd with varying flock sizes and densities. We then examine the effect of adding multiple shepherds, with *SI* alone and *SI* plus a small quasi-arc-formation, upon the dynamics and completion time.

3.2 Experimental and Computational Details

3.2.1 Problem Formulation

Consider an unbounded, continuous-space environment of length and width L, containing N sheep agents with positions $P^t_{\pi_i}$ for agent π_i at time t, where $i \in \{1,\ldots,N\}$. M shepherd agents with positions $P^t_{\beta_j}$ for agent β_j at time t, where $j \in \{1,\ldots,M\}$. The sheep are initialised in the top-right quadrant of the $L \times L$ environment and the shepherds' initial positions are evenly spaced along the vertical axis, centred at the midpoint, as a function of M. There is a static goal position P^t_G at time t, with the maximum allowable distance $\mathbb{D}$ required between any sheep and the goal. This will be known as the target region T_D (3.1), and represents the acceptable final region for the sheeps' collective Global Centre of Mass (GCM) $\Gamma^t_{\pi_i}$. Throughout this discussion, the centre of the goal region will be positioned at [0, 0] to align our coordinate system origin to the target goal point. To be considered as successfully shepherded, the GCM must be within T_D and the sheep clustered within the distance $f(N)$; detailed by Eq. (3.7).

$$T_D\left(P^t_G\right) = \left\{\left\|\Gamma^t_{\pi_i} - P^t_G\right\| \leq D\right\} \tag{3.1}$$

Definition 3.1 (Multi-Shepherd Herding) Develop a multi-shepherd swarm guidance algorithm for M_j shepherds with dynamics detailed by Eq. (3.8) to collect and drive a flock of N_i sheep with dynamics detailed by Eq. (3.6) to a predetermined target region detailed by Eq. (3.1).

Definition 3.2 (Intelligent Shepherd Positioning) Develop an extension to the multi-shepherd swarm guidance algorithm to encourage a more intelligent method for shepherds to position themselves.

3.2.2 Sheep Agent Model

The sheep agent model mimics that developed by Strömbom et al. [15] and does not restrict the form our sheep agents may take. For example, the same agent model may be used to model the movement of other entities such as fish, birds, sheep, cattle, robots, and humans. By simply tuning the sheep agent's characteristic parameter weights, this model may be utilised for simulating experiments in a number of engineered "*shepherding*" problems. Here, the sheep are modelled using the potential field technique that Cameron and Probert [2] theorised and proved effective for flocking by Vaughan et al. [16]. Analogous to herding behaviour [7], the sheep are governed by five spatial rules [11]: separation, attraction, alignment, grazing/jittering, and threat repulsion.

The separation rule is enforced to avoid inter-agent collisions when the distance between sheep π_i and any other sheep in its flock $\pi_k \neq i$ that is within $R_{\pi\pi}$ distance, by a short but strong repulsive force $W_{\pi\pi}$ in the direction shown in Eq. 3.2.

$$F^t_{\pi_i\pi_k} = \sum_{k\neq i}^{N} \frac{P^t_{\pi_i} - P^t_{\pi_k}}{\| P^t_{\pi_i} - P^t_{\pi_k} \|} \forall k \text{ where } \| P^t_{\pi_i} - P^t_{\pi_k} \| \leq R_{\pi\pi} \tag{3.2}$$

Due to the nature of their flocking dynamics, sheep also experience inter-agent attraction forces when a threat is detected within a distance $R_{\pi\beta}$. They are attracted to the Local Centre of Mass (LCM) $\Lambda^t_{\pi_i}$ of a number $\Omega_{\pi_i\pi}$ of their nearest neighbours, with an attraction force $W_{\pi\Lambda}$ in the direction shown in Eq. 3.3.

$$F^t_{\pi_i\Lambda^t_{\pi_i}} = \frac{\Lambda^t_{\pi_i} - P^t_{\pi_i}}{\| \Lambda^t_{\pi_i} - P^t_{\pi_i} \|} \tag{3.3}$$

Sheep π_i motion is influenced by its inertia through a weighted multiple W_{π_υ} of its previous direction of travel $F^{t-1}_{\pi_i}$; it grazes the environment by being typically motionless with a sporadic movement by a small percentage $W_{e\pi_i}$ of random movement $F^t_{\pi_i\epsilon}$. Once sheep *i* detects a shepherd within $R_{\pi\beta}$, they are repelled in the opposite direction to that of the location of the threat. The direction of this repulsive force applied to sheep π_i by shepherd β_j is shown in Eq. 3.4.

$$F^t_{\pi_i\beta} = \sum_{j} \frac{P^t_{\pi_i} - P^t_{\beta_j}}{\| P^t_{\pi_i} - P^t_{\beta_j} \|}, \forall j \text{ where } \| P^t_{\pi_i} - P^t_{\beta_j} \| \leq R_{\pi\beta} \tag{3.4}$$

When the sheep agent model characteristics (force vectors) and behavioural traits (strengths) are combined, they create the sheep's total force vector. This details each sheep agent's next movement (direction and magnitude). Summarising, this force vector is a weighted function of the sheep's previous step, attraction to its local centre of mass, repulsion from a threat, repulsion from an agent that is too close, and its grazing movement as shown in Eq. 3.5.

$$F^t_{\pi_i} = W_{\pi_\upsilon} F^{t-1}_{\pi_i} + W_{\pi\Lambda} F^t_{\pi_i \Lambda^t_{\pi_i}} + W_{\pi\beta} F^t_{\pi_i \beta_j} + W_{\pi\pi} F^t_{\pi_i \pi_{-i}} + W_{e\pi_i} F^t_{\pi_i \epsilon} \tag{3.5}$$

Through empirical data collected from sheep-sheepdog interactions, Strömbom et al. [15] calculated the weighting such that: $W_{\pi\pi} > W_{\pi\Lambda} > W_{\pi\beta} > W_{\pi_\upsilon}$. They attribute the inequality to real world sheep behaviour [4]. Sheep-to-sheep repulsion dominates in strength but is shortest in distance. Sheep tend to aggregate rather than immediately disperse when a threat is detected and therefore $W_{\pi\Lambda}$ is weighted greater than $W_{\pi\beta}$. The speed of the sheep is denoted by $S^t_{\pi_i}$. Lastly, W_{π_υ} has been included to promote a smooth simulation trajectory by preventing sharp turns and is therefore subordinate to all other forces. As a result, sheep i moves to its new position using Eq. 3.6.

$$P^{t+1}_{\pi_i} = P^t_{\pi_i} + S^t_{\pi_i} F^t_{\pi_i} \tag{3.6}$$

3.2.3 *Shepherd Agent Model*

The shepherd's task is to collect all sheep present in the environment and drive them to the target location. To achieve this, each shepherd is programmed with Strömbom et al.'s biologically sound switching algorithm [15]. That is, at each moment, a shepherd follows one of two behaviours, dictated by the relative position of the sheep, see Fig. 3.1. These behaviours are:

Behaviour 1 (σ_1): If all sheep are located within a distance $f(N)$ of the flock's *GCM*, the shepherd aims to position itself directly behind the flock's *GCM* in relation to the target region. This location is known as the driving position $P^t_{\beta_j\sigma_1}$.

Behaviour 2 (σ_2): If at least one sheep is further than $f(N)$ from the flock's *GCM*, it is referred to as a separated sheep. The shepherd aims to position itself directly behind the separated sheep in relation to the flock's *GCM*. This location is known as the collecting position $P^t_{\beta_j\sigma_2}$.

Additionally, if a shepherd finds itself within $3R_{\pi\pi}$ of any sheep π_i, it reduces its speed $S^t_{\beta_j}$ at time t to 0 for that time step. This is due to an observation made during Strömbom et al.'s study that showed shepherds rarely approach flocks at this range as it caused the flock to rapidly disperse [15]. To allow for the natural asymmetry of sheep flocks, the allowable separation distance is calculated as:

$$f(N) = R_{\pi\pi} N^{\frac{2}{3}} \tag{3.7}$$

Shepherds β_j are governed by basic swarm intelligence behaviour which elicits decentralised guidance and mitigates the need for communication with other shepherds. As such, each shepherd is repelled from $\beta_{y\neq j}$ other shepherds within distance $R_{\beta\beta}$, calculated by substituting β for π in Eq. (3.2), and attracted to the shepherds *LCM* within distance $3R_{\beta\beta}$, calculated by substituting β for π in

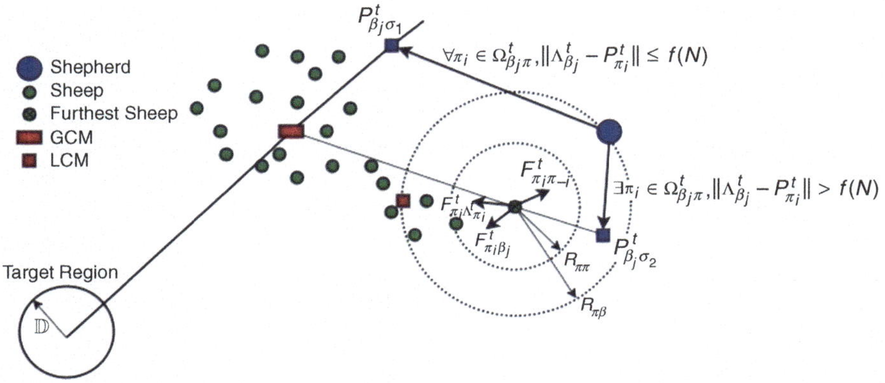

Fig. 3.1 The multi-shepherd swarm system model's governing rules, positions, influence distances, and key interaction forces. Modified from [15]

Eq. (3.3). When these *SI* forces are combined with either of the two shepherding behaviour forces $F^t_{\beta_j\sigma_1\sigma_2}$, the resultant force for shepherd β_j's is shown in Eq. 3.8.

$$F^t_{\beta_j} = F^t_{\beta_j\sigma_1\sigma_2} + W_{e\beta_j} F^t_{\beta_j\epsilon} \tag{3.8}$$

A shepherd's new location is determined by its current position, $P^t_{\beta_j}$ and the force vector calculated by (3.8), weighted by the speed factor $S^t_{\beta_j}$ as shown in Eq. 3.9.

$$P^{t+1}_{\beta_j} = P^t_{\beta_j} + S^t_{\beta_j} F^t_{\beta_j} \tag{3.9}$$

The multi-shepherd swarm system model is depicted in Fig. 3.1. It shows the system's governing rules, key interaction forces, positions, and influence distances. As we assume no communication between shepherds, each shepherd's behaviour is influenced only by the sheep's positions and the positions of the other shepherds.

3.2.4 Swarm Guidance Algorithm Design

We presented the geometric and vector descriptions of the system, which allows for mapping of the multi-shepherd swarm kinematics to a quasi-arc-formation. It represents the synergising of shepherding with a swarm intelligence behaviour in the form of formation control.

The goal, as stated in problem 3.2, is to develop a guidance scheme to promote the shepherds to position themselves in a more intelligent manner to improve the overall system performance. To achieve this, we propose a geometric transformation about the sheep *GCM* in relation to the target region at a radius $R_{\pi\beta}$ from the *GCM* as shown in Fig. 3.2. Given the multi-shepherd swarm's desired position as a result of solving Problem 3.1, the quasi-arc-formation position for each shepherd can be calculated along the circumference of a partly enclosing circle. These formation positions require a weighted force vector in the shepherd's next step direction. To design the additional force vector, consider the new desired position of each shepherd as $P^{*t}_{\beta_j}$, on the circular radius $R_{\pi\beta}$, with equal spacing Δ_j, from $0 \rightarrow \pi$. Let the original environment axis be $\emptyset$, and the rotated axis be $\emptyset^*$, and the rotation angle be $\Delta\emptyset$. The original shepherd positions can be given by

$$x = R_{\pi\beta}\cos(\emptyset) \qquad y = R_{\pi\beta}\sin(\emptyset) \tag{3.10}$$

By applying the rotation matrix to the original axis, the resultant shepherd positions are given by

$$\begin{bmatrix} x^* \\ y^* \end{bmatrix} = \begin{bmatrix} \cos(\Delta\emptyset) & -\sin(\Delta\emptyset) \\ \sin(\Delta\emptyset) & \cos(\Delta\emptyset) \end{bmatrix} \begin{bmatrix} x \\ y \end{bmatrix} \tag{3.11}$$

With the new rotated axis and shepherd position identifiers transposed, the equal spacing Δ_j can be calculated. To obtain an even Δ_j, θ is calculated by

$$\theta_j = \frac{\pi}{M+1} \tag{3.12}$$

Therefore, the shepherd's new desired position or formation aware desired position $P^{*t+1}_{\beta_j}$ becomes its calculated swarm position $P^{*t}_{\beta_j}$, plus a small weighted

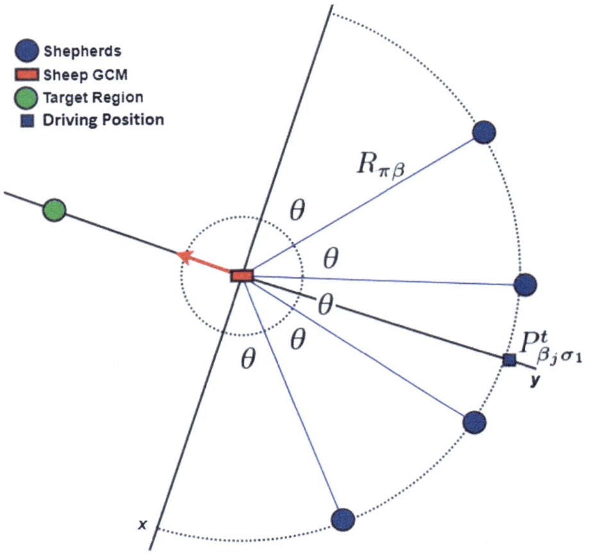

Fig. 3.2 The rotated axis about the sheep *GCM* in relation to the target region from 0 to π and the desired quasi-arc shepherd driving positions

quasi-arc-formation direction, $F^{*t}_{\beta_j}$:

$$P^{*t+1}_{\beta_j} = P^{*t}_{\beta_j} + S^{t}_{\beta_j} F^{*t}_{\beta_j} \tag{3.13}$$

The multi-shepherd swarm plus quasi-arc-formation stimulus shepherding guidance algorithm can then be described by Algorithm 1. All shepherds will attempt to move towards the collecting position if required. This is due to the lack of explicit shepherd-to-shepherd communication, but could be fixed easily by adding a repulsion force among the shepherds.

Algorithm 1 Multi-shepherd swarm plus quasi-arc-formation

```
while sheep GCM not within target region do
    if shepherd j's distance to any other shepherd is < threshold then
        Repel in opposite direction
    end if
    if shepherd j's distance to any sheep is < threshold then
        Stop moving
    else
        if all sheep are within the cohesive flock threshold then
            Move towards driving position
            if closest arc-position vacant then
                Move towards it*
            else
                Head towards next arc-position*
            end if
        else
            Move towards the collecting position
        end if
    end if
end while
* Additional algorithm steps to promote intelligent positioning
```

3.2.5 *Experimental Design*

Experiments were conducted in Matlab R2016b simulation environment using the developed constant speed model outlined above. To achieve baseline results, we first simulated the single shepherd scenario, $M = 1$, herding a small flock of sheep, $N = 50$, over thirty trials. The two measures of performance were task completion times and sheep distances from their *GCM*. The flock size incrementally increased from $N = \{50$, to 100, 150, and lastly to $200\}$ with varying degrees of density. These flock sizes are termed: small; small-medium; medium-large; and large, respectively. As a means to progressively reduce the density of the sheep and promote a greater initial dispersion from their *GCM*, a saved, repeatable pseudo-random sheep position initiator for each sheep π_i was used. The flocks were initiated in five initial *cluster configurations*: one tight cluster; two clusters; five clusters; ten clusters; and finally, dispersed over the entire paddock (termed the whole cluster), as seen in Fig. 3.3. The maximum number of simulation steps increased approximately linearly with the number of sheep in the flock, and was calculated as $T_{sim} = 20N + 630$. This formula follows Strömbom et al.'s approach [15].

Following the baseline simulations, the number of shepherds in the multi-shepherd swarm is scaled to $M = \{2, 4, 5, 10, 20\}$. For each group-size of shepherds in the swarm, thirty simulation trials were conducted for each combination of flock size and density, replicating those used for the single shepherd scenario. Furthermore, we investigate the effects of including a small stimulus to promote a quasi-semi-circular arc-formation behind the flock while the multi-shepherd swarm is in a *driving* mode.

3.3 Simulation Results

3.3.1 Herding with a Single Shepherd

The baseline simulations show the case of $M = 1$ shepherd and $N = \{50, 100, 150, 200\}$ sheep, dispersed as one, two, five, ten, and whole clusters around the environment achieves a combined mean successful task completion rate

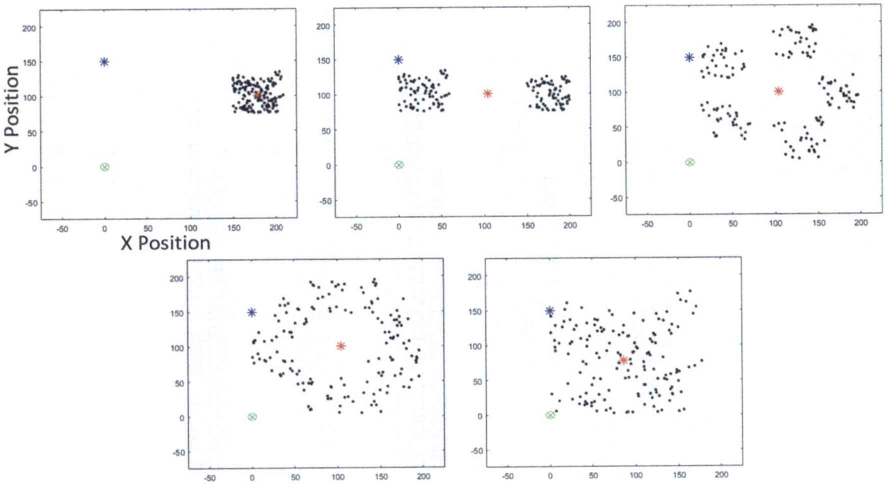

Fig. 3.3 The five simulated initial sheep cluster variations: one tight cluster (top-left); two clusters (top-center); five clusters (top-right); ten clusters (bottom-left); and, one dispersed cluster (bottom-right). Shown within the environment are the shepherd ($*$), sheep ($\bullet$), sheep *GCM* ($*$), and the target region ($*$) with a flock size of $N = 150$

of 71.5%, as seen in Fig. 3.4, and a mean time to completion of 2250 steps (71.9% of the maximum allowed), shown in Fig. 3.5. Additionally, the mean number of sheep separation occurrences per simulation was 1777. The mean distance of the furthest sheep from the *GCM* to the *GCM* increased approximately linearly from 50 units in the case of a small flock size to 70 units in the case of a larger flock size, as shown in Fig. 3.6.

There was little to no change in task completion times between herding a small flock and herding a small-medium flock, at 1200 and 1240 steps, respectively (Fig. 3.5). However, a dramatic increase in completion times is evident when the flock size increases beyond medium-large, with the simulation taking an average of 2450 and 4200 steps, respectively, as shown in Fig. 3.7.

Figure 3.8 (L) illustrates a simulation dynamic movement plot of $M = 1$ and $N = 100$. It shows the shepherd's trajectory in blue, sheep's *GCM* in red, sheep i's trajectory in grey, and the target region denoted by the green circle.

3.3.2 *Herding with a Multi-Shepherd Swarm*

For the multi-shepherd swarm case of $M = \{2, 4, 5, 10, 20\}$ shepherds and all combinations of flock sizes and dispersion densities, the simulation achieves a combined mean successful task completion rate of 88.9% as shown in Fig. 3.4, and a mean time to completion of 1181 steps (just 37.7% of the maximum allowed) as shown in Fig. 3.5. The mean distance of the furthest sheep from the *GCM* to the

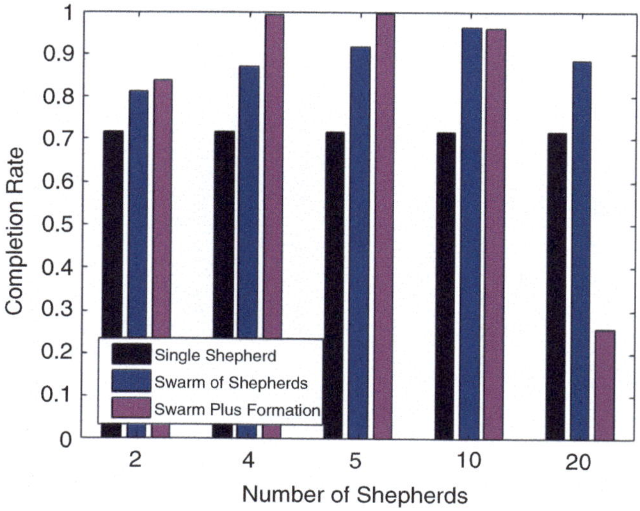

Fig. 3.4 The mean successful completion rates of the single shepherd, multi-shepherd swarm, and multi-shepherd swarm plus quasi-arc formation

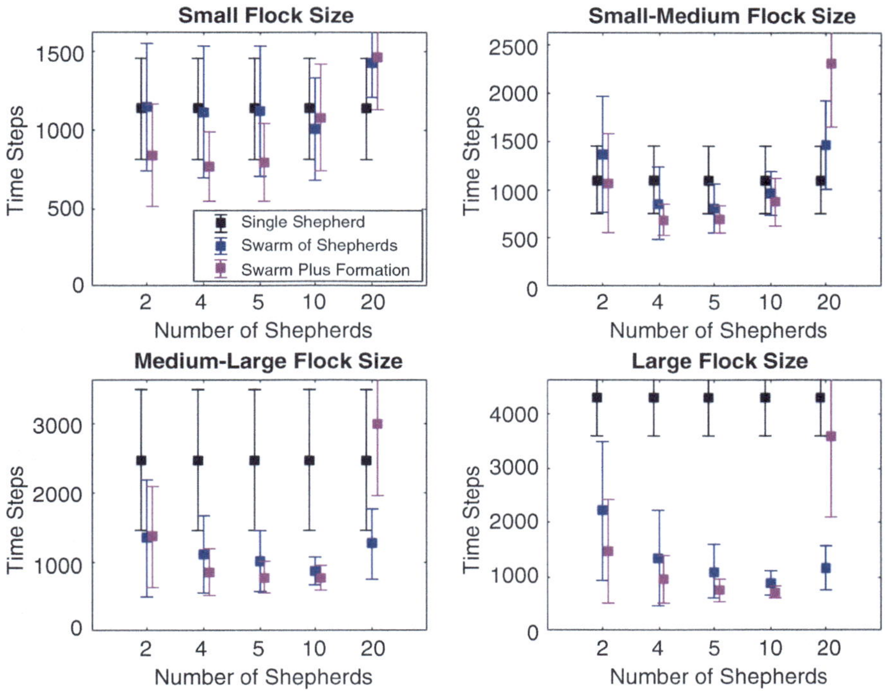

Fig. 3.5 The mean simulation time steps (including standard deviation) for task completion achieved for the single shepherd, multi-shepherd swarm, and multi-shepherd swarm plus quasi-arc formation

GCM reduced to 38 units in the case of a small flock size and 66 units in the case of the larger flock size as shown in Fig. 3.6.

The average task completion times remained relatively consistent for all flock sizes, with 1150 time steps for the small flock size, 1130 for the small-medium flock, 1100 for the medium-large flock, and 1360 for the large flock as shown in Fig. 3.5. Additionally, the recorded mean number of sheep separation occurrences per simulation was 1178. Figure 3.8 (C) illustrates a simulation dynamic movement plot of $M = 4$ and $N = 100$. It shows the shepherd's trajectory in blue, sheep's *GCM* in red, sheep i's trajectory in grey, and the target region denoted by the green circle. The resultant trajectories are smoother and more direct with a consistently tighter flock when compared to that of the single shepherd system.

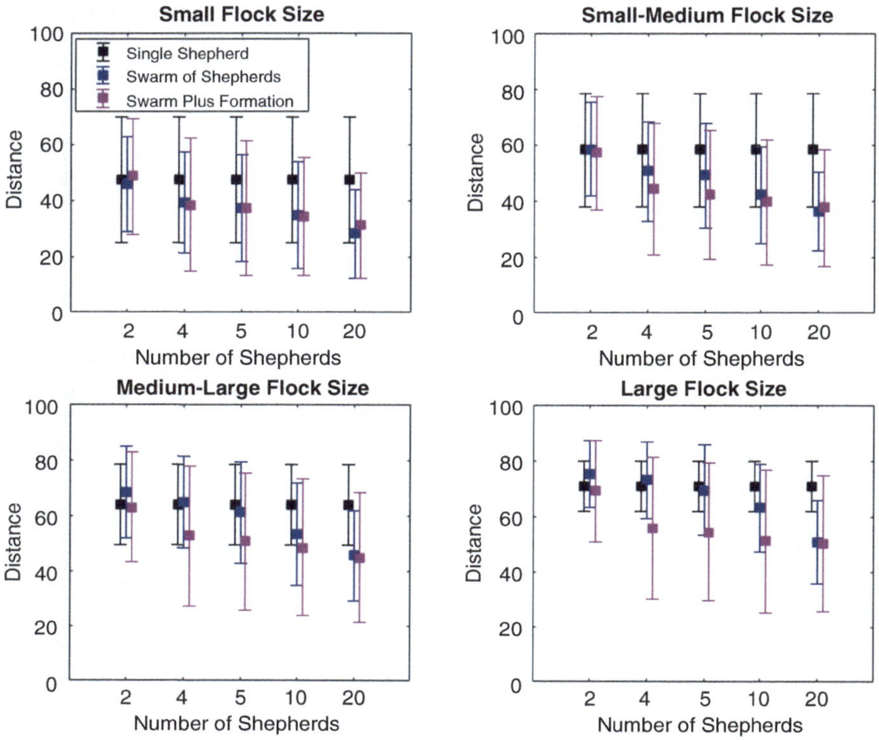

Fig. 3.6 The mean distance of the furthest sheep from the *GCM* to the *GCM* (including standard deviation) over the duration of the task for the single shepherd, multi-shepherd swarm, and multi-shepherd swarm plus quasi-arc formation

3.3.3 *Herding with a Multi-Shepherd Swarm Plus Formation*

For the multi-shepherd swarm plus formation stimulus case of $M = \{2, 4, 5, 10, 20\}$ shepherds and all combinations of flock size and dispersion densities, the simulation achieves a combined mean successful task completion rate of 80.9%, Fig. 3.4, and mean time to completion of 1181 steps (39.7% of the maximum allowed), Fig. 3.5. Additionally, the recorded mean number of sheep separation occurrences per simulation was 954. The mean distance of the furthest sheep from the *GCM* to the *GCM* has reduced further to 36 units for the small flock size and 58 units for the large flock size, respectively, as shown in Fig. 3.6. The average task completion times are further reduced with the addition of the formation stimulus, down to 990 steps, 1060 steps, 1090 steps, and 1200 steps for small to large flock sizes, respectively, as shown in Fig. 3.5. It should be noted that, within Fig. 3.8, the inclusion of formation guidance maintains a greater shepherd-to-shepherd separation over time compared to the multi-shepherd case without formation.

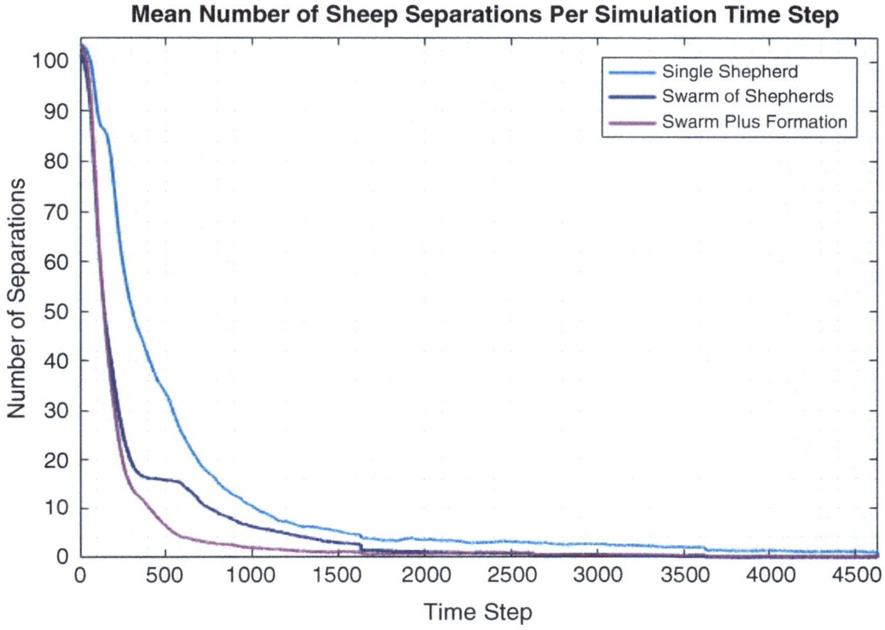

Fig. 3.7 The exponential decay relationship between the number of sheep separations from the flock as a function of simulation time step

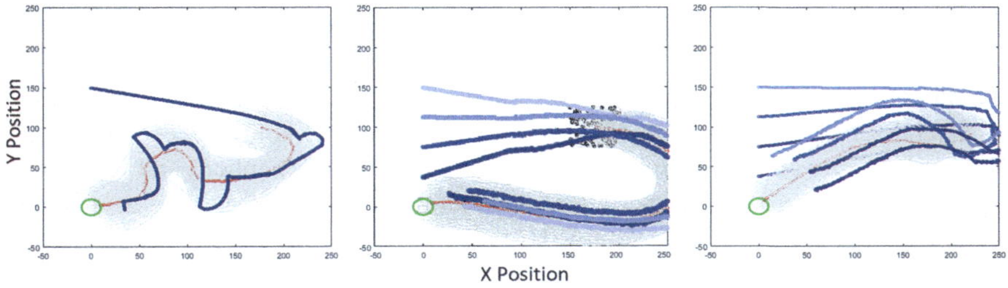

Fig. 3.8 From left to right: Dynamic step plot of the single shepherd, multi-shepherd swarm (4 shepherds), and multi-shepherd swarm task with formation stimulus (4 shepherds) denoted in shades of blue, with small-medium flock size of 100 sheep (denoted in grey). The sheep *GCM* highlighted in red and the goal region by a green circle

3.3.4 Analysis of the Shepherding Task as a Function of Guidance Scheme

To complete the shepherding task, the shepherd starts by firstly collecting the sheep into a cohesive flock prior to driving them toward the target region. The shepherd's constant influence against the centre of the flock causes elongation of the flock, resulting in sheep separations. At separation, the shepherd switches from a driving mode to a collecting mode to re-establish a cohesive flock prior to switching back to a driving mode. This process continued until the sheep *GCM* was within the target region and all sheep were considered *collected*, signaling the completion of the task and thereby confirming the effectiveness of the shepherding algorithm, Fig. 3.8. The shepherd's performance in *time to completion* and *flock cohesiveness* were tested for each combination of flock size and density under pseudo-random initial sheep conditions over thirty trials for each configuration. The measured mean simulation time and sheep distance from their *GCM* are shown in Figs. 3.5 and 3.7, respectively. The results show a marked improvement following the inclusion of multiple shepherds programmed with *SI* showing an increase in task completion rates by 18%. This may be attributed to reduction of up to 33% in the number of sheep separated from the flock, which reduced the mean distance of the furthest sheep from the *GCM* by approximately 15%. These two factors combined resulted in a more compact and cohesive flock.

Further improvements were made to these metrics with the additional quasi-arc-formation stimulus by way of two additional steps in the guidance algorithm. Interestingly, the results show a dramatic decrease in performance when the number of shepherds in the swarm was increased to 20. The task completion rate drops to just 20% and an increase mean time to completion for flock sizes of small to medium-large of up to 200%. The decrease in performance is attributed to the high volume of shepherds in a limited space, resulting in obscured trajectories. We postulate that a large and well spread out shepherd formation causes problems when there are many shepherds attempting to collect a single wayward sheep. The stalling effect of the sheep-shepherd separation distance has likely resulted in a large number of shepherds unduly impacting the flock formation, while they all attempt to collect the wayward sheep. Without formation control (or with a smaller number of sheepdogs in the shepherd swarm), there is a lower likelihood of the sheep-shepherd stalling effect occurring. With a larger number of shepherd and formation control, the shepherds exert undue influence on the flock shape as soon as a single sheep starts to wander. By rejecting the results obtained using 20 shepherds, the improvement in performance is as hypothesised. The task completion rate increased to 94.3%, 24% above that achieved from the single shepherd and 6% above the no formation stimulus system. The number of sheep separations was reduced by approximately 60%, which attains a 32% reduction in the mean distance of the furthest sheep from the *GCM* to the *GCM*.

3.4 Conclusions

In summary, through synergising shepherding and a simple formation control swarming behaviour, we have developed a multi-shepherd guidance algorithm that outperforms the single shepherd in task completion rates and the time required to complete the task while maintaining a more compact and cohesive flock. The results obtained throughout our experimental study show that there is no linear relationship between the number of shepherds and task performance. Several combinations of sheep flock size and density were tested, from which we have identified conditions that result in both desired and undesired performance outcomes. Using the proposed guidance scheme, we achieved a high increase in performance in the shepherding task for the use of between and including 4 to 10 shepherds. However, an increase from ten to twenty shepherds resulted in a dramatic decrease in shepherding performance. Investigation into this performance decrease and identification of alternative approaches to rectify it are earmarked as areas for future work.

References

1. ATSB: A safety analysis of remotely piloted aircraft systems. Technical Report, Australian Transport Safety Bureau (2017). https://www.atsb.gov.au/media/5772450/ar-2017-016-final.pdf
2. Cameron, S., Probert, P.: Advanced Guided Vehicles: Aspects of the Oxford AGV Project, vol. 9. World Scientific, Singapore (1994)
3. Deakin, S.: Wise men and shepherds: a case for taking non-lethal action against civilians who discover hiding soldiers. J. Military Ethics **10**(2), 110–119 (2011)
4. Dudek, G., Jenkin, M., Milios, E., Wilkes, D.: A taxonomy for swarm robots. In: Proceedings of the 1993 IEEE/RSJ International Conference on Intelligent Robots and Systems' 93, IROS'93, vol. 1, pp. 441–447. IEEE, Piscataway (1993)
5. Fingas, M.: The Basics of Oil Spill Cleanup. CRC Press, Boca Raton (2012)
6. Gade, S., Paranjape, A.A., Chung, S.J.: Herding a flock of birds approaching an airport using an unmanned aerial vehicle. In: AIAA Guidance, Navigation, and Control Conference. AIAA, Kissimmee (2015)
7. Hamilton, W.D.: Geometry for the selfish herd. J. Theoret. Biol. **31**(2), 295–311 (1971)
8. Hughes, R.L.: A continuum theory for the flow of pedestrians. Transp. Res. Part B Methodol. **36**(6), 507–535 (2002)
9. Hughes, R.L.: The flow of human crowds. Ann. Rev. Fluid Mech. **35**(1), 169–182 (2003)
10. Kim, D.H., Han, S.H.: Robust self-organization for swarm robots. In: International Conference on Control, Automation and Systems, 2008. ICCAS 2008, pp. 1228–1233. IEEE, Piscataway (2008)
11. Lakshika, E., Barlow, M., Easton, A.: Co-evolving semi-competitive interactions of sheepdog herding behaviors utilizing a simple rule-based multi agent framework. In: 2013 IEEE Symposium on Artificial Life (ALIFE), pp. 82–89. IEEE, Piscataway (2013)
12. Lien, J.-M., Rodriguez, S., Malric, J., Amato, N.M.: Shepherding behaviors with multiple shepherds. In: Proceedings of the 2005 IEEE International Conference on Robotics and Automation, pp. 3402–3407 (2005). https://doi.org/10.1109/ROBOT.2005.1570636
13. Safe Work Australia: Summary of quad bike fatalities. Canberra: Safe Work Australia (2013). https://www.safeworkaustralia.gov.au/quad-bike-fatality-data#2011
14. Safe Work Australia: Work-related injuries and fatalities on Australian farms. Canberra: Safe Work Australia (2013). https://www.safeworkaustralia.gov.au/system/files/documents/1702/work-related-injuries-fatalities-farms.pdf
15. Strömbom, D., Mann, R.P., Wilson, A.M., Hailes, S., Morton, A.J., Sumpter, D.J.T., King, A.J.: Solving the shepherding problem: heuristics for herding autonomous, interacting agents. J. R. Soc. Interf. **11**(100) (2014). https://browzine.com/articles/52614503
16. Vaughan, R., Sumpter, N., Henderson, J., Frost, A., Cameron, S.: Experiments in automatic flock control. Rob. Auton. Syst. **31**(1), 109–117 (2000)

Chapter 4
The Influence of Stall Distance on Effective Shepherding of a Swarm

Anthony Perry

Abstract In the shepherding problem, an external agent (the shepherd) attempts to influence the behaviour of a swarm of agents (the sheep) by steering them towards a goal that is known to the shepherd but not the sheep. This chapter outlines some of the dynamics inside a shepherding task known as herding, focusing on what we call the stall distance. We describe the stall distance as the minimum distance that the shepherd must maintain from all members of the swarm while carrying out the primary task of herding the swarm to a particular location. This chapter shows how herding performance is influenced by setting the value for stall distance to an appropriate level in a herding model. A connection is made between the concept of stall distance for shepherds herding sheep agents in the model and real-world shepherds herding a group of sheep.

Keywords Human–Swarm interaction · Influence · Shepherding · Swarm robotics

In the shepherding problem, an external agent (the shepherd) attempts to influence the behaviour of a swarm of agents (the sheep) by steering them towards a goal that is known to the shepherd but not the sheep. This chapter outlines some of the dynamics inside a shepherding task known as herding, focusing on what we call the stall distance. We describe the stall distance as the minimum distance that the shepherd must maintain from all members of the swarm while carrying out the primary task of herding the swarm to a particular location. This chapter shows how herding performance is influenced by setting the value for stall distance to an

A. Perry (✉)
Advanced Vehicle Systems, Land Vehicles and Systems, Land Division, Defence Science and Technology Group, Edinburgh, SA, Australia
e-mail: Anthony.perry@dst.defence.gov.au

H. A. Abbass, R. A. Hunjet (eds.), *Shepherding UxVs for Human-Swarm Teaming*, Unmanned System Technologies,
https://doi.org/10.1007/978-3-030-60898-9_4

appropriate level in a herding model. A connection is made between the concept of stall distance for shepherds herding sheep agents in the model and real-world shepherds herding a group of sheep.

4.1 Introduction

The shepherding problem for multi-agent systems has been addressed by a number of authors with a great deal of variation in models for both the shepherds and sheep. For models of sheep, the most common approach is BOIDS [1, 2, 7, 8, 11, 12]. This approach sees the sheep as a BOIDS swarm with the three classic rules of separation, cohesion and alignment.

However, the BOIDS model does not describe the behaviour of the sheep once they reach a destination, leaving the interpretation/choice of this behaviour up to the individual researchers. This results in variations in successful shepherding behaviour between approaches. For example, Lien et al. [4] assume that once individual sheep enter the target area, they will not leave. This allows for behaviours where shepherds can steer small groups or individuals to the destination and then leave to collect another group without worrying about dispersion of previously collected groups. In contrast in Lee and Kim [3], there is no such assumption resulting in shepherds being required to stay with sheep that have entered the target area or risk them wandering out.

The environments modelled by the authors vary greatly between different studies. Some works model obstacles in the environment [3, 6], while others do not [2, 12]. The representation of target areas also varies, from points [13], to circular areas [9] or simply the corner of a 'paddock' [12].

The complexity of the problem influences the development of the shepherding behaviour. Linder and Nye [5] investigated the implications of the problem design on the utility of the shepherding behaviour solution. They concluded that oversimplification of the problem, especially in learning systems, can result in shepherding behaviours that have a little generalisation value.

In this chapter, we explore a dimension of shepherding called stall distance, namely the closest distance that a shepherd may move towards a swarm or flock member before it must reduce its velocity to zero. A sheepdog moves a flock by changing its position relative to the members of the flock; however, if it comes too close to any single member of the flock, it reaches a point where it must stop in order to prevent its position generating an inappropriate influence (which would cause flock members to scatter). We call this reduction to zero velocity under these circumstances, 'stalling', and the distance where velocity is reduced to zero the 'stalling distance'. We hypothesise that it has a profound impact on task success and task performance. In the real world (biological shepherding), a shepherd that attempts to herd sheep by moving very close might be able to generate a strong repulsive effect but has risk of scattering the flock. Likewise, a shepherd that attempts to herd sheep while keeping at a greater distance might have less risk of scattering the flock, but might generate less repulsive effect, taking more time to achieve its task. The remainder of the chapter is organised as follows. We start with a literature review of the shepherding to define key terms in shepherding, showcase the variety of approaches used and establish the research gap that the current chapter addresses in Sect. 4.2. I present the simulation model and introduce the challenge of setting the stall distance and associated stall factor for the shepherd in Sect. 4.3. Experimental design is presented in Sect. 4.4 with the results in Sect. 4.5. Conclusions are drawn, and potential future extensions to the shepherding model are in Sect. 4.6.

4.2 Background

Shepherding literature in the multi-agent system domain is limited to a dozen or so papers, with the work diverse due to the multi-faceted nature of simultaneous control of multiple agents with one or a small number of operators/shepherds.

Most shepherding models in the literature employ a single shepherd to control the sheep swarm or flock. Scripted approaches to the shepherding tasks are generally formulated around a single shepherd, either as a stepping stone to multi shepherd systems [2, 4] or due to the complexity of real-world robotic systems as described in [11]. In the pre-programmed shepherding literature, the goal is in defining and demonstrating optimal shepherding behaviour considering metrics such as task completion time, task completion rate, distances covered and spread of the sheep flock or swarm.

Lee and Kim [3] and Lien et al. [4] describe herding, covering, patrolling and collecting as common types of shepherding behaviours. Lien et al. [4] (pp. 4 and 5) describe these behaviours as follows:

- In **herding** tasks, the 'shepherd needs to move all the flock members from a start region to a goal region'.
- In **covering** tasks, the 'shepherd guides a flock to the areas of the environment that have not been visited'.
- In **patrolling** tasks, the 'shepherd needs to guard a designated region called the forbidden area or FA. Once the intrusion of the flock is discovered, the shepherd will chase the flock until it vacates the FA'.
- In **collecting** tasks, the shepherd 'gathers initially scattered flock members into a designated region'.

Strömbom et al. describe algorithms for herding behaviours 'based on adaptive switching between collecting the agents when they are too dispersed and driving them once they are aggregated' [12] (p. 4). These two modes are being described as

- the **driving mode**, where the shepherd 'is directly behind or in front of the flock relative to the target';

- the **collecting mode**, where the shepherd 'is directly behind, or on the opposite side of the flock, relative to the furthest agent'.

When comparing the herding behaviours of Strömbom et al. [12] with those described by Lee and Kim and Lien et al. [3, 4], we agree that Lee and Kim and Lien et al.'s **herding** task might be equivalent to a combination of both **collecting** and **driving** from Strömbom's work.

The literature has rich models to describe shepherding tasks including driving straight at individual sheep [8], towards the centre of mass of the flock [12], by moving side to side [4], or by moving in a V-formation [1, 2]. Collecting appears to be much more limited in terms of available strategies, with behaviour dictated by the ratio of shepherds to sheep [3].

Fujioka et al. [1, 2] and Singh et al. [10] use the model based on Strömbom et al. [12] with herding composed of both collecting and driving. These authors set a driving point based on the centre of mass of the flock and the location of the goal; similarly, they set a collecting point based on the size of the flock and the location of the sheep furthest from the flock's centre of mass. The shepherd heads towards the driving point when carrying out a driving task and a collecting point when carrying out a collecting task. Strömbom et al. [12] require that the shepherd changes from a velocity of 1.5–0 m/s if it comes too close to any single sheep. This stalling distance (distance at which the shepherd's velocity must be set to zero) is empirically set to three times that of the agent-to-agent activation distance, 6 m in their work. Fujioka et al. [1, 2] take a slightly different approach, using the same distance of 6 m, but rather than switching the shepherd velocity from 1.5 m/s down to 0 m/s, the velocity is reduced to 0.6 m/s. This means that for Singh et al. [10] and Strömbom et al. [12], the shepherd may move towards a driving point or collecting point as necessary, but never actually reach it due to the stalling effect triggered by a single sheep being too close to the intended trajectory of the shepherd, unless a small random force is added to the shepherd velocity vector that allows it to escape this state.

The anticipated interaction of the stalling behaviour and driving or collecting behaviour implies that within Strömbom's model, the shepherd does not actually drive or collect from the locations defined by the driving point or the collecting point, rather these are points that define the direction of travel for the shepherd's movement (i.e. the shepherd moves towards these points but will not reach them). While not actually reaching the collecting or driving point might not necessarily result in failure of the task, we hypothesise that it may have an effect on task performance.

4.2.1 Driving Interactions

When carrying out a driving task as shown in Fig. 4.1, the shepherd (black diamond) is able to move towards the driving point (red diamond) to drive the flock of sheep

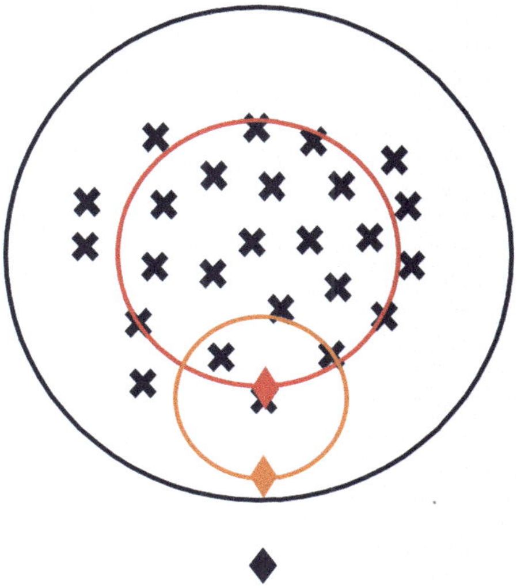

Fig. 4.1 Driving mode (part 1): the shepherd (black diamond) is able to move towards the driving point (red diamond) to drive the flock of sheep (black crosses) towards the top of the page because they are sufficiently clustered within the black circle. The shepherd will stall at the stalling radius (orange circle) due to the proximity of the nearest sheep

(black crosses) towards the top of the page. The shepherd is driving the flock because the sheep are sufficiently clustered within the black circle. The size of this black circle is a function of the flock size and centred on the flock's centre of mass. The shepherd would be unable to reach the driving point because it would stall at the stalling radius (orange circle) due to the proximity of the nearest sheep. The influence of the shepherd on the flock of the sheep is likely to produce sheep movement in a useful direction.

Continuing the driving task in Fig. 4.2, the shepherd (black diamond) is unable to move any further towards the driving point (red diamond) as it has reached the stalling radius (orange circle). As the sheep move away from the stalled shepherd, the shepherd will cease to be stalled, heading towards the new driving point, or collection point as appropriate. Even from this stalled location, the shepherd's position is still sufficient to continue the driving task, driving the flock upwards page. If a sheep exceeds a threshold distance from the flock's centre of mass (i.e. begins to stray too far from the flock), as shown by the bottom left sheep in the figure, the shepherd will change its behaviour to collecting.

4.2.2 Collecting Interactions

The collecting task is shown in Fig. 4.3 where the wandering sheep (black cross) in the bottom left corner has exceeded the maximum allowable radius (black circle) from the flock's centre of mass triggering the collecting behaviour of the shepherd. The shepherd (black diamond) will move left towards the collecting point (yellow

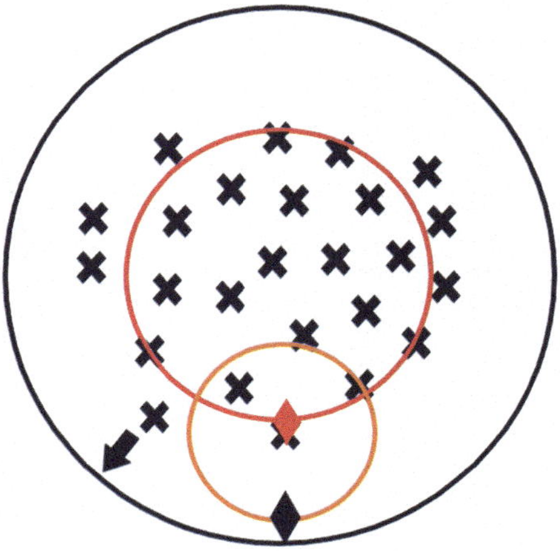

Fig. 4.2 Driving mode (part 2): the shepherd (black diamond) is unable to move any further towards the driving point (red diamond) as it has reached the stalling radius (orange circle). Note that one sheep is moving towards the bottom left, and this will trigger a change in behaviour from the shepherd if it exceeds the threshold distance from the centre of mass defined by the black circle

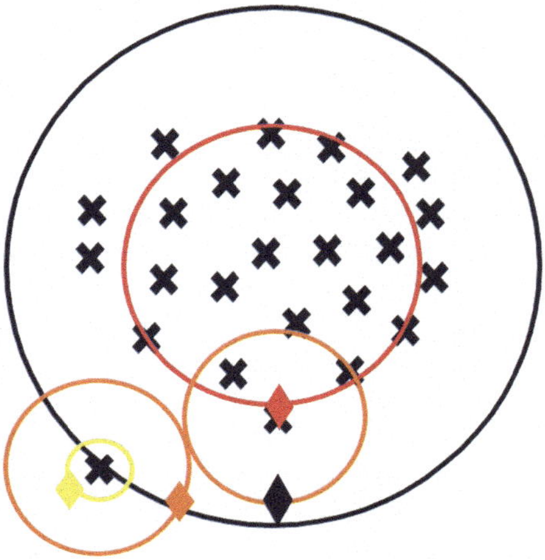

Fig. 4.3 Collecting mode: the sheep (black cross) in the bottom left corner has exceeded the maximum allowable radius (black circle) from the flock's centre of mass triggering the collecting behaviour of the shepherd. The shepherd (black diamond) will move left towards the collecting point (yellow diamond). The shepherd would stall at the stalling radius (lower left orange circle) due to the proximity of the nearest sheep

diamond). The shepherd will not reach this collection point but instead would stall at the stalling radius (lower left orange circle) due to the proximity of the nearest sheep. Again, as the sheep move, the stalled state of the shepherd will cease.

It is arguable that in the situation shown in Fig. 4.3, the collecting task might not be carried out effectively as the repulsive action of the shepherd on the wandering sheep is not directed towards the centre of the flock. If the stalling radius were smaller, then the shepherd would move much closer to this wandering sheep, generating a repulsive force more appropriately directed.

4.3 Methodology

In this section, we first describe the Strömbom et al.'s [12] model for herding and then explore the implications of the effect of the stall distance on the shepherd's behaviour. While the model itself is from Strömbom et al. [12], the notation used here was first described by Singh et al. [10].

We denote the set of sheep agents with $\Pi = \{\pi_1, \ldots, \pi_i, \ldots, \pi_N\}$, where the letters Π and π are chosen as the first character of the Greek word for sheep $\Pi\rho\acute{o}\beta\alpha\tau o$ and denote the set of shepherd agents with $B = \{\beta_1, \ldots, \beta_j, \ldots, \beta_M\}$, where the letters B and β are chosen as the first character of the Greek word for shepherds $B o\sigma\kappa\acute{o}\varsigma$. We denote the set of behaviours in the simulation with $\Sigma = \{\sigma_1, \ldots, \sigma_K\}$, where the letters Σ and σ are chosen as the first character of the Greek word for behaviour $\sigma\upsilon\mu\pi\epsilon\rho\iota\varphi o\rho\acute{\alpha}$.

Agents are initialised in a square area. We use u to denote the unit, where u is a metre in Strömbom et al.'s [12] original model but could equally generalise to other units per the application. The agents adopt different behaviours as described below:

1. Shepherd's β_j driving behaviour: When the sheep are clustered in one group, the shepherd drives the sheep towards the goal by moving towards a driving point that is situated behind the sheep on the ray between the sheep's centre of mass and the goal. The shepherd moves towards the driving point with normalised force vector, $F^t_{\beta_j cd}$.
2. Shepherd's β_j collecting behaviour: If one of the sheep is further away from the sheep's centre of mass than a nominated threshold, the shepherd moves in a straight line to a collection point behind this sheep to move it towards the centre of mass of the herd, in other words, to collect it. The shepherd moves to the collection point with a normalised force vector $F^t_{\beta_j cd}$. A visualisation of driving and collecting behaviour discussed above can be found in [12].
3. Shepherd β_j adds a random force, $F^t_{\beta_j \epsilon}$, at each time step to help resolving deadlocks. The strength of this angular noise is denoted by $W_{e\beta_j}$.
4. Shepherd's β_j total force $F^t_{\beta_j}$ is then calculated as

$$F^t_{\beta_j} = F^t_{\beta_j cd} + W_{e\beta_j} F^t_{\beta_j \epsilon}. \tag{4.1}$$

5. Sheep π_i repulses from β_j using a force $F^t_{\pi_i \beta}$.
6. Sheep π_i repulses from other sheep $\pi_{i1}, i1 \neq i$ using a force $F^t_{\pi_i \pi_{i1}}$.
7. Sheep π_i is attracted to the centre of mass of its neighbors $\Lambda^t_{\pi_i}$ using a force $F^t_{\pi_i \Lambda^t_{\pi_i}}$.
8. Sheep's π_i angular noise uses a force $F^t_{\pi_i \epsilon}$.
9. Sheep's π_i total force is calculated as

$$F^t_{\pi_i} = W_{\pi_\upsilon} F^{t-1}_{\pi_i} + W_{\pi\Lambda} F^t_{\pi_i \Lambda^t_{\pi_i}} + W_{\pi\beta} F^t_{\pi_i \beta_j} + W_{\pi\pi} F^t_{\pi_i \pi_{-i}} + W_{e\pi_i} F^t_{\pi_i \epsilon}, \tag{4.2}$$

where each W representing the weight of the corresponding force vector.

The total force of each agent is used to update the agent's position as depicted in Eqs. 4.3 and 4.4.

In Strömbom's standard model, if there is a sheep within three times the sheep-to-sheep interaction radius, the agent will stop; thus, it will set its speed to zero, $S^t_{\beta_j} = 0$, otherwise, it will use its default speed, $S^t_{\beta_j} = S_{\beta_j}$. The speed of a sheep is assumed constant; that is, $S^t_{\pi_i} = S_{\pi_i}$

$$P^{t+1}_{\pi_i} = P^t_{\pi_i} + S^t_{\pi_i} F^t_{\pi_i} \tag{4.3}$$

$$P^{t+1}_{\beta_j} = P^t_{\beta_j} + S^t_{\beta_j} F^t_{\beta_j}. \tag{4.4}$$

Both π and β agents in Strömbom's model move with fixed speed or are stationary. The standard model uses a fixed minimum distance between the shepherd and the nearest sheep; this chapter will be exploring the effects of varying this distance at which the speed of the shepherd is set to zero, and we call this value the *stall distance*. Through simulation, we analyse the effect of changes to the stall distance on the effectiveness and efficiency of successful herding in terms of the success rate, time taken and the distance covered by the shepherd. A genetic algorithm was used to find appropriate settings for the stall distance.

4.4 Experimental Design

The distance between the shepherd and the members of the swarm has an effect on the direction and velocity of the movement of the swarm as well as the rate at which the swarm members might spread out.

Several performance measures are used to analyse shepherd performance of the shepherd completing the task. These include the total distance travelled by the shepherd, the number of time steps and the success rate of the shepherd in completing the task. All simulations are run for a number of steps per the equation suggested by Strömbom et al. [12].

The shepherd's initial location is 50 m south of the goal. Each initial sheep cluster is initialised 75 m away from the goal, with all sheep initialised within a radius of 30 m of the centre of the cluster. The shepherd's initial location is 50 m south of the centre of the sheep cluster. The environment size is set to 150 × 150 m area with the goal located at the origin in the lower left corner. The task is assumed to be completed when all sheep are within the 20×20 m box in the lower left corner of the area.

The parameters of the Strömbom model were set to their default values as presented in the original paper. Strömbom experimentally induced that all runs will be successful if $0.53N \leq \Omega_{\pi_i\pi} \leq N$, where $\Omega_{\pi_i\pi}$ represents the number of π agents (neighborhood) a π_i agent operates on. Accordingly, we use $\Omega_{\pi_i\pi} = 0.8N$.

Two sets of experiments are carried out to analyse the effect of the stalling distance on the herding task. The first uses a genetic algorithm to determine an optimal stalling distance for the shepherd with sheep swarm sizes ranging from 10 to 120 agents. The second experiment examines the effect of the stall distance on success rate, the number of time steps for task completion, and on the total distance covered by the shepherd completing the task.

4.4.1 *Genetic Algorithm Exploration of Stall Distance*

The genetic algorithm experiment was carried out using the Global Optimisation Toolbox of Matlab 2018b. Minimisation of time steps to complete the herding task was used as the single objective. A series of runs were carried out in this phase using a range flock sizes. The smallest flock size used contained 10 agents, and each of the subsequent runs incremented the flock size by 10 up to a maximum flock size of 120. Fifty generations were used along with a population size of 150. The lower bound for the stalling distance multiplier was 1, making it equal to the sheep-to-sheep avoidance distance of 2 m. The upper bound of the stalling distance multiplier was set to 35 giving an effective stalling distance upper limit of 70 m, 5 m more than the sheep-to-shepherd avoiding distance of 65 as used by Strömbom. This was to ensure that all distances up to a maximum range that the sheep could detect the shepherd were explored.

4.4.2 *Systematic Analysis of Stall Distance*

The systematic stall distance analysis used the average of 30 runs for flock sizes ranging between 10 and 120 sheep. For each of these flock sizes, the stall distance multiplier was varied between 0 and 10. This multiplier when combined with the sheep activation distance of 2 m gave a stalling distance ranging between 0 and 20 m. The zero-metre stalling distance configuration effectively allowed the shepherd to proceed directly to a collecting point regardless of the position of any sheep, and the larger distance causes the shepherd to stall at a much greater distance from the nearest sheep when heading towards a driving point or collecting point.

The performance metrics recorded for each flock size included

- time steps for completion vs. stall distance multiplier,
- shepherd distance for task completion vs. stall distance multiplier and
- task completion success performance vs. stall distance multiplier.

4.5 Results

4.5.1 Results of Genetic Algorithm Exploration of Stall Distance

A genetic algorithm was used to determine the best setting for the stall factor using minimisation of the number of time steps as the single objective. As shown in Fig. 4.4, there is a general trend towards smaller stall factors for small flock sizes (10–30 sheep) and larger stall factors for larger flock sizes (90–120 sheep). Strömbom et al. [12] suggested that the use of a stall factor of 3 for all flock sizes produced reasonable behaviour that was well matched to field experiments with real animals; however, no claim of optimality was made for this setting.

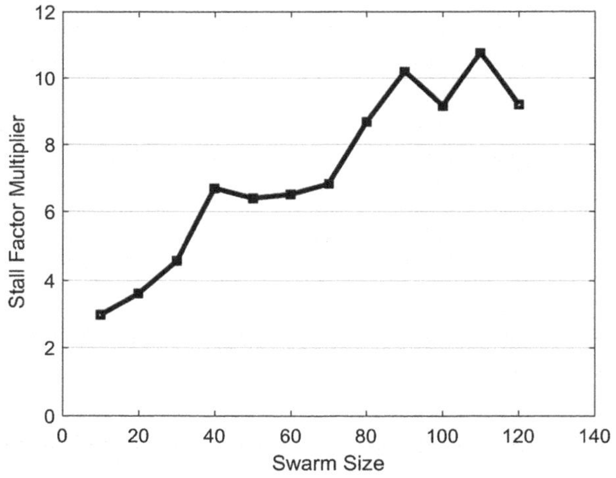

Fig. 4.4 Stall factor vs. swarm size (minimising time steps for task completion), showing lower stall factors for small swarm/flock size and generally larger stall factors for larger swarm/flock size

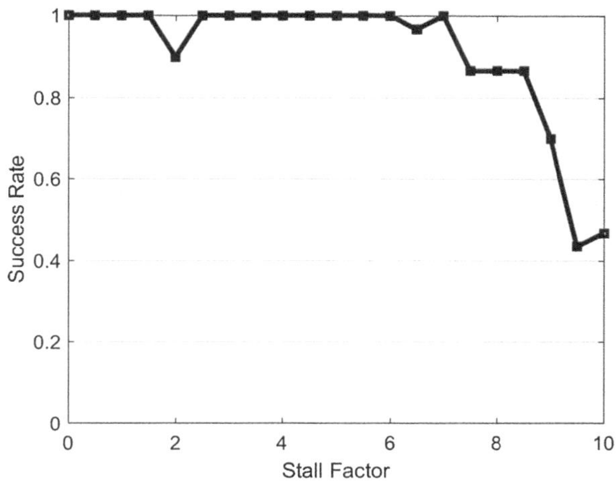

Fig. 4.5 Success rate vs. stall factor for a flock of 10 sheep, with a drop in success rate at stall factor of 2.0, 6.5 and between 7.5 and 10.0

4.5.2 Systematic Analysis of Stall Distance

4.5.2.1 Success Rates for Herding

The analysis of the success rate for the different stall factors for the small (10 sheep) flock can be seen in Fig. 4.5. While most stall factors have a success rate of 1.0, the success rate for the stall factor of 2.0 is 0.9, and for stall factors 6.5 and above, there is a downward trend in success rate. The failures at the stall factor of 2.0 occurred when the shepherd found its way to the centre of the sheep flock; the attraction force vectors of the sheep to the centre of the flock and the repulsive force vectors from the shepherd balancing out to an extent that the shepherd was not able to drive the flock successfully. The failures for the higher stall factors were different; although the shepherd occasionally reached the centre of the flock, the main contribution to failure was the high stalling factors preventing the shepherd from performing an effective collect manoeuvre.

For flock sizes of between 20 and 110 sheep, the herding remained stable with a success rate of 1.0 for all stall factors tested between 0.0 and 10.0. This indicates that herding is likely to be reasonably stable for a wide range of stall factors for these herds or swarm sizes. When the flock size was increased to 120 sheep, the success factor was generally 1.0, only dropping to 0.97 for stall factors of both 7.5 and 8.5. Once again, these herding failures were brought about by the high stalling factor occasionally preventing the shepherd from being able to perform an effective collect manoeuvre; the large stall distance prevented the shepherd from reaching the collect point and, in turn, prevented the application of an appropriate repulsive effect onto wandering sheep.

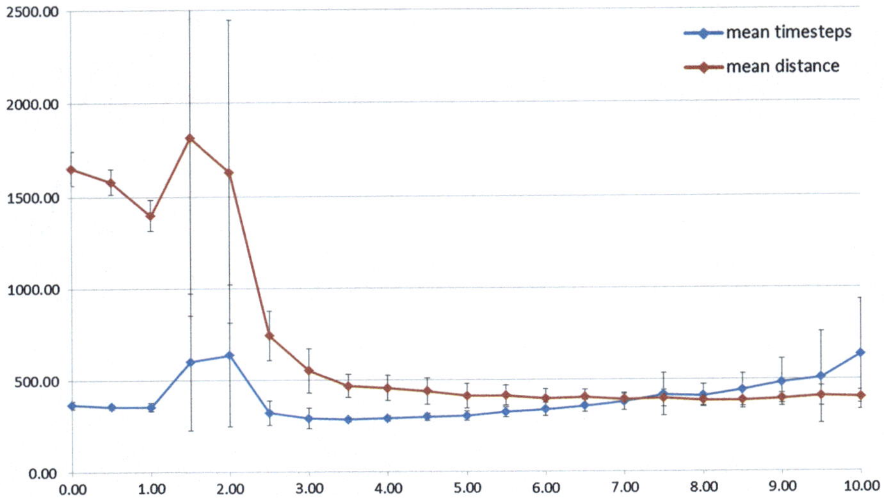

Fig. 4.6 Time steps and distance travelled by shepherd herding 10 sheep

4.5.2.2 Herding Time Steps and Distances

For a herd size of 10 sheep, Fig. 4.6 shows that as the stall factor increases, the distance required by the shepherd to complete the task gradually decreases (noting some problematic herding behaviour at a stall factor of 2.0). In fact, the greater the stall factor, the shorter total distance that the shepherd covers during the herding task. The mean fastest time achieved for the shepherd carrying out the task occurs at a stall factor of 3.0. Incidentally, this is the stall factor used by Strömbom et al. [12] for all flock sizes. Increasing the stall factor beyond 3.0 for the flock size of 10 does decrease the distance covered by the shepherd; however, the time taken increases.

For the herd of 20 sheep, Fig. 4.7 shows smoother curves for both time steps and distance covered by the shepherd. Again, as the stall factor increases, the distance required by the shepherd to complete the task gradually decreases. The stall factor with the lowest time steps is about 3.5, but the slight dip in the values for the required time steps for task completion is subtle.

For herd sizes of 40, 60, 80, 100 and 120, Figs. 4.8, 4.9, 4.10, 4.11, 4.12 show a similar pattern for the effect of the increasing stall factor leading to reduced distance travelled for the shepherd. The stall factors resulting in the smallest number of time steps across the various herd sizes are summarised in Table 4.1. The reader will note that stall factor increases near linearly with the flock size (the correlation between the two is 0.97).

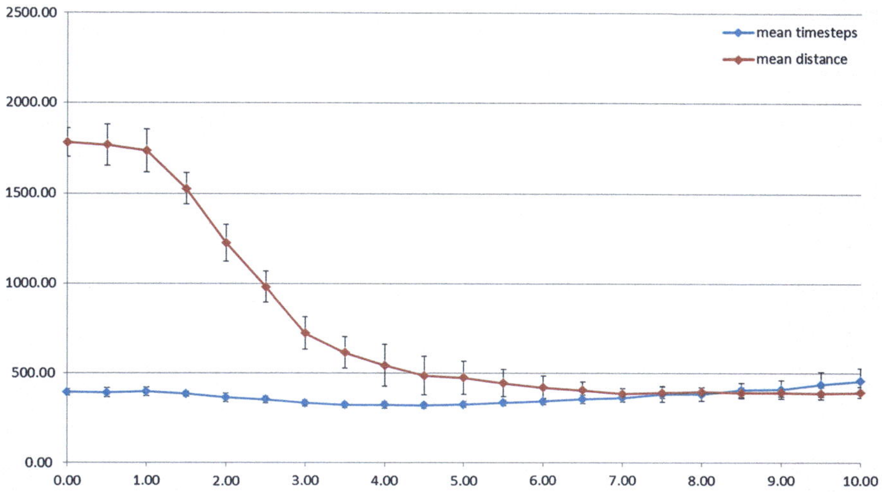

Fig. 4.7 Time steps and distance travelled by shepherd herding 20 sheep

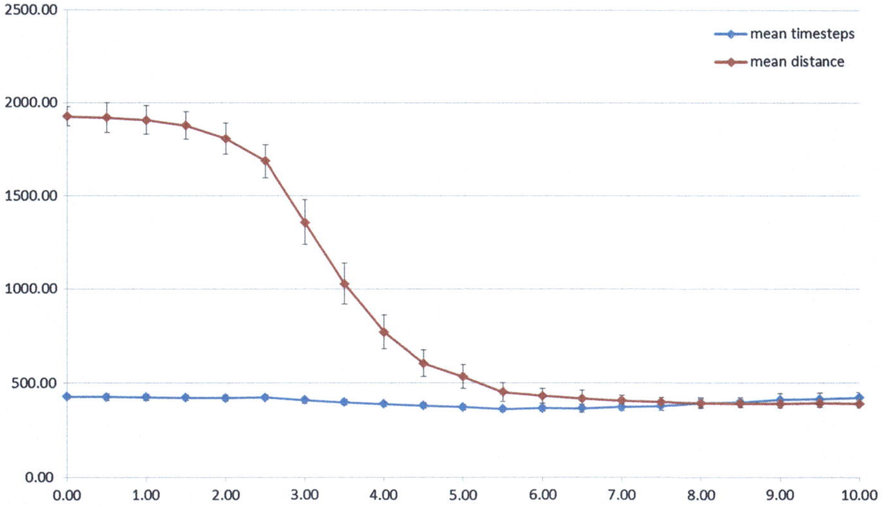

Fig. 4.8 Time steps and distance travelled by shepherd herding 40 sheep

4.6 Conclusion

In this chapter, we have explored the effect of the stall factor on the performance of the single shepherd herding task in terms of success rate, time steps required for task completion and the distance covered by the shepherd. Strömbom et al. [12] did not claim that the standard setting for the stall factor of 3.0 was optimal, rather that it agreed with the observations of real animals.

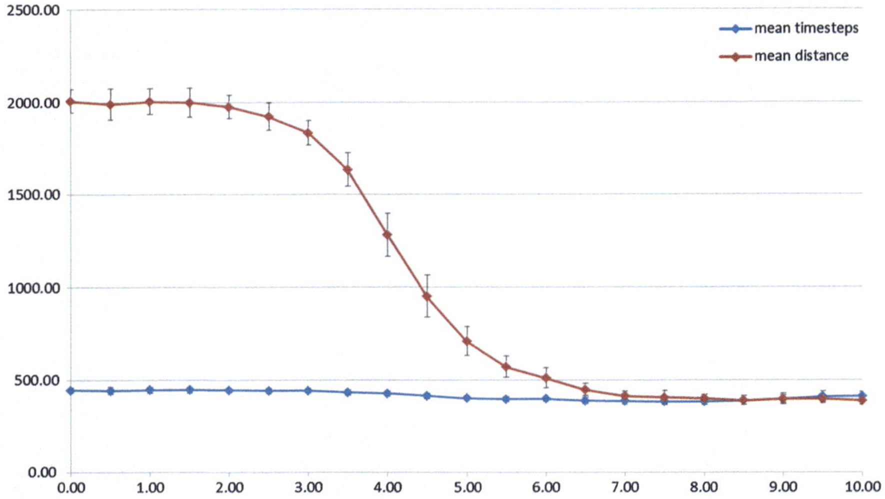

Fig. 4.9 Time steps and distance travelled by shepherd herding 60 sheep

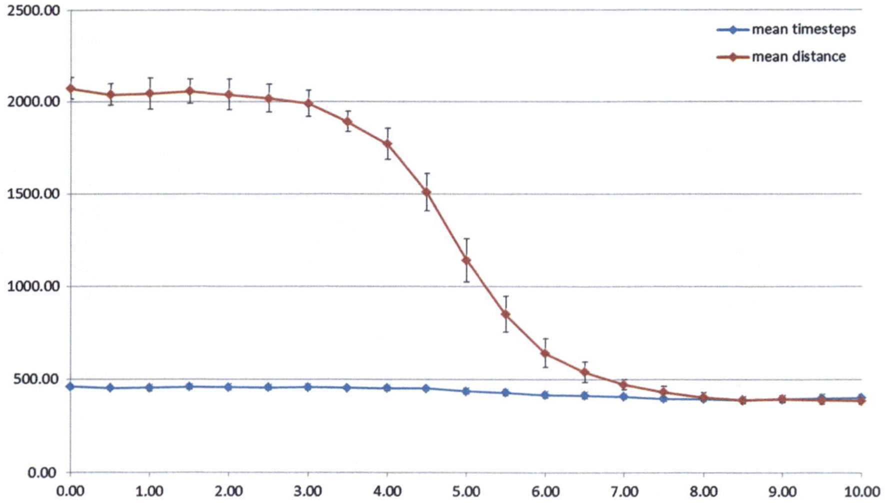

Fig. 4.10 Time steps and distance travelled by shepherd herding 80 sheep

The analysis in this chapter shows that a lower stall factor describes a more active shepherd that gets much closer to the individual sheep in the flock while carrying out the herding task. This more active kind of shepherd needs to carry out more frequent collection tasks between a series of driving tasks. Shepherd's behaviour with a lower stall factor (and shorter associated stalling distance) can lead to the shepherd invading the flock, causing the flock to scatter.

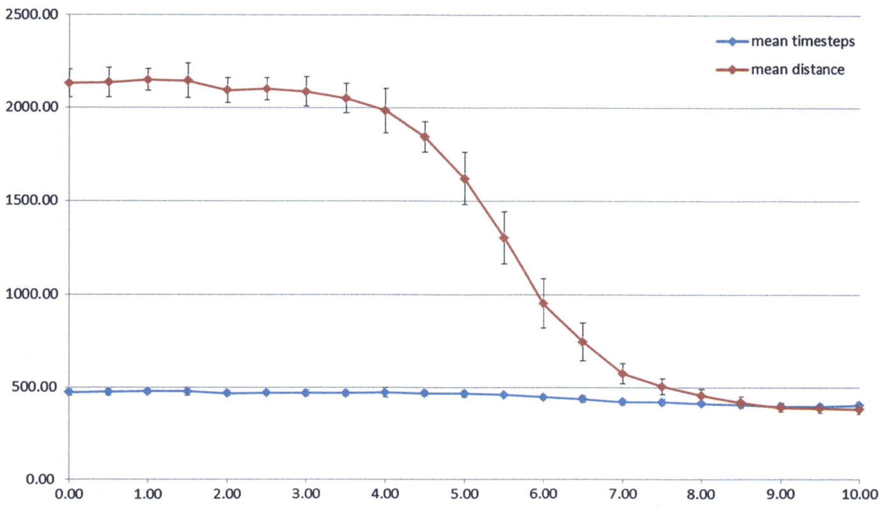

Fig. 4.11 Time steps and distance travelled by shepherd herding 100 sheep

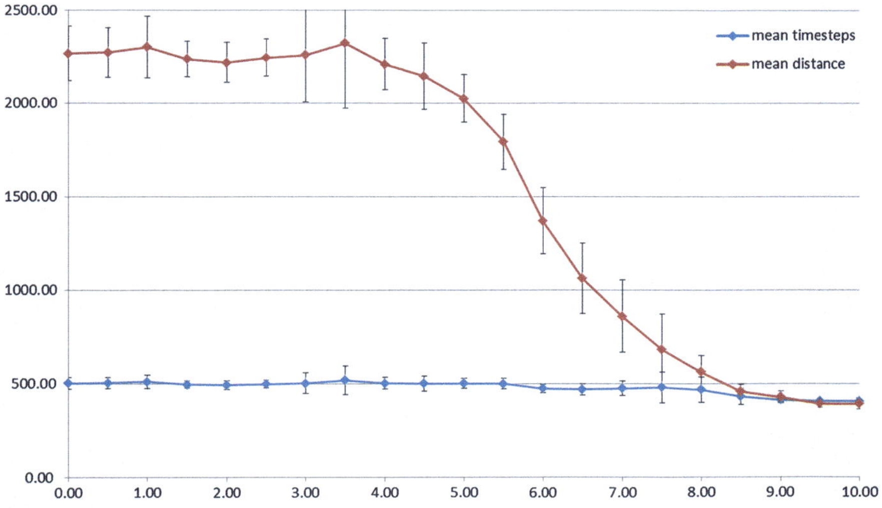

Fig. 4.12 Time steps and distance travelled by shepherd herding 120 sheep

Conversely, a shepherd using a higher stall factor keeps a greater distance from individual sheep in the flock. This higher stall factor produces shepherd's behaviour that keeps the shepherd further from the flock, leading to fewer collection tasks and shorter total distance travelled for the shepherd. Occasionally, even with a higher stall factor, there is some evidence that for flock sizes, both small (10 sheep or fewer) and large (120 sheep), the shepherd has difficulty of getting to a collecting

Table 4.1 Stall factor that results in the minimum time steps for task completion

Flock size	Stall factor	Stall distance (m)
10 sheep	3.0	6.0
20 sheep	4.5	9.0
40 sheep	5.5	11.0
60 sheep	8.0	16.0
80 sheep	8.5	17.0
100 sheep	9.5	19.0
120 sheep	10.0	20.0

point when required, leading to slow herding performance or a failure to complete the task.

In our experiments, we observe that there is a relationship between the most efficient (in terms of time taken for task completion) stall factor setting and the herd size.

In the real world, a more experienced shepherd may tend to patiently drive the flock from a greater distance, minimising the distance it needs to traverse to complete the task (this shepherd's behaviour uses a larger stalling distance). Conversely, a less experienced shepherd gets much closer to the flock in an attempt to drive it more forcefully, and however, this behaviour causes the sheep to scatter more often, requiring more frequent collection tasks, the shepherd covering a greater distance (this shepherd's behaviour uses a smaller stalling distance).

Strömbom et al. did suggest that the stalling distance was selected 'because of our observations of our sheepdog in the field, where the dog would rarely approach the flock at close range (since this causes the flock to fission)' [12] (p. 7). Exercising Strömbom's model has demonstrated that a very low stall distance does make task completion difficult sometimes, although not necessarily always, generating fission.

In the field, if a shepherd gets too close to a sheep, the sheep's response may be to run away from the shepherd at a much higher speed than normal. The sheep might ignore the location of the rest of the flock entirely, causing other sheep to follow it, breaking up the dense flock. This behaviour is not currently a part of Strömbom's standard model, but this kind of flight behaviour could be useful when extending Strömbom's style of herding in more detail. More explicitly, capturing flight behaviour of the sheep agents would also be useful when modelling the other two types of shepherding behaviour, namely covering and patrolling.

References

1. Fujioka, K.: Effective herding in shepherding problem in V-formation control. Trans. Inst. Syst. Control Inf. Eng. **31**(1), 21–27 (2018)
2. Fujioka, K., Hayashi, S.: Effective Shepherding Behaviours Using Multi-Agent Systems, pp. 3179–3182. Institute of Electrical and Electronics Engineers, Piscataway (2017)
3. Lee, W., Kim, D.: Autonomous shepherding behaviors of multiple target steering robots. Sensors **17**(12), 2729 (2017)

4. Lien, J.M., Bayazit, O.B., Sowell, R.T., Rodriguez, S., Amato, N.M.: Shepherding behaviors. In: IEEE International Conference on Robotics and Automation, vol. 4, pp. 4159–4164. Citeseer (2004)
5. Linder, M.H., Nye, B.: Fitness, environment and input: Evolved robotic shepherding, pp. 1–10 (2010)
6. Masehian, E., Royan, M.: Cooperative control of a multi robot flocking system for simultaneous object collection and shepherding. In: Computational Intelligence, pp. 97–114. Springer, Berlin (2015)
7. Miki, T., Nakamura, T.: An Effective Simple Shepherding Algorithm Suitable for Implementation to a Multi-Mmobile Robot System. pp. 161–165. Institute of Electrical and Electronics Engineers, Piscataway (2006)
8. Miki, T., Nakamura, T.: An effective rule based shepherding algorithm by using reactive forces between individuals. Int. J. InnovativeComput. Inf. Control **3**(4), 813–823 (2007)
9. Razali, S., Meng, Q., Yang, S.H.: Immune-inspired cooperative mechanism with refined low-level behaviors for multi-robot shepherding. Int. J. Comput. Intell. Appl. **11**(01), 1250007 (2012)
10. Singh, H., Campbell, B., Elsayed, S., Perry, A., Hunjet, R., Abbass, H.: Modulation of force vectors for effective shepherding of a swarm: A bi-objective approach. In: 2019 IEEE Congress on Evolutionary Computation (CEC), pp. 2941–2948. IEEE, Piscataway (2019). https://doi.org/10.1109/CEC.2019.8790228
11. Strömbom, D., King, A.J.: robot collection and transport of objects: a biomimetic process. Front. Rob. AI **5**, 1–7 (2018)
12. Strömbom, D., Mann, R.P., Wilson, A.M., Hailes, S., Morton, A.J., Sumpter, D.J.T., King, A.J.: Solving the shepherding problem: heuristics for herding autonomous, interacting agents. J. R. Soc. Interf. **11**(100) (2014). https://browzine.com/articles/52614503
13. Sueoka, Y., Ishitani, M., Osuka, K.: Analysis of sheepdog-type robot navigation for goal-lost-situation. Robotics **7**(2), 21 (2018)

Part II
Learning and Optimisation for Shepherding

Chapter 5
Mission Planning for Shepherding a Swarm of Uninhabited Aerial Vehicles

Jing Liu, Sreenatha Anavatti, Matthew Garratt, and Hussein A. Abbass

Abstract Uninhabited aerial vehicles (UAVs) are widely used in many areas for completing complex missions such as tracking targets, search and rescue, farming (shepherding in the traditional sense) and mapping. Mission planning for shepherding a UAV swarm is advantageous for Human-Swarm Teaming. While most research on shepherding see the shepherd as a simple reactive agent, a smart shepherd in a complex environment will need to consider many dimensions and sub-decisions to successfully guide a swarm through complex environment and towards a goal.

In this chapter, we review and offer formal definitions for the sub-problems required for a shepherd to complete a mission successfully. The swarm mission planning system needs to have decision modules capable of solving four main problems: task decomposition, task assignment, path planning and trajectory generation. These sub-problems are coupled differently depending on the scenario. This chapter defines these sub-problems in their general form and gives UAV swarm shepherding problem as a specific application. A brief review of the widely used algorithms for tackling these problems and the state of art of mission planning are also given in this chapter.

Keywords Uninhabited aerial vehicles · Mission planning · Task assignment · Evolutionary algorithms

Uninhabited aerial vehicles (UAVs) are widely used in many areas for completing complex missions such as tracking targets, search and rescue, farming (shepherding in the traditional sense) and mapping. Mission planning for shepherding a UAV

J. Liu (✉) · S. Anavatti · M. Garratt · H. A. Abbass
School of Engineering and Information Technology, University of New South Wales, Canberra, ACT, Australia
e-mail: jing.liu5@student.adfa.edu.au; s.anavatti@adfa.edu.au; m.garratt@adfa.edu.au; h.abbass@adfa.edu.au

H. A. Abbass, R. A. Hunjet (eds.), *Shepherding UxVs for Human-Swarm Teaming*, Unmanned System Technologies,
https://doi.org/10.1007/978-3-030-60898-9_5

swarm is advantageous for Human-Swarm Teaming. While most research on shepherding see the shepherd as a simple reactive agent, a smart shepherd in a complex environment will need to consider many dimensions and sub-decisions to successfully guide a swarm through complex environment and towards a goal.

In this chapter, we review and offer formal definitions for the sub-problems required for a shepherd to complete a mission successfully. The swarm mission planning system needs to have decision modules capable of solving four main problems: task decomposition, task assignment, path planning and trajectory generation. These sub-problems are coupled differently depending on the scenario. This chapter defines these sub-problems in their general form and gives UAV swarm shepherding problem as a specific application. A brief review of the widely used algorithms for tackling these problems and the state of art of mission planning are also given in this chapter.

5.1 Introduction

Uninhabited aerial vehicles (UAVs) are being increasingly used in a wide range of applications such as infrastructure inspection [56] and traffic monitoring [73] due to their autonomy, manoeuvrability, low cost and accessibility to dangerous areas. Mission planning for shepherding a swarm of UAVs aims to generate a plan for shepherding the UAV swarm to reach the goal in an optimal manner by means of a number of shepherding agents. The shepherding agents could be a number of specific UAVs or some other intelligent agents with special capabilities. This allows humans to shepherd a swarm through operation of the shepherding agents. In this chapter, specific UAVs with high-level capabilities are considered as the shepherding agents (called shepherd UAVs), which are used to shepherd a swarm of low-level UAVs (called sheep UAVs).

During UAVs mission planning, there are multiple objectives that need to be achieved by accomplishing some tasks with a group of resources over a period of time. A number of sub-problems (e.g., task assignment, path planning) are involved in mission planning, which increases the difficulty and complexity of the optimisation problem. In realistic application scenarios, different sub-problems are coupled with specific optimisation objectives and constraints. Additionally, UAV planning becomes more complicated due to the increasingly challenging environment that UAVs face, such as the dynamic environment with disturbances, failures and uncertainties. Therefore, UAV mission planning in complex environments has become an important and challenging research area.

In this chapter, the definitions of the major sub-problems involved in mission planning are presented, and the mathematical formulation of mission planning for shepherding a UAV swarm is derived. This chapter also presents a brief review of the well-known methods for addressing sub-problems involved in mission planning. The combinatorial optimisation problems in mission planning are also reviewed to provide an overview of the state of art in mission planning and some directions for future research.

The rest of this chapter is organised as follows. Section 5.2 presents the overview of mission planning for shepherding UAV swarm. The major sub-problems involved in mission planning are described in Sects. 5.3 and 5.4. Section 5.5 presents some typical coupled planning problems that are subsets of mission planning, followed by the conclusion in Sect. 5.6.

5.2 Overview of Mission Planning for Shepherding UAV Swarm

Mission planning offers an effective and efficient way to shepherd a swarm of UAVs towards mission completion. Figure 5.1 presents a simple UAV swarm shepherding problem where three shepherding agents are required to control the UAV swarm to move to the goal area without colliding with one another or violating the no-flight zone.

Figure 5.2 presents the major sub-problems involved in UAV mission planning. Specifically, in the UAV swarm shepherding problem, the mission is to guide a UAV swarm using force vectors that arrive to the UAV from behind towards a pre-determined goal area that is known only to the shepherds.

Task decomposition is to take the mission as a whole and divide into a series of tasks that need to be performed sequentially or in parallel to complete the mission. Task decomposition may happen at the start of the mission as a one-off decision or continuously as the mission gets executed. The latter normally takes place in highly uncertain environments.

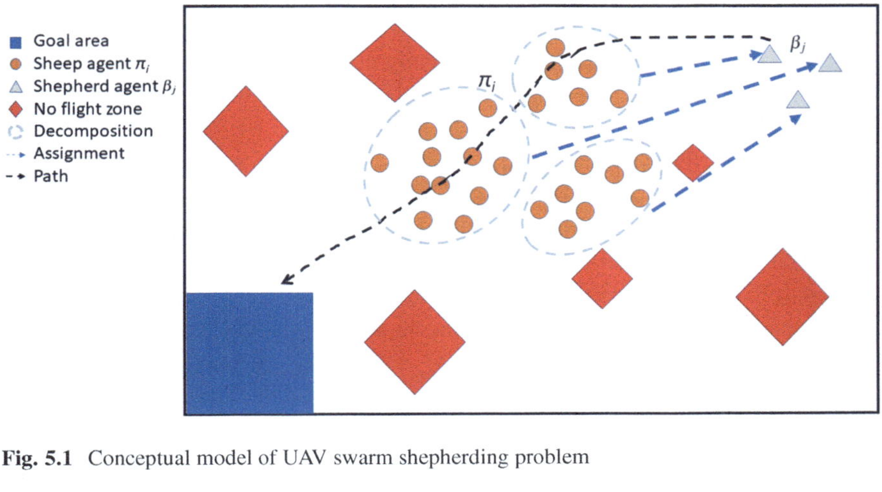

Fig. 5.1 Conceptual model of UAV swarm shepherding problem

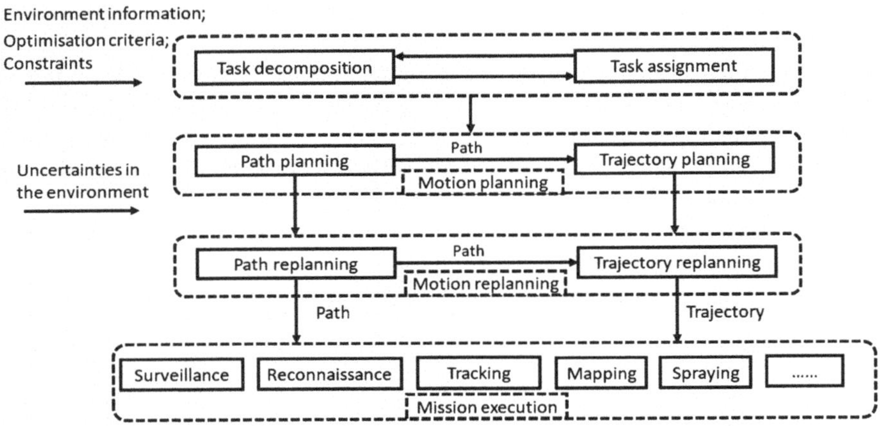

Fig. 5.2 Illustration of UAVs mission planning

Task assignment is to split the mission among the different shepherds so that each shepherd is assigned a subset of the swarm. This may happen as a one-off decision at the start of the mission or dynamically as the mission progresses.

In order to accomplish the assigned tasks, the planning system needs to find the path that each shepherd needs to follow and generate the trajectory for each shepherding UAV so that the shepherd and the UAVs can reach the goal safely and efficiently. This form of motion planning may need to be done continuously if the shepherd is operating in an uncertain and/or partially observable environment.

These sub-problems are optimised simultaneously to achieve a set of objectives. According to different scenarios and preferences, the optimisation objectives could take the form of minimising the total cost, minimising the total flight time, minimising system-level risks, minimising a mission's makespan, and maximising the success rate of mission, to name but a few. In complex environments with dangerous zones and obstacles, the commonly considered optimisation objectives include *operational cost minimisation*, *time cost minimisation*, *risk minimisation* and *completion rate maximisation*.

Operational Cost Minimisation Operational cost (C_{oper}) for shepherding UAV swarm considers factors related to the operation of the platform such as costs associated with fuel C_{fuel} and length of flight route C_{length}. The total fuel cost is the total amount of fuel consumed by M shepherding agents during the flight, denoted as $\sum_{j=1}^{M} C_{fuel}^{j}$, which is related to several issues such as the flight time and UAV velocity. The flying length cost is the sum of the flying length L_{path} of each shepherding agent, computed by $\sum_{j=1}^{M} L_{path}^{j}$. Sometimes turning cost $C_{turning}$ and climbing cost C_{height} are also considered (see [89]).

Time Cost Minimisation Time cost (C_{time}) is usually evaluated by the makespan of the mission, which is the time span between the start time of the first task and the end time of all tasks to be completed. The makespan can be measured by the maximum time among the total time $Time^j$ required by each shepherding agent to complete the assigned tasks, denoted as $\max_{j=1,\ldots,M}(Time^j)$.

Risk Minimisation In complex environments, there are areas that pose threats to UAVs, such as those with excessive radio frequency interference that may interrupt the UAV's control channel. UAVs should try to avoid such areas.

Completion Rate Maximisation The completion rate Rate_{comp} is the amount of the mission that can be completed, which is used to evaluate the reliability of mission execution. It can be computed as the percentage of completed tasks in a mission.

It is hard to satisfy all these objectives simultaneously since some of them are conflicting. For example, the reduction of the operational cost may increase risk and reduce the completion rate. Therefore, solutions to multiple objective problems are usually presented as a set of non-dominated solutions (called Pareto-optimal solutions) across the various objectives; for a discussion on Pareto optimality, the reader is referred to [57].

There are two main classes of approaches to solve multi-objective optimisation problems. One is to convert the multi-objective problem into a single objective problem, e.g., via a weighted linear combination of the individual objectives [89]. A single objective function with weighted sum for minimisation is given as an example in the following Eq. 5.1.

$$\begin{aligned} F_{obj}(\mathscr{S}) = w_1 \cdot C_{fuel}(\mathscr{S}) + w_2 \cdot C_{length}(\mathscr{S}) + w_3 \cdot C_{time}(\mathscr{S}) \\ + w_4 \cdot \text{Risk}(\mathscr{S}) - w_5 \cdot \text{Rate}_{comp}(\mathscr{S}) \end{aligned} \tag{5.1}$$

where $\mathscr{S}$ is the complete solution of mission planning consisting of a set of sub-solutions $[\Phi, \mathscr{A}, \mathscr{B}, \mathscr{C} \ldots]$ that are defined in the following sections, and w_i, $i = 1, 2, \ldots, 5$ are the weight coefficients. However, this kind of formulation needs to define the weight coefficients that need some domain knowledge including the decision maker's priorities and preferences. This formulation can only discover one Pareto-optimal solution at a time and need to run the optimisation many times with different weight coefficients to sample a subset of the set of Pareto-optimal solutions.

Another method is to optimise the multiple objective functions simultaneously by using specialised multi-objective methods such as Non-dominated Sorting Genetic Algorithm II (NSGA II) [15] and Multi-Objective Particle Swarm Optimisation (MOPSO) [14]. These methods are population-based heuristics that are able to obtain a set of near Pareto-optimal solutions in a single simulation.

Constraints vary between scenarios. The differential constraints in the UAVs mission planning system are represented by $\Upsilon = \{\upsilon', \ldots, \upsilon_o, \ldots, \upsilon_O\}$, which include physical constraints, kinematic constraints, temporal constraints, and so on. O is the total number of constraints. For example, the flight time and length for each UAV cannot exceed their maximum flight time and length. Considering the above, the mission planning problem can be generalised as follows:

$$
\begin{aligned}
\mathit{Minimise} \quad & F_{obj}(\mathscr{S}) = (C_{fuel}(\mathscr{S}),\ C_{length}(\mathscr{S}),\ C_{time}(\mathscr{S}), \\
& \qquad \text{Risk}(\mathscr{S}),\ \text{Rate}_{comp}(\mathscr{S}) \ldots) \\
\text{Subject to} \quad & \Upsilon = \{\upsilon'(\mathscr{S}), \ldots, \upsilon_o(\mathscr{S}), \ldots, \upsilon_O(\mathscr{S})\}
\end{aligned}
\tag{5.2}
$$

More details of the major sub-problems in UAVs mission planning system and specific examples for the UAV swarm shepherding problem are described in the following sections.

5.3 Task Planning

Task decomposition and task assignment in a multi-agent system are referred to as task planning [84]. Task assignment seeks to assign tasks to the agents such that they are performed in a feasible (and ideally optimal) manner; this process is also known as task allocation, where an assignment is a special case of allocation. In assignment, one agent is assigned a single task and one task is assigned to a single agent. In allocation, the mapping is not one to one. Task decomposition involves breaking down a complex mission into a set of tasks. More details of task decomposition and task assignment problems are described in Sects. 5.3.1 and 5.3.2. Then the algorithms for solving them are reviewed in Sect. 5.3.3.

5.3.1 Task Decomposition

Task decomposition seeks to divide an overall mission into smaller and simpler sub-tasks that can be executed separately, allowing for a better mission efficiency. The general problem has multiple names in the literature including partitioning, as in graph partitioning problems, split and merge algorithms, divide and conquer and clustering and decomposition.

Task decomposition can be used for a wide variety of problems such as modular neural network design [6], image segmentation[69], pattern classification[52] and multi-agent system operation[1]. For example, Yazdani et al. [87] decomposed a large-scale dynamic optimisation problem into a number of lower dimensional sub-problems by using a variable interaction analysis algorithm and then optimised these sub-problems concurrently. For dealing with multi-target tracking and surveillance problem, Adamey et al.[1] split the environment into a set of regions based on the probability distributions of the target tracks and allocated these regions to specific agents. In [12], Chen et al. used task decomposition to divide the global optimal path into a number of local optimal paths for reducing the time cost of path planning.

By generalising special cases of task decomposition, the general definition of this problem is presented in the following.

Denoting the overall mission as Θ and the number of decomposed tasks as Q, the solution of task decomposition can be generally stated as follows:

$$\Phi = \{\phi_1, \ldots, \phi_q, \ldots, \phi_Q\} \tag{5.3}$$

where Φ is the set of decomposed tasks ϕ_q that constitute the overall mission Θ. The task decomposition problem is defined as:

Definition 5.1 (Task Decomposition) Given an overall mission Θ and a set of constraints $\Upsilon = \{\upsilon', \upsilon'', \upsilon''' \ldots\}$, task decomposition is to decompose the overall mission into several relative subsets $\phi_1, \ldots, \phi_q, \ldots, \phi_Q$, where $\bigcup_{q=1}^{Q} \phi_q = \Theta$, $\bigcap_{q=1}^{Q} \phi_q = \emptyset$ and $\phi_q \neq \emptyset$, $\forall q \in \{1, \ldots, Q\}$, contributing to the achievement of the overall mission separately.

The mathematical formulation of task decomposition can be generalised as follows:

$$\begin{aligned}
&\text{Minimise} \quad F_{obj}(\Phi), \quad \forall \Phi = \{\phi_1, \ldots, \phi_q, \ldots, \phi_Q\} \\
&\text{Subj. to} \quad \bigcup_{q=1}^{Q} \phi_q = \Theta, \ \bigcap_{q=1}^{Q} \phi_q = \emptyset \\
&\qquad \phi_q \neq \emptyset, \ \forall q \in \{1, \ldots, Q\} \\
&\qquad \Upsilon = \{\upsilon'(\Phi), \upsilon''(\Phi), \upsilon'''(\Phi) \ldots\}
\end{aligned} \tag{5.4}$$

where the objective $F_{obj}(\Phi)$ is one of the objectives we discussed before such as cost minimisation. The first three constraints ensure that the subsets of the mission are exclusive and nonempty and constitute the overall mission. These first three constraints represent the classic conditions to define a partition for a set, while $\Upsilon = \{\upsilon'(\Phi), \upsilon''(\Phi), \upsilon'''(\Phi) \ldots\}$ represents the set of some other possible constraints related to the specific scenarios. For example, a specific part of the mission might be required to be undivided.

In the mission planning problem for UAV swarm shepherding shown in Fig. 5.1, specifically, an approach to task decomposition is to divide the shepherding mission into several smaller tasks of shepherding different subsets of the UAVs. We divide the UAV swarm into several subsets shown as dotted ellipses in Fig. 5.1. Let the UAV swarm be $\Pi = \{\pi_1, \ldots, \pi_i, \ldots, \pi_N\}$, where π_i denotes a UAV (sheep agent) and N is the number of sheep agents in the swarm. Task decomposition divides the UAV swarm Π into Q subsets. Since task decomposition is usually coupled with task assignment, contributing to the same optimisation objectives, the mathematical formulation of task decomposition for UAV swarm shepherding problem is presented along with that of task assignment as Eq. 5.7 in the following Sect. 5.3.2.

5.3.2 Task Assignment

Assuming that the set of tasks and required associated information (e.g., capabilities of agents, environment information) are given, the task assignment problem is to assign the tasks to those agents that will carry them out. There are many variants of task assignment problems based on different scenarios and constraints. Gerkey and Mataric [30] proposed a taxonomy of multi-robot task allocation problems based on three dimensions: single-task (ST) robots versus multi-task (MT) robots, single-robot (SR) tasks versus multi-robot (MR) tasks and instantaneous assignment (IA) versus time-extended assignment (TA). Bertuccelli et al. [8] addressed a MT-SR-TA task assignment problem where each UAV can be assigned to deal with multiple tasks, while the task assignment solved in [67] is MT-MR-TA. A time window-based cooperative task assignment model, which is also MT-SR-TA, was presented by Chen et al. [11], and a heuristic algorithm named Multi-Objective Symbiotic Organisms Search algorithm (MOSOS) was proposed to assign necessary tasks and determine the optimal task sequence to UAVs. The general definition of task assignment is presented in the following.

Denoting the set of tasks as $\Phi = \{\phi_1, \ldots, \phi_q, \ldots, \phi_Q\}$ and the set of agents as $B = \{\beta_1, \ldots, \beta_j, \ldots, \beta_M\}$, the solution of task assignment problem can be generally stated as follows:

$$\mathscr{A} : \Phi \to B \tag{5.5}$$

where $\mathscr{A}$ is a mapping function indicating the assignment of task $\phi_q,\ q \in \{1, \ldots, Q\}$ to a specific agent $\beta_j,\ j \in \{1, \ldots, M\}$. $\mathscr{A}^r_{jq} = 1$ denotes that task ϕ_q is assigned to agent β_j and is the rth task to be executed by β_j; $\mathscr{A}^r_{jq} = 0$, otherwise.

Let the cost function of each agent β_j specifying the cost of performing each task ϕ_q be $Cost^q_j$, and then the optimal task assignment problem can be generally defined as:

Definition 5.2 (Optimal Task Assignment) Given a task assignment problem (Φ, B, $Cost^q_j$), the optimisation objective function $F_{obj}(\mathscr{A})$ and a set of constraints $\Upsilon = \{\upsilon', \upsilon'', \upsilon''' \ldots\}$, optimal task assignment is to find the feasible matching $\mathscr{A}^*$ of tasks Φ to agents B with the optimal $F_{obj}(\mathscr{A}^*)$.

The mathematical formulation of optimal task assignment can be generalised as follows:

$$\begin{array}{ll} \text{Minimise} & F_{obj}(\mathscr{A}), \quad \forall \mathscr{A} : \Phi \to B \\ \text{Subj. to} & \Upsilon = \{\upsilon'(\mathscr{A}), \upsilon''(\mathscr{A}), \upsilon'''(\mathscr{A}) \ldots\} \end{array} \tag{5.6}$$

where the objective F_{obj} usually considers minimising the cost (e.g., operational cost, time cost), which is consistent to the overall objective of mission planning.

The set of constraints Υ are enforced to task assignment problem for finding feasible matching of tasks Φ to agents B and the appropriate task execution sequence when it is required. One example of the constraints in this problem is that the number of tasks assigned to each agent should be less than the maximum number of tasks that can be performed by the agent, that is, an agent capacity constraint.

Applying this to the UAV swarm shepherding problem shown in Fig. 5.1 as an example, task assignment seeks to assign decomposed subsets of the UAV swarm to the shepherding agents to control in order to maximise the overall performance. In this specific problem, $B = \{\beta_1, \ldots, \beta_j, \ldots, \beta_M\}$ is the set of shepherding agents (shepherd UAVs), and $\Phi = \{\phi_1, \ldots, \phi_q, \ldots, \phi_Q\}$ is the set of tasks (to shepherd UAV swarm subsets); the subsets are formed through the aforementioned task decomposition approach. One shepherding agent may be assigned to multiple tasks, but each task can only be performed by one agent. The mathematical formulation of this MT-SR-TA task assignment with its associated task decomposition can be described as:

$$
\begin{aligned}
\text{Minimise } F_{obj}(\mathscr{A}) &= \sum_{j=1}^{M}\sum_{q=1}^{Q}\sum_{r=1}^{N_r} Cost_j^q \cdot \mathscr{A}_{jq}^r, \quad \forall \Phi,\ \mathscr{A} : \Phi \rightarrow B \\
\text{Subj. to } \upsilon' &: \bigcup_{q=1}^{Q} \phi_q = \Pi;\ \bigcap_{q=1}^{Q} \phi_q = \emptyset;\ \phi_q \neq \emptyset,\ \forall q \in \{1, \ldots, Q\} \\
\upsilon'' &: \sum_{q=1}^{Q}\sum_{r=1}^{N_r} \mathscr{A}_{jq}^r \leq N_r, \quad \forall j \in \{1, \ldots, M\} \\
\upsilon''' &: \sum_{j=1}^{M}\sum_{r=1}^{N_r} \mathscr{A}_{jq}^r = 1, \quad \forall q \in \{1, \ldots, Q\} \\
\upsilon'''' &: \mathscr{A}_{jq}^r \in \{0, 1\}, \quad \forall j \in \{1, \ldots, M\};\ q \in \{1, \ldots, Q\}; \\
&\quad r \in \{1, \ldots, N_r\}
\end{aligned}
\tag{5.7}
$$

where N_r is the maximum number of tasks that one agent can be assigned. The objective is to minimise the total cost of completing all the decomposed tasks. Constraint υ' ensures that the overall mission is decomposed into a set of tasks. Constraint υ'' ensures that the number of tasks assigned to each shepherding agent is less than the maximum number of tasks that the agent can perform. Constraint υ''' ensures that each task is assigned to one and only one shepherding agent. Constraint υ'''' guarantees the value of logical variables. The algorithms for optimising task decomposition and task assignment, which can be applied for UAV swarm shepherding problem, are reviewed in the following Sect. 5.3.3.

5.3.3 Algorithms for Task Planning

Task planning algorithms comprise both algorithms for task decomposition and task assignment. Various task decomposition methods have been developed. A mission can be divided into a set of sub-tasks by the designer who has knowledge concerning the problem domain and decomposition [27]. This kind of method is called explicit decomposition [52] and requires that the prior knowledge concerning the mission is sufficient. If only the inherent class relations are known, class decomposition method can be applied [5]. When the prior knowledge about the problem is absent, the automatic decomposition through learning is applied. Lu and Ito [52] proposed a new decomposition method based on the idea of using class relations of agents and learning framework. In the UAV swarm shepherding problem, task decomposition may be regarded as a clustering problem that partitions the swarm consisting of N UAV agents into Q exclusive and nonempty clusters. Therefore, the clustering methods can be used for task decomposition in the UAV swarm shepherding problem. As a famous clustering method, k-means clustering is widely used and studied[34, 36, 54]. For example, Kanungo et al. [38] presented a filtering algorithm to implement Lloyd's k-means clustering algorithm efficiently. Wagstaff et al. [81] presented a constrained k-means clustering algorithm by utilising the background knowledge about the problem domain to improve the clustering accuracy. The pseudocode of the k-means clustering algorithm [51] where $k = Q$ is presented in Algorithm 2 as an example for task decomposition algorithm in this chapter. The algorithm terminates when it converges to a local optimum where no further change in clusters can be made.

Algorithm 2 The pseudocode of k-means algorithm

Require: $\Pi = \{\pi_1, \ldots, \pi_j, \ldots, \pi_M\}$
1: Selecting k initial centres of clusters $\phi_1, \ldots, \phi_q, \ldots, \phi_Q$
2: **while** The termination condition is not met **do**
3:     **for** each π_j in Π **do**
4:         Assign π_j to the close t cluster ϕ_q
5:     **end for**
6:     **for** each cluster ϕ_q in Φ **do**
7:         Update its centre by averaging all of the π_j that have been assigned to it
8:     **end for**
9: **end while**
Ensure: $\Phi = \{\phi_1, \ldots, \phi_q, \ldots, \phi_Q\}$

Within the MAS literature, there is a larger body of research on task assignment than task decomposition. Some works [70, 74] formulated the task assignment problem as a mixed integer linear programming (MILP) problem, which allows the optimal solution to be found while satisfying all timing constraints. A cooperative UAV task assignment problem with uncertainty was solved as a single MILP in [7]. When MT agents and MR tasks are involved, and complex constraints such as those caused by the cooperation among agents are considered, the general task assignment problem becomes much more complex and falls into the class of NP-hard problems; that is, the problem cannot be solved in a polynomial time. Many efforts have been made to solve the task assignment problem, and various approaches have been proposed, such as *market-based methods* [60], *threshold-based algorithms* [48] and *bioinspired algorithms* [11].

In *market-based methods*, where auctions are the most commonly used mechanisms, agents compete through bidding to win those tasks that they are capable to perform in low cost and then minimise their cost. Generally, agents place bids on tasks, and the highest bid wins the assignment. An auction-based task assignment system named MURDOCH, which is built on a resource centric publish communication model, was presented and implemented by Gerkey and Mataric[29]. Choi, et al. [13] presented the consensus-based auction algorithm (CBAA) and generalised it to the multi-assignment problem as the consensus-based bundle algorithm (CBBA). The market-based decision strategy was utilised for decentralised task selection, and a consensus routine was used as the conflict resolution mechanism to achieve agreement on the winning bid values in [13]. CBBA was also extended to handle real-time multi-UAV task assignment in dynamic environment by introducing a churning-mitigation formulation [8]. An auction-based method was proposed by Landén et al. [44] to solve the complex task allocation with task dependencies, temporal and spatial requirements considered. A market-based assignment algorithm was proposed to determine task assignment and execution sequence while considering the estimated time of arrival in [60].

In *threshold-based algorithms* [10], agents have activation thresholds for each task that represent the reference of the agent to the task. For each task, each agent continuously receives stimulus; once the stimulus for a particular task exceeds its threshold, the agent starts to perform the task. Based on the threshold model, Agassounon and Martinoli [2] proposed three distributed algorithms for task allocation in multi-agent systems and analysed the efficiency and robustness of these algorithms. A task allocation algorithm for swarm robotic system was presented based on the response threshold model for the foraging mission [85]. Kim et al. [42] proposed a distributed method by using a probabilistic decision-making protocol based on the response threshold model for addressing UAV search planning and task allocation problem. *Bioinspired algorithms* are approaches that take inspiration from biological evolution or natural phenomena. They overcome the weaknesses where general mathematical model-based algorithms have difficulties when addressing NP-hard problem with a large number of variables and nonlinear objective functions. The random search element of bioinspired algorithms helps them to escape local optima and find better solutions. Meanwhile, the majority of these algorithms do not require any special assumption to be made about the optimisation problem. These algorithms usually perform well in finding good solutions, but it should be noted that optimality cannot be guaranteed.

Some typical bioinspired algorithms include Genetic Algorithm (GA), Particle Swarm Optimisation (PSO) [41] and Ant Colony Optimisation (ACO) [17]. For example, GA [35] is one of the most famous population-based optimisation algorithms, where a group of solutions are manipulated by several operators to progress towards an optimal solution over a series of generations. These operators include selection according to a fitness function, crossover to create new solutions and random mutation of new solutions. GA was applied for assigning cooperating UAVs to simultaneous tasks on consecutive targets and has shown its viability for different sized scenarios of the assignment problem [75]. Salman et al. [72] introduced PSO for general task assignment problem in 2002. An extension of ACO was applied to solve UAV cooperative multi-assignment problem that considers the dynamic time window constraints and the capabilities of UAVs [22].

The pseudocode of basic auction algorithm [29] is given in Algorithm 3 as an example for task assignment algorithm.

Algorithm 3 The pseudocode of auction algorithm

Require: $\Phi = \{\phi_1, \ldots, \phi_q, \ldots, \phi_Q\}$, $B = \{\beta_1, \ldots, \beta_j, \ldots, \beta_M\}$, $Cost_j^q$

for each task ϕ_q in Φ **do**
  Task announcement:
  Publish the details of task and select the set of candidate agents CB from B that have capabilities to perform the task
  for each agent β_j in CB **do**
    Bid submission which is correlated to $Cost_j^q$
  end for
  Winner determination
  Select the winner β_i based on the bids
  $\mathscr{A}_{iq}^r = 1$ if it is the rth task assigned to β_i
end for

Ensure: $\mathscr{A}$

5.4 Motion Planning

In the planning system, motion planning is a crucial problem that should be resolved before each stage of mission can be executed. Motion planning can be loosely defined as how to determine the required movement of agents to accomplish tasks[45]. It produces paths or trajectories for agents to follow from start states to goal states. A basic motion planning task could be formulated as a path planning problem where motion is restricted to paths with collision avoidance and the dynamic properties of agents are ignored. The basic problem could be defined as generating a path consisting of a continuous sequence of locations and orientations of agents (e.g., UAVs, robots), which is collision-free with obstacles, from the initial location to the goal location[45]. When considering the time along the path, motion planning refers to trajectory planning that couples the spatial and temporal elements [61]. Motion planning with dynamic elements may be more suitably transferred into a trajectory planning problem. Note that path and trajectory are not always distinguished strictly in the literature such as in the work of [31, 53].

Here, we distinguish them to avoid misunderstanding. More details of UAV path planning and trajectory planning problems are discussed in Sects. 5.4.1 and 5.4.2, respectively. Then, the methods for solving them are reviewed in Sect. 5.4.3.

5.4.1 Path Planning

Path planning seeks to find a path in the configuration space (C-space) from the start location to the destination for each agent to follow, while satisfying given constraints and optimising the quality criteria of solution. The C-space is a set of all possible configurations of the agent, which represents its pose or position in two dimensions (2-D) [59, 92] or three dimension (3-D) space [24, 63], while the set of all feasible configurations of the agent is called free C-space. Path planning problems are formulated as an extension of the Travelling Salesman Problem (TSP) in some works [47, 64] or more generally a Vehicle Routing Problem (VRP) [65] where some constraints (e.g., coverage range, collision avoidance) are considered.

The consideration of different constraints and objectives gives rise to many possible variations of the path planning problem. For example, the constraint of maintaining a fixed flight level over the ground in a non-uniform terrain was added to the planning problem by Gonzalez et al. [32], and the Fast Marching Square (FM^2) method was proposed to compute an obstacles-free path for UAVs. Ergezer and Leblebicioglu [20, 21] focused on maximising the collected information (CI) from desired regions (DR) when designing an evolution-based algorithm to find the collision-free path to the destination in 2D and 3D environments, respectively. The general definition of path planning is presented below.

Denoting the C-space as $\mathbb{C} = \mathbb{R}^2$ or $\mathbb{R}^3$, the free C-space as $\mathbb{C}_{free}$, the initial configuration of the agent as x_{init} and the goal configuration of the agent as x_{goal}, the solution of the path planning problem is a curve that can be represented as a continuous function:

$$\mathscr{B} : [0, 1] \rightarrow \mathbb{C} \tag{5.8}$$

where $\mathscr{B}(0) = x_{init}$, $\mathscr{B}(1) = x_{goal}$, $\mathscr{B}(\omega) \in \mathbb{C}_{free}$ for all $\omega \in [0, 1]$. The optimal path planning can be generally defined as [61]:

Definition 5.3 (Path Planning) Given a path planning problem $(x_{init}, x_{goal}, \mathbb{C}_{free})$ and the optimisation objective functions $F_{obj}(\mathscr{B})$, we seek to find the feasible path $\mathscr{B}$, where $\mathscr{B}(0) = x_{init}$ and $\mathscr{B}(1) = x_{goal}$, subject to differential constraints $\Upsilon = \{\upsilon', \upsilon'', \upsilon''' \ldots\}$. Optimal path planning is to find the optimal path $\mathscr{B}^*$ with optimal $F_{obj}(\mathscr{B}^*)$ from all the feasible paths.

The mathematical formulation of optimal path planning for each agent β can be generalised as follows:

$$\begin{aligned} \text{Minimise } & F_{obj}(\mathscr{B}), \quad \forall \mathscr{B} : [0, 1] \to \mathbb{C} \\ \text{Subj. to } & \mathscr{B}(0) = x_{init}, \; \mathscr{B}(1) = x_{goal} \\ & \mathscr{B}(\omega) \in \mathbb{C}_{free}, \; \forall \omega \in [0, 1] \\ & \Upsilon = \{\upsilon'(\mathscr{B}), \upsilon''(\mathscr{B}), \upsilon'''(\mathscr{B}) \ldots\} \end{aligned} \tag{5.9}$$

For example, in the UAV swarm shepherding problem, path planning seeks to find the feasible paths for shepherding agents to follow from their initial positions to goal positions. These are likely to be collecting or driving positions for a subset of the UAV swarm assigned to each shepherding agent through task assignment. Considering Fig. 5.1, the red no-flight zones should be avoided by shepherding agents and sheep agents during mission execution. The black dotted line, which is collision-free with respect to no-flight zones and other agents, is given as an example of a feasible path for the shepherding agent β. Optimal path planning is to generate a feasible path that has optimal quality that is defined by the objective functions. The mathematical formulation of the optimal path planning for shepherding agents $B = \{\beta_1, \ldots, \beta_j, \ldots, \beta_M\}$ in the UAV swarm shepherding problem described above can be stated as

$$\begin{aligned} \text{Minimise } F_{obj}(\mathscr{B}) = & \Bigg(\sum_{j=1}^{M} UC_{fuel}^{j} \cdot Time^{j}, \\ & \sum_{j=1}^{M} L_{path}^{j}, \; \max(Time^{j}, j = 1, \ldots, M) \Bigg), \\ & \forall \mathscr{B} : [0, 1] \to \mathbb{C} \\ \text{Subj. to } & \mathscr{B}^{j}(0) = x_{init}^{j}, \; \mathscr{B}^{j}(1) = x_{goal}^{j}, \; \forall j \in \{1, \ldots, M\} \\ & \mathscr{B}^{j}(\omega) \in \mathbb{C}_{free}, \; \forall j \in \{1, \ldots, M\}, \; \omega \in [0, 1] \\ & \mathscr{B}^{j}(\omega) = x_{\phi}^{j}, \; \forall j \in \{1, \ldots, M\}, \; \exists \omega \in [0, 1] \\ & \upsilon' : Time^{j} \le Time^{j\_max}, \; \forall j \in \{1, \ldots, M\} \\ & \upsilon'' : L_{path}^{j} \le L_{path}^{j\_max}, \; \forall j \in \{1, \ldots, M\} \end{aligned} \tag{5.10}$$

where UC_{fuel}^{j} is the unit fuel cost of agent β_j; $Time^{j}$ is the total flight time of agent β_j for completing the assigned tasks; L_{path}^{j} is the length of the generated path for agent β_j; $\max(Time^{j}, j = 1, \ldots, M)$ is the makespan of the mission; x_{ϕ}^{j} is the configurations of the UAV swarm subsets assigned to agent β_j; $Time^{j\_max}$ is the maximum flight time for each agent β_j; $L_{path}^{j\_max}$ is the maximum path length that each agent β_j can travel. The objective is to minimise the total fuel cost of the shepherding agents, the total distance travelled by the shepherding agents and the makespan of the mission. The first three constraints assure that the shepherding agents depart from their initial positions, pass by the positions of UAV swarm subsets assigned to them and finally reach the goal positions. Constraints υ' and υ'' ensure that the flight time and length for each agent are less than their maximum flight time and length, respectively.

5.4.2 *Trajectory Planning*

Trajectory planning and path planning are two distinct but related parts of the planning system; trajectory is a path that is associated with time and could describe how the configuration of the UAV evolves over time [61]. Trajectory planning is to compute a trajectory for the agent from the start location to a destination while considering geometrical, physical and temporal constraints, which is usually related to path planning. Trajectory planning can be seen as a generalisation of path planning from static to dynamic environment.

Denoting T as the planning horizon and the free C-space at time $t \in [0, T]$ as $\mathbb{C}_{free}$, the solution of the trajectory planning problem can be represented as a time-parameterised function:

$$\mathscr{C}(t) : [0, T] \rightarrow \mathbb{C} \tag{5.11}$$

where $\mathscr{C}(0) = x_{init}$, $\mathscr{C}(T) = x_{goal}$, $\mathscr{C}(t) \in \mathbb{C}_{free}(t)$ for all $t \in [0, T]$. Then the optimal trajectory planning can be defined as [61]:

Definition 5.4 (Trajectory Planning) Given a trajectory planning problem $(x_{init}, x_{goal}, \mathbb{C}_{free}, T)$, the optimisation objective functions $F_{obj}(\mathscr{C})$, trajectory planning is to find the feasible trajectory $\mathscr{C}$, where $\mathscr{C}(0) = x_{init}$ and $\mathscr{C}(T) = x_{goal}$, that subjects to differential constraints $\Upsilon = \{\upsilon', \upsilon'', \upsilon''' \ldots\}$. Optimal trajectory planning is to find the optimal trajectory $\mathscr{C}^*$ with optimal $F_{obj}(\mathscr{C}^*)$ from all the feasible trajectories.

The mathematical formulation of optimal trajectory planning for each agent can be generalised as follows:

$$\begin{aligned} \text{Minimise } & F_{obj}(\mathscr{C}), \quad \forall \mathscr{C} : [0, T] \rightarrow \mathbb{C} \\ \text{Subj. to } & \mathscr{C}(0) = x_{init}, \ \mathscr{C}(T) = x_{goal} \\ & \mathscr{C}(t) \in \mathbb{C}_{free}(t), \ \forall t \in [0, T] \\ & \Upsilon = \{\upsilon'(\mathscr{C}), \upsilon''(\mathscr{C}), \upsilon'''(\mathscr{C}) \ldots\} \end{aligned} \tag{5.12}$$

For example, in the UAV swarm shepherding problem shown in Fig. 5.1, trajectory planning is necessary in order to avoid collision with other UAVs or moving obstacles since positions of these are changing over time. T is the total planning horizon. UAVs trajectory planning is to generate safe and short trajectories for the shepherd UAVs to follow from the start positions to the destinations over time. Here, we consider a trajectory planning problem that is based on the path planning problem described in Sect. 5.4.2 but is associated with time. The mathematical formulation of the optimal trajectory planning for shepherding agents $B = \{\beta_1, \ldots, \beta_j, \ldots, \beta_M\}$ in UAV swarm shepherding problem can be stated as

$$\begin{aligned}
\text{Minimise } F_{obj}(\mathscr{C}) = \Bigg(&\sum_{j=1}^{M} UC^j_{fuel} \cdot Time^j, \\
&\sum_{j=1}^{M} L^j_{traj}, \ \max(Time^j, j = 1, \ldots, M)\Bigg), \\
\forall \mathscr{C} &: [0, T] \rightarrow \mathbb{C} \\
\text{Subj. to } \mathscr{C}^j(0) &= x^j_{init}, \ \mathscr{C}^j(T) = x^j_{goal}, \ \forall j \in \{1, \ldots, M\} \\
\mathscr{C}^j(t) &\in \mathbb{C}_{free}(t), \ \forall j \in \{1, \ldots, M\}, \ t \in [0, T] \\
\mathscr{C}^j(t) &= x^j_{\phi}, \ \forall j \in \{1, \ldots, M\}, \ \exists t \in [0, T] \\
\upsilon' &: Time^j \leq Time^{j_max}, \ \forall j \in \{1, \ldots, M\} \\
\upsilon'' &: L^j_{traj} \leq L^{j_max}_{traj}, \ \forall j \in \{1, \ldots, M\} \\
\upsilon''' &: \max(Time^j, j = 1, \ldots, M) <= T
\end{aligned} \tag{5.13}$$

where L^j_{traj} is the length of the generated trajectory for agent β_j; $L^{j\_max}_{traj}$ is the maximum trajectory length that each agent β_j can travel. The optimisation objectives and the main constraints are consistent with those described in the path planning problem. Constraint υ''' assures that the set of tasks can be completed with the planning horizon T. The algorithms for optimising path planning and trajectory planning problems, which can be applied for UAV swarm shepherding problem, are introduced in the following Sect. 5.4.3.

5.4.3 Algorithms for Motion Planning

Algorithms for motion planning consist of both algorithms for path planning and trajectory planning. The complexity of path planning problem has been well studied and is known to be P-SPACE-hard [68]. This means that it is at least as hard as solving any NP-complete problem and there is no efficient (polynomial-time) algorithm that is able to solve all instances of the problem [61]. Many efforts have been undertaken, and a wide variety of algorithms have been proposed to solve path planning problems. Mac et al. [53] classified path planning algorithms into two major categories: classical approaches (e.g., cell decomposition method, potential field method, sampling-based methods) and heuristic approaches (e.g., neural network, fuzzy logic, bioinspired algorithms, hybrid algorithms). Yang et al. [86] reviewed path planning algorithms and classified them into five categories from the perspective of explore mechanism, including sampling-based algorithms, node-based optimal algorithms, mathematic model, bioinspired algorithms and multifusion-based algorithms. Some typical path planning algorithms are briefly introduced in the following, including *Sampling-based algorithms*, *Node-based algorithms*, *Bioinspired algorithms* and *Hybrid planning algorithms*.

Sampling-based algorithms require some initial information of the C-space and usually sample the environment as a set of nodes, cells or other forms[86]. This algorithm type maps the environment or searches randomly to find a feasible path. Rapidly exploring Random Trees (RRT) [46] is one of the most famous sampling-based algorithms and was first proposed by Lavalle. In each step of RRT, a new node is sampled; if the extension from the sampled to the nearest node succeeds, a new node will be added. RRT explores based on the Monte Carlo random sampling, which can guarantee completeness. However, always biases explore regions and are time-consuming when the environment is complex. A number of variants of RRT have been presented such as Dynamic Domain RRT (DDRRT) [88] for fast local escaping, Rapidly Exploring Random Graph (RRG) for ensuring asymptotic optimality [39] and RRT* for addressing differential constraints[39].

Probabilistic Road Map (PRM) [40] is another typical sampling-based algorithm. PRM samples a random node in feasible space $\mathbb{C}_{free}$ after the definition of $\mathbb{C}$ and $\mathbb{C}_{free}$. Then PRM tries to connect it to nearest k neighbours that should be collision-free. PRM only generates the road map from the start to the goal; hence, a search algorithm (e.g., A*, D*) is usually used to find the path with optimal objective value from all the feasible paths after the formation of road map. For example, PRM used a random sampling method to more efficiently distribute nodes of road map in 3D space, and then A* was used to generate a feasible path in [83]. Artificial potential field methods are classified as sampling-based algorithm as they need the whole configuration sampling information to escape from local minima [86].

Node-based algorithms explore through the decomposed graph. They deal with nodes' and arcs' weight information and calculate the cost by exploring among a set of nodes in the map. The typical representation of node-based algorithms is Dijkstra's algorithm [16], A* [33], D* [77], etc. Dijkstra's algorithm[16] aims to find the shortest path in a graph where edges' or arcs' weights are already known depending on the local path cost. A* [33] extended Dijkstra's algorithm by introducing a heuristic estimation of the cost from the current state to the goal state for reducing the total number of states. D* [77] is short for dynamic A* and can deal with dynamic obstacles by changing in real time its edge's weights to form a temporal map. Then D* moves the agent from its current position to a goal position in the shortest feasible path. The pseudocode of A* is given in Algorithm 4 as an example of an algorithm for path planning. Let $g(n)$ be the cost from the start node to the current node n, and $h(n)$ be the heuristic function for estimating the cost from the current node n to the goal. Then the evaluation function $f(n)$ can be calculated as

$$f(n) = g(n) + h(n) \tag{5.14}$$

In Algorithm 4, $Open$ is the set of nodes that can be considered for expansion. $Closed$ is the set of nodes that have been expanded, which makes sure that each node can be travelled at most once. $c(n, i)$ is the cost from the node n to node i. $Parent(i)$ is to record the path with the lowest cost.

Algorithm 4 The pseudocode of A*

```
Require: g(x_init) = 0; Open = ∅; Closed = ∅; Parent(x_init) = x_init
Inset x_init into Open with g(x_init) + h(x_init);
while Open ≠ ∅ do
    Select the node n in Open with the smallest value of g(n) + h(n) ;
    if n = X_goal then
        return Path;
    end if
    Remove the node n from Open;
    Add the node n to Closed;
    for each neighbour i of n do
        if i ∉ Closed then
            if i ∉ Open then
                g(i) = ∞;
                Parent(i) = NULL;
            end if
            if g(n) + c(n, i) < g(i) then
                g(i) = g(n) + c(n, i);
                Parent(i) = n;
                Inset i into Open or update i in Open with g(i) + h(i);
            end if
        end if
    end for
end while
Ensure: Parent
```

Bioinspired algorithms also provide an effective way for solving motion planning problems. They are flexible and promising when addressing NP-hard problem with an increasing number of variables and complex objective functions as well as constraints as they do not have special requests for the optimisation problems. A wide variety of bioinspired algorithms have been applied for path planning. For example, based on the improved GAs, an offline path planner in known environment and an online path planner in completely unknown environment are designed for UAV navigation [58]. Allaire et al.[4] presented Field Programmable Gate Array (FPGA) implementation to reduce the computational time of GA for designing a path planning generator that can recompile a path in real time, as unforeseen events are met by the UAV. A novel cooperative path planning algorithm based on the parallel GA was also proposed in [76] for multi-UAV system. Gautam and Verma[28] proposed a novel method for path planning by combining GA and Artificial Neural Network (ANN) where the output generated from GA is used to train the network of ANN. Fu et al. [26] designed a new route planner for unmanned aerial vehicle (UAV) to generate a safe and flyable path in the presence of different threat environments based on the phase angle-encoded and quantum-behaved PSO (θ-QPSO). Pigeon-inspired optimisation (PIO), another bioinspired algorithm, was improved by integrating the concept of predator-prey and was applied to solve the UAV 3D path planning problem in dynamic environment by Zhang and Duan [90, 91].

Hybrid planning algorithms are increasingly popular for generating optimal paths as they can utilise the advantages of different algorithms. Algorithms are usually combined in a multi-layer way for achieving better performance. For example, Yu et al. [89] presented a two-layer algorithm called ACO-A*, which combines Ant Colony Optimisation (ACO) with A* search for finding an optimal path to traverse multiple targets in complex environments. In ACO-A*, ACO takes responsibility for determining the optimal travelling order of targets, while A* plans paths based on a fine-grained modelling. In [55], a Voronoi diagram, visibility graph and potential fields were integrated for solving the path planning problem, and this algorithm can obtain a better trade-off between risk minimisation and length minimisation.

The trajectory planning problem is also PSPACE-hard since it is a generalisation of path planning by changing the problem definition from a static environment to a dynamic one. Trajectory planning algorithms are even more difficult to address due to their need to consider both time and C-space simultaneously. Path planning algorithms can be applied to find the solution since trajectory planning problems can be transformed to path planning problems in the state-time space that combines C-space with an added time dimension [25, 80]. For example, Suzuki et al. [79] used A* path search algorithm followed by direct optimisation of the trajectory to generate real-time 4-D flight trajectory with time as the extra dimension.

Decoupled trajectory planning is another efficient way to deal with trajectory planning problems. For example, the trajectory planning problem can be decomposed as a static path planning problem and a temporal velocity planning problem, where time is presented as an extra dimension, to avoid collision with static and moving obstacles, respectively [37]. In [71], Richter et al. utilised RRT* algorithm first to find a collision-free path without considering the dynamics and then to generate optimal polynomial trajectories for quadrotors in cluttered indoor environments by allocating time along the path. Foirini and Shiller[23] used velocity information to calculate the potential collisions directly and presented a first-order method by selecting avoidance manoeuvres to decide the trajectory of a robot in the dynamic environment.

Bioinspired algorithms are also applied for trajectory planning. A novel Max–Min adaptive Ant Colony Optimisation (ACO) approach was proposed by Duan et al. [18] to handle the multi-UAV coordinated trajectory replanning in dynamic and uncertain environments. Wang et al.[82] proposed an online trajectory planning algorithm based on receding horizon control and GA for the UAVs cooperative target tracking in the urban area. Alejo et al. [3] presented a new system for collision-free 4D trajectory planning, which automatically identifies conflicts between multiple UAVs and proposes the most effective solution considering the available computation time. In this system, a simple one-at-a-time strategy was used to quickly compute a feasible but non-optimal initial solution that will be continuously evolved based on PSO. Zhang et al.[93] used PSO to optimise the trajectory planning problem for improving the target localisation accuracy using UAV by improving estimation accuracy of heading angle bias.

The pseudocode of a road map-based trajectory planning algorithm [80] is given in Algorithm 5 as an example algorithm for trajectory planning. In this algorithm, a two-level search is performed to find the trajectory in the state-time space. The local level is for finding trajectories on single edges of the road map by using a depth-first search, while the global level is for finding the global trajectory by using an A*-like search. A road map is first constructed, consisting of a set of collision-free vertices and edges from static obstacles. Then x_{init} and x_{goal} are the start and goal vertices in the road map. Given the length of the edge be l and a state time (x, t), the new state time of the agent can be calculated as follows:

$$
\begin{aligned}
&v \in \{v_p, 0, -v_p\} \\
&\Delta x = x(t + \Delta t) - x(t) = v \cdot \Delta t / l
\end{aligned} \tag{5.15}
$$

where v is the velocity, v_p is the principal velocity within the velocity bound v_{max}, Δt is a small time step and l is the length of the edge. In the pseudocode, u is a vertex in the road map; $t_0 = 0$ is the start time; $cf(c, t)$ is a Boolean function denoting that whether an agent at state c and time t collides with any moving obstacles.

5.5 Mission Planning

Although mission planning problem can be decomposed into a number of sub-problems (e.g., task decomposition, task assignment, motion planning), considering each sub-problem independently may result in inefficient solutions since the variables involved in each sub-problem may have influence on other sub-problems. The interaction among variables in mission planning for the UAV swarm shepherding problem is shown in Fig. 5.3. It is essential to consider several sub-problems of mission planning simultaneously in some scenarios to improve the overall performance of multi-agent systems. A number of existing works discussed various scenarios of mission execution where combinatorial optimisation problems with different optimisation objectives and constraints were considered.

Algorithm 5 The pseudocode of road map-based trajectory planning algorithm

Require: All advancing probes on all the outgoing edges (x_{init}, u) of x_{init} with state time $(0, t_0)$ and return probes on incoming edges (u, x_{init}) of x_{init} with state time $(1, t_0)$, store them in the priority queue

while the priority queue is not empty **do**
    $p \leftarrow$ top element of the priority queue
    $(u_s, u_d) \leftarrow$ edge on which p is active
    if $f(p) > t_{max}$ **then**
        Terminate algorithm, report failure
    end if
    $(x, t) \leftarrow p.DOSTEP()$
    if $x = 1$ and $u_d = x_{goal}$ **then**
        Success
    end if
    if $x = 1$ and t is in an unvisited interval on u_d **then**
        Initialise advancing probes on all the outgoing edges (u_d, u) of u_d with state time $(0, t)$ and returning probes on incoming edges (u, u_d) of u_d with state time $(1, t)$, add them to the priority queue
    end if
    if p.STACKEMPTY() **then**
        Delete p and remove it from the priority queue
    end if
    Report no trajectory exists
end while

FUNCTION DOSTEP
$(x, t) \leftarrow STACKPOP()$
if not $(x, t).visited$ and $cf(x, t)$ **then**
    STACKPUSH($x - \Delta x, t + \Delta t$)
    $(x - \Delta x, t + \Delta t)$.backpointer $\leftarrow (x, t)$
    STACKPUSH($x, t + \Delta t$)
    $(x, t + \Delta t)$.backpointer $\leftarrow (x, t)$
    STACKPUSH($x + \Delta x, t + \Delta t$)
    $(x + \Delta x, t + \Delta t)$.backpointer $\leftarrow (x, t)$
    $(x, t).visited \leftarrow$ true
    return (x, t)
else
    $(x, t).visited \leftarrow$ true
    return NULL
end if

For example, task decomposition and assignment are coupled in some works. The mission might be decomposed first and then be allocated to the agents [78], while in some methods, complex missions are first allocated to agents and then each agent locally decomposes its assigned mission [9]. Task decomposition and assignment are coupled in multi-target tracking and surveillance problem and are tackled simultaneously by utilising region allocation trees in [1]. Liu et al.[50] proposed a motif-based rescue mission planning method for UAV swarm while considering dynamic reconfiguration. Motif was introduced as the basic unit of configuration, which only requires a few communication connections among UAVs.

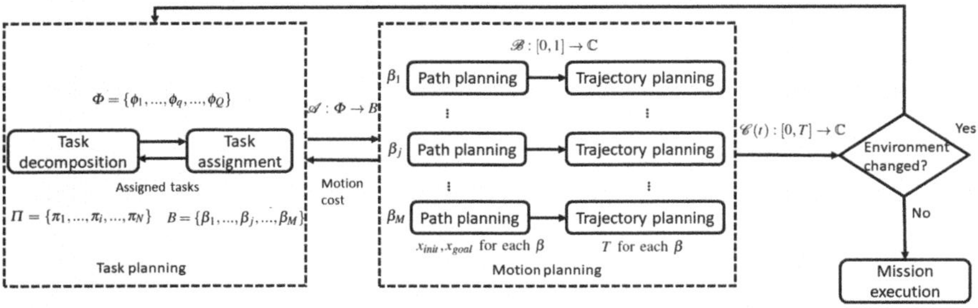

Fig. 5.3 Illustration of the interaction among variables in mission planning for shepherding UAV swarm

The mission is decomposed into logically related tasks that are assigned to motifs, and the NSGA-III was applied to select the optimal task execution sequence in this work.

Some researchers investigated the combinatorial problem of task assignment and path planning. Edison and Shima [19] applied GA for the integrated task assignment and path planning in order to reduce the computational complexity of the resulting combinatorial optimisation problem where multiple UAVs are required to perform multiple consecutive tasks cooperatively on each ground target. Lamont[43] presented a mission planning system consisting of a vehicle route planner for assigning a set of vehicles to a set of targets and a path planner to generate a feasible path for each vehicle based on a parallel multi-objective evolutionary algorithm. The task assignment problem was joined with the route planning problem in reference [94]. The route was planned first to obtain the sensing latency and the sensing cost (i.e., the energy consumption) of each UAV, based on which the task assignment strategies are optimised.

Real-time planning or replanning in uncertain or dynamic environments is another area of research in the literature. It usually involves new planning for the previous mission plan due to uncertain factors, such as a vehicle or sensor failure, moving obstacles or a new task arrival during the real-time execution of the mission. Ramirez and Camacho [66] extended QGroundControl, an open-source simulation environment for controlling multi-vehicle, for automated mission planning and replanning. Model Predictive Control (MPC) was applied in [62] to continuously update the environmental information for the dynamic multi-objective path planner. Duan et al. [18] investigated multi-UAV coordinated trajectory replanning to generate new trajectories for UAVs from their current sites to the desired destinations after the environment changed. A task priority execution order was given as an input to automatically generate the mission planning in [49], and an improved NSGA-III with adaptive genetic operators was used to optimise the mission planning problem with changing task priority execution orders.

5.6 Conclusion and Discussion

UAVs can be utilised as effective tools to complete missions in a wide range of areas, ranging from infrastructure monitoring and inspection, precision agriculture, surveillance to terrain mapping. Mission planning is essential to shepherding UAV swarms in complex environments.

The reliance of optimisation methods to replace reactive models offers opportunities. First, the optimisation methods get integrated in the user interface as recommender systems to offer a human operator with a number of options to select from. Second, the optimisation methods can quickly narrow down a search space that a human may not be able to do cognitively. Third, as recommender systems, they reduce human workload by looking after some of the most complex and demanding tasks on the behaviour of the human.

The mission planning for shepherding UAV swarm can be decomposed into a number of sub-problems, such as task decomposition, task assignment and path planning, where the solution of one sub-problem is the input to the next one. The definitions of the major sub-problems involved in mission planning were introduced in the general form, and how those can be used for shepherding UAV swarm is also presented in this chapter. This chapter also reviewed the well-known algorithms for sub-problems involved in mission planning, as well as the state of art of combinatorial mission planning.

There exists efficient algorithms for relatively simpler sub-problems, and the decomposition of mission planning allows the application of well-developed sub-problem approaches. Although significant progress has been made towards the optimisation of sub-problems, the trade-off between the search accuracy and the computational complexity has not been completely solved yet. This might cause the difficulty of real-time planning. The algorithms that can optimise the planning problems accurately are usually too computationally expensive to be applied in real time; when they have reduced complexity, the solutions are not proven to be satisfactory.

One area that needs more attention is the optimisation of comprehensive mission planning problems with most or all of its sub-problems considered simultaneously. As mentioned before, addressing each sub-problem independently might cause inefficient solutions as the solution of one sub-problem is the input to the next one. More efforts should be made to integrate and tailor the sub-problems approaches or to present new carefully designed approaches, so that the comprehensive mission planning problem can be optimised in an efficient way.

Another issue is the modelling of complicated environments. The increasingly challenging environment that UAVs face (e.g., uncertainties arising from the environment or from the failures of agents, large-scale environment) increases the complexity of modelling, which further increases the complexity of the mission planning optimisation problem. One example of this in UAV swarm shepherding is that the shepherding agent might fail to collect or drive one or more sheep UAV in the swarm. The uncontrolled sheep UAV leaves the UAV mass so that the planning system needs to do the replanning.

Scalability also remains a significant challenge for a UAV swarm mission planner as it takes time to address for a large set of agents, a large set of tasks, a number of sub-problems and complex environments where a number of complex and dynamic constraints need to be considered. Little work has been done on the UAV swarm shepherding modelling and optimisation. The application of mission planning techniques to shepherding should receive more attention as these techniques are promising ways to enhance the shepherding efficiency.

When considering the above challenges, there is a trade-off that needs to be made between the quality of solutions obtained by algorithm and the computational complexity. Designing a comprehensive mission planning optimisation algorithm for complicated environments remains as a significant challenge in the field of optimisation. This challenge causes gaps between simulation, where researchers have the time to run and tweak their algorithms, and the real world, where an agent is faced with novel situations and needs to make a decision to navigate and sustain its performance during a mission. Future work on optimisation and shepherding should focus on some of these issues to advance the design of smarter shepherd capable to operate safely and near-optimally in complex dynamic environments.

References

1. Adamey, E., Oğuz, A.E., Özgüner, Ü.: Collaborative multi-MSA multi-target tracking and surveillance: a divide & conquer method using region allocation trees. J. Intell. Rob. Syst. **87**(3–4), 471–485 (2017)
2. Agassounon, W., Martinoli, A.: Efficiency and robustness of threshold-based distributed allocation algorithms in multi-agent systems. In: Proceedings of the First International Joint Conference on Autonomous Agents and Multiagent Systems: Part 3, pp. 1090–1097. ACM, New York (2002)
3. Alejo, D., Cobano, J.A., Heredia, G., Ollero, A.: Collision-free 4d trajectory planning in unmanned aerial vehicles for assembly and structure construction. J. Intell. Rob. Syst. **73**(1–4), 783–795 (2014)
4. Allaire, F.C., Tarbouchi, M., Labonté, G., Fusina, G.: FPGA implementation of genetic algorithm for UAV real-time path planning. In: Unmanned Aircraft Systems, pp. 495–510. Springer, Berlin (2008)
5. Anand, R., Mehrotra, K., Mohan, C.K., Ranka, S.: Efficient classification for multiclass problems using modular neural networks. IEEE Trans. Neural Netw. **6**(1), 117–124 (1995)
6. Auda, G., Kamel, M.: Modular neural networks: a survey. Int. J. Neur. Syst. **9**(02), 129–151 (1999)
7. Bertuccelli, L., Alighanbari, M., How, J.: Robust planning for coupled cooperative UAV missions. In: 2004 43rd IEEE Conference on Decision and Control (CDC)(IEEE Cat. No. 04CH37601), vol. 3, pp. 2917–2922. IEEE, Piscataway (2004)
8. Bertuccelli, L., Choi, H.L., Cho, P., How, J.: Real-time multi-UAV task assignment in dynamic and uncertain environments. In: AIAA Guidance, Navigation, and Control Conference, p. 5776 (2009)

9. Botelho, S.C., Alami, R.: M+: A scheme for multi-robot cooperation through negotiated task allocation and achievement. In: Proceedings 1999 IEEE International Conference on Robotics and Automation (Cat. No. 99CH36288C), vol. 2, pp. 1234–1239. IEEE, Piscataway (1999)
10. Campos, M., Bonabeau, E., Theraulaz, G., Deneubourg, J.L.: Dynamic scheduling and division of labor in social insects. Adaptive Behav. **8**(2), 83–95 (2000)
11. Chen, H.X., Nan, Y., Yang, Y.: Multi-UAV reconnaissance task assignment for heterogeneous targets based on modified symbiotic organisms search algorithm. Sensors **19**(3), 734 (2019)
12. Chen, Y., Xie, L., He, W., Jiang, Q., Xu, J.: An improved A* algorithm based on divide-and-conquer method for golf unmanned cart path planning. In: International Conference on Artificial Intelligence for Communications and Networks, pp. 497–505. Springer, Berlin (2019)
13. Choi, H.L., Brunet, L., How, J.P.: Consensus-based decentralized auctions for robust task allocation. IEEE Trans. Rob. **25**(4), 912–926 (2009)
14. Coello, C.A.C., Pulido, G.T., Lechuga, M.S.: Handling multiple objectives with particle swarm optimization. IEEE Trans. Evol. Comput. **8**(3), 256–279 (2004)
15. Deb, K., Pratap, A., Agarwal, S., Meyarivan, T.: A fast and elitist multiobjective genetic algorithm: NSGA-II. IEEE Trans. Evol. Comput. **6**(2), 182–197 (2002)
16. Dijkstra, E.W.: A note on two problems in connexion with graphs. Numer. Math. **1**(1), 269–271 (1959)
17. Dorigo, M., Stützle, T.: The ant colony optimization metaheuristic: Algorithms, applications, and advances. In: Handbook of Metaheuristics, pp. 250–285. Springer, Berlin (2003)
18. Duan, H.b., Zhang, X.y., Wu, J., Ma, G.j.: Max-min adaptive ant colony optimization approach to multi-UAVs coordinated trajectory replanning in dynamic and uncertain environments. J. Bionic Eng. **6**(2), 161–173 (2009)
19. Edison, E., Shima, T.: Integrated task assignment and path optimization for cooperating uninhabited aerial vehicles using genetic algorithms. Comput. Operat. Res. **38**(1), 340–356 (2011)
20. Ergezer, H., Leblebicioglu, K.: Path planning for UAVs for maximum information collection. IEEE Trans. Aerosp. Electro. Syst. **49**(1), 502–520 (2013)
21. Ergezer, H., Leblebicioğlu, K.: 3D path planning for multiple UAVs for maximum information collection. J. Intell. Rob. Syst. **73**(1–4), 737–762 (2014)
22. Fei, S., Yan, C., Lin-Cheng, S.: UAV cooperative multi-task assignment based on ant colony algorithm. Acta Aeronautica et Astronautica Sinica **29**, 188–s189 (2008)
23. Fiorini, P., Shiller, Z.: Motion planning in dynamic environments using velocity obstacles. Int. J. Rob. Res. **17**(7), 760–772 (1998)
24. Foo, J.L., Knutzon, J., Oliver, J., Winer, E.: Three-dimensional path planning of unmanned aerial vehicles using particle swarm optimization. In: 11th AIAA/ISSMO Multidisciplinary Analysis and Optimization Conference, p. 6995 (2006)
25. Fraichard, T.: Trajectory planning in a dynamic workspace: a' state-time space'approach. Adv. Rob. **13**(1), 75–94 (1998)
26. Fu, Y., Ding, M., Zhou, C.: Phase angle-encoded and quantum-behaved particle swarm optimization applied to three-dimensional route planning for UAV. IEEE Trans. Syst. Man Cybern. Part A Syst. Humans **42**(2), 511–526 (2012)
27. Gallinari, P.: Training of modular neural net systems. In: The Handbook of Brain Theory and Neural Networks, pp. 582–585. MIT Press, Cambridge (1998)
28. Gautam, S.A., Verma, N.: Path planning for unmanned aerial vehicle based on genetic algorithm & artificial neural network in 3D. In: 2014 International Conference on Data Mining and Intelligent Computing (ICDMIC), pp. 1–5. IEEE, Piscataway (2014)
29. Gerkey, B.P., Mataric, M.J.: Sold!: Auction methods for multirobot coordination. IEEE Trans. Rob. Autom. **18**(5), 758–768 (2002)
30. Gerkey, B.P., Matarić, M.J.: A formal analysis and taxonomy of task allocation in multi-robot systems. Int. J. Rob. Res. **23**(9), 939–954 (2004)

31. Goerzen, C., Kong, Z., Mettler, B.: A survey of motion planning algorithms from the perspective of autonomous UAV guidance. J. Intell. Rob. Syst. **57**(1–4), 65 (2010)
32. González, V., Monje, C., Moreno, L., Balaguer, C.: UAVs mission planning with flight level constraint using fast marching square method. Rob. Auton. Syst. **94**, 162–171 (2017)
33. Hart, P.E., Nilsson, N.J., Raphael, B.: A formal basis for the heuristic determination of minimum cost paths. IEEE Trans. Syst. Sci. Cybern. **4**(2), 100–107 (1968)
34. Hartigan, J.A., Wong, M.A.: Algorithm as 136: A k-means clustering algorithm. J. R. Stat. Soc. Ser. C (Appl. Stat.) **28**(1), 100–108 (1979)
35. Holland, J.H., et al.: Adaptation in Natural and Artificial Systems: An Introductory Analysis with Applications to Biology, Control, and Artificial Intelligence. MIT Press, Cambridge (1992)
36. Jain, A.K.: Data clustering: 50 years beyond k-means. Pattern Recogn. Lett. **31**(8), 651–666 (2010)
37. Kant, K., Zucker, S.W.: Toward efficient trajectory planning: The path-velocity decomposition. Int. J. Rob. Res. **5**(3), 72–89 (1986)
38. Kanungo, T., Mount, D.M., Netanyahu, N.S., Piatko, C.D., Silverman, R., Wu, A.Y.: An efficient k-means clustering algorithm: Analysis and implementation. IEEE Trans. Pattern Anal. Mach. Intell. **24**(7), 881–892 (2002)
39. Karaman, S., Frazzoli, E.: Optimal kinodynamic motion planning using incremental sampling-based methods. In: 49th IEEE Conference on Decision and Control (CDC), pp. 7681–7687. IEEE, Piscataway (2010)
40. Kavralu, L.E., Svestka, P., Latombe, J.C., Overmars, M.H.: Probabilistic roadmaps for path planning in high-dimensional configuration spaces. IEEE Trans. Rob. Autom. **12**(4), 566–580 (1996)
41. Kennedy, J., Eberhart, R.: Particle swarm optimization. In: Proceedings., IEEE International Conference on Neural Networks, 1995, vol. 4, pp. 1942–1948. IEEE, Piscataway (1995)
42. Kim, M.H., Baik, H., Lee, S.: Response threshold model based UAV search planning and task allocation. J. Intell. Rob. Syst. **75**(3–4), 625–640 (2014)
43. Lamont, G.B., Slear, J.N., Melendez, K.: UAV swarm mission planning and routing using multi-objective evolutionary algorithms. In: 2007 IEEE Symposium on Computational Intelligence in Multi-Criteria Decision-Making, pp. 10–20. IEEE, Piscataway (2007)
44. Landén, D., Heintz, F., Doherty, P.: Complex task allocation in mixed-initiative delegation: a UAV case study. In: International Conference on Principles and Practice of Multi-Agent Systems, pp. 288–303. Springer, Berlin (2010)
45. Latombe, J.C.: Robot Motion Planning, vol. 124. Springer Science & Business Media, Berlin (2012)
46. LaValle, S.M.: Rapidly-Exploring Random Trees: A New Tool for Path Planning. CiteSeer (1998)
47. Lim, K.K., Ong, Y.S., Lim, M.H., Chen, X., Agarwal, A.: Hybrid ant colony algorithms for path planning in sparse graphs. Soft Comput. **12**(10), 981–994 (2008)
48. Lin, W., Wang, J.Z., Liang, C., Qi, D.: A threshold-based dynamic resource allocation scheme for cloud computing. Procedia Eng. **23**, 695–703 (2011)
49. Liu, J., Wang, W., Li, X., Wang, T., Bai, S., Wang, Y.: Solving a multi-objective mission planning problem for UAV swarms with an improved nsga-iii algorithm. Int. J. Comput. Intell. Systems **11**(1), 1067–1081 (2018)
50. Liu, J., Wang, W., Li, X., Wang, T., Wang, T.: A motif-based mission planning method for UAV swarms considering dynamic reconfiguration. Defence Sci. J. **68**(2), 159–166 (2018)
51. Lloyd, S.: Least squares quantization in PCM. IEEE Trans. Inf. Theory **28**(2), 129–137 (1982)
52. Lu, B.L., Ito, M.: Task decomposition and module combination based on class relations: a modular neural network for pattern classification. IEEE Trans. Neur. Netw. **10**(5), 1244–1256 (1999)
53. Mac, T.T., Copot, C., Tran, D.T., De Keyser, R.: Heuristic approaches in robot path planning: a survey. Rob. Auton. Syst. **86**, 13–28 (2016)

54. MacQueen, J., et al.: Some methods for classification and analysis of multivariate observations. In: Proceedings of the Fifth Berkeley Symposium on Mathematical Statistics and Probability, vol. 1, pp. 281–297. Oakland (1967)
55. Masehian, E., Amin-Naseri, M.: A voronoi diagram-visibility graph-potential field compound algorithm for robot path planning. J. Rob. Syst. **21**(6), 275–300 (2004)
56. Máthé, K., Buşoniu, L.: Vision and control for UAVs: a survey of general methods and of inexpensive platforms for infrastructure inspection. Sensors **15**(7), 14,887–14,916 (2015)
57. Miettinen, K.: Nonlinear Multiobjective Optimization, vol. 12. Springer Science & Business Media, Berlin (1999)
58. Nikolos, I.K., Valavanis, K.P., Tsourveloudis, N.C., Kostaras, A.N.: Evolutionary algorithm based offline/online path planner for UAV navigation. IEEE Trans. Syst. Man Cybern. Part B (Cybern.) **33**(6), 898–912 (2003)
59. Nikolos, I.K., Zografos, E.S., Brintaki, A.N.: UAV path planning using evolutionary algorithms. In: Innovations in Intelligent Machines-1, pp. 77–111. Springer, Berlin (2007)
60. Oh, G., Kim, Y., Ahn, J., Choi, H.L.: Market-based distributed task assignment of multiple unmanned aerial vehicles for cooperative timing mission. J. Aircraft **54**(6), 2298–2310 (2017)
61. Paden, B., Čáp, M., Yong, S.Z., Yershov, D., Frazzoli, E.: A survey of motion planning and control techniques for self-driving urban vehicles. IEEE Trans. Intell. Veh. **1**(1), 33–55 (2016)
62. Peng, X., Xu, D.: Intelligent online path planning for UAVs in adversarial environments. Int. J. Adv. Rob. Syst. **9**(1), 3 (2012)
63. Peng, Z.h., Wu, J.p., Chen, J.: Three-dimensional multi-constraint route planning of unmanned aerial vehicle low-altitude penetration based on coevolutionary multi-agent genetic algorithm. J. Cent. South Univ. Technol. **18**(5), 1502 (2011)
64. Phung, M.D., Quach, C.H., Dinh, T.H., Ha, Q.: Enhanced discrete particle swarm optimization path planning for UAV vision-based surface inspection. Autom. Constr. **81**, 25–33 (2017)
65. Pohl, A.J., Lamont, G.B.: Multi-objective UAV mission planning using evolutionary computation. In: 2008 Winter Simulation Conference, pp. 1268–1279. IEEE, Piscataway (2008)
66. Ramirez-Atencia, C., Camacho, D.: Extending QGroundcontrol for automated mission planning of UAVs. Sensors **18**(7), 2339 (2018)
67. Ramirez-Atencia, C., R-Moreno, M.D., Camacho, D.: Handling swarm of UAVs based on evolutionary multi-objective optimization. Prog. Artif. Intell. **6**(3), 263–274 (2017)
68. Reif, J.H.: Complexity of the mover's problem and generalizations. In: 20th Annual Symposium on Foundations of Computer Science (SFCS 1979), pp. 421–427. IEEE, Piscataway (1979)
69. Ren, X., Zhang, L., Ahmad, S., Nie, D., Yang, F., Xiang, L., Wang, Q., Shen, D.: Task decomposition and synchronization for semantic biomedical image segmentation (2019). Preprint arXiv:1905.08720
70. Richards, A., Bellingham, J., Tillerson, M., How, J.: Coordination and control of multiple UAVs. In: AIAA Guidance, Navigation, and Control Conference and Exhibit, p. 4588 (2002)
71. Richter, C., Bry, A., Roy, N.: Polynomial trajectory planning for aggressive quadrotor flight in dense indoor environments. In: Robotics Research, pp. 649–666. Springer, Berlin (2016)
72. Salman, A., Ahmad, I., Al-Madani, S.: Particle swarm optimization for task assignment problem. Microprocess. Microsyst. **26**(8), 363–371 (2002)
73. Salvo, G., Caruso, L., Scordo, A.: Urban traffic analysis through an UAV. Procedia-Soc. Behav. Sci. **111**, 1083–1091 (2014)
74. Schumacher, C., Chandler, P., Pachter, M., Pachter, L.: Constrained optimization for UAV task assignment. In: AIAA Guidance, Navigation, and Control Conference and Exhibit, p. 5352 (2004)
75. Shima, T., Schumacher, C.: Assigning cooperating UAVs to simultaneous tasks on consecutive targets using genetic algorithms. J. Oper. Res. Soc. **60**(7), 973–982 (2009)
76. Shorakaei, H., Vahdani, M., Imani, B., Gholami, A.: Optimal cooperative path planning of unmanned aerial vehicles by a parallel genetic algorithm. Robotica **34**(4), 823–836 (2016)
77. Stentz, A.: Optimal and efficient path planning for partially known environments. In: Intelligent Unmanned Ground Vehicles, pp. 203–220. Springer, Berlin (1997)

78. Stone, P., Veloso, M.: Task decomposition, dynamic role assignment, and low-bandwidth communication for real-time strategic teamwork. Artif. Intell. **110**(2), 241–273 (1999)
79. Suzuki, S., Komatsu, Y., Yonezawa, S., Masui, K., Tomita, H.: Online four-dimensional flight trajectory search and its flight testing. In: AIAA Guidance, Navigation, and Control Conference and Exhibit, p. 6475 (2005)
80. Van Den Berg, J.P., Overmars, M.H.: Roadmap-based motion planning in dynamic environments. IEEE Trans. Rob. **21**(5), 885–897 (2005)
81. Wagstaff, K., Cardie, C., Rogers, S., Schrödl, S., et al.: Constrained k-means clustering with background knowledge. In: Proceedings of the Eighteenth International Conference on Machine Learning, vol. 1, pp. 577–584 (2001)
82. Wang, L., Su, F., Zhu, H., Shen, L.: Active sensing based cooperative target tracking using UAVs in an urban area. In: 2010 2nd International Conference on Advanced Computer Control, vol. 2, pp. 486–491. IEEE, Piscataway (2010)
83. Yan, F., Liu, Y.S., Xiao, J.Z.: Path planning in complex 3D environments using a probabilistic roadmap method. Int. J. Autom. Comput. **10**(6), 525–533 (2013)
84. Yan, Z., Jouandeau, N., Cherif, A.A.: A survey and analysis of multi-robot coordination. Int. J. Adv. Rob. Syst. **10**(12), 399 (2013)
85. Yang, Y., Zhou, C., Tian, Y.: Swarm robots task allocation based on response threshold model. In: 2009 4th International Conference on Autonomous Robots and Agents, pp. 171–176. IEEE, Piscataway (2009)
86. Yang, L., Qi, J., Song, D., Xiao, J., Han, J., Xia, Y.: Survey of robot 3D path planning algorithms. J. Control Sci. Eng. **2016**, 5 (2016)
87. Yazdani, D., Omidvar, M.N., Branke, J., Nguyen, T.T., Yao, X.: Scaling up dynamic optimization problems: A divide-and-conquer approach. IEEE Trans. Evol. Comput. **24**, 1–15 (2019)
88. Yershova, A., Jaillet, L., Siméon, T., LaValle, S.M.: Dynamic-domain RRTs: Efficient exploration by controlling the sampling domain. In: Proceedings of the 2005 IEEE International Conference on Robotics and Automation (2005)
89. Yu, X., Chen, W.N., Gu, T., Yuan, H., Zhang, H., Zhang, J.: Aco-A*: Ant colony optimization plus A* for 3d traveling in environments with dense obstacles. IEEE Trans. Evol. Comput. **23**, 617–631 (2018)
90. Zhang, B., Duan, H.: Predator-prey pigeon-inspired optimization for UAV three-dimensional path planning. In: International Conference in Swarm Intelligence, pp. 96–105. Springer, Berlin (2014)
91. Zhang, B., Duan, H.: Three-dimensional path planning for uninhabited combat aerial vehicle based on predator-prey pigeon-inspired optimization in dynamic environment. IEEE/ACM Trans. Comput. Biol. Bioinform. **14**(1), 97–107 (2015)
92. Zhang, S., Zhou, Y., Li, Z., Pan, W.: Grey wolf optimizer for unmanned combat aerial vehicle path planning. Adv. Eng. Softw. **99**, 121–136 (2016)
93. Zhang, L., Chen, J., Deng, F., Bi, Y., Phang, S.K., Chen, X.: Trajectory planning for improving vision-based target geolocation performance using a quad-rotor UAV. IEEE Trans. Aerosp. Electron. Syst. **55**, 2382–2394 (2018)
94. Zhou, Z., Feng, J., Gu, B., Ai, B., Mumtaz, S., Rodriguez, J., Guizani, M.: When mobile crowd sensing meets UAV: energy-efficient task assignment and route planning. IEEE Trans. Commun. **66**(11), 5526–5538 (2018)

Chapter 6
Towards Ontology-Guided Learning for Shepherding

Benjamin Campbell

Abstract Shepherding offers an exciting application for machine learning research. Shepherding tasks are scalable in terms of both complexity and dimension. This scalability supports investigations into the generality of learned multi-agent solutions. Shepherding is also valuable for the study of how multi-agent learning systems transition from simulation to physical systems. This chapter reviews previous learning strategies for shepherding and highlights the advantages of applying prior knowledge to the design of learning systems for shepherding. It presents ontology guided learning, a hybrid learning approach to learning. Ontology guided learning will enable the application of symbolic prior knowledge to non-symbolic learning systems. This will allow a non-symbolic system to reason on abstract concepts, reduce dimensionality by partitioning the state and action space, increase transparency and allow learning to focus on the parametric rather than semantic parts of the problem, where it will likely be most effective. This chapter concludes by describing how ontology guided learning could be applied to the shepherding problem.

Keywords Guided learning · Hybrid learning · Shepherding

Shepherding offers an exciting application for machine learning research. Shepherding tasks are scalable in terms of both complexity and dimension. This scalability supports investigations into the generality of learned multi-agent solutions. Shepherding is also valuable for the study of how multi-agent learning systems transition from simulation to physical systems. This chapter reviews previous learning strategies for shepherding and highlights the advantages of applying prior knowledge to the design of learning systems for shepherding. It presents ontology guided learning,

B. Campbell (✉)
Advanced Vehicle Systems, Land Vehicles and Systems, Land Division, Defence Science and Technology Group, Edinburgh, SA, Australia
e-mail: benjamin.campbell@dst.defence.gov.au

H. A. Abbass, R. A. Hunjet (eds.), *Shepherding UxVs for Human-Swarm Teaming*, Unmanned System Technologies,
https://doi.org/10.1007/978-3-030-60898-9_6

a hybrid learning approach to learning. Ontology guided learning will enable the application of symbolic prior knowledge to non-symbolic learning systems. This will allow a non-symbolic system to reason on abstract concepts, reduce dimensionality by partitioning the state and action space, increase transparency and allow learning to focus on the parametric rather than semantic parts of the problem, where it will likely be most effective. This chapter concludes by describing how ontology guided learning could be applied to the shepherding problem.

6.1 Introduction

It has been suggested that shepherding approaches can be applied to many applications; it has been suggested as a potential approach to crowd control [27], robotic tour guides [4], environmental cleanup [4, 27] and manipulating micro-organisms and nanomedicine [19].

Shepherding also offers an interesting and complex problem to the machine learning field of research. Applications that allow systems to be trained in simulation before transitioning to the real world are important. Without them, learning systems in robotics would remain confined to theory and simulations. Learning shepherding systems have previously been used in this manner [23]. Shepherding has a number of advantages in this capacity compared to more widely known robotics problems such as driverless cars. Shepherding is easily scalable in terms of both dimension and complexity [17, 21]. This allows for incremental development as well as investigation of the generalizability of learned solutions. Shepherding environments are controlled and relatively predictable (sheep behaviour has not changed much in the last several 1000 years) and verified and widely accepted models of the controlled objects exist [27]. Additionally, compared to other robotics applications in which the consequences of poor performance are significant, in shepherding the stakes are relatively low. This means that trust is not as great an obstacle to adoption in physical systems.

The following sections of this chapter review previous learning approaches to shepherding. Based on this review, it is proposed that applying domain knowledge to learning systems has the potential to significantly improve the performance of learning shepherding systems. This is followed by a justification for applying domain knowledge using hybrid symbolic/non-symbolic learning and a review of hybrid learning techniques. The chapter concludes by presenting a hybrid approach to learning, called ontology guided learning, explaining how it could be applied to the shepherding problem and describing future work. Applying ontology guided learning and transitioning from simulation to robotics would provide a valuable demonstration of a learning system capable of reasoning on abstract concepts controlling a multi-agent robotic system.

6.2 Learning Shepherding Systems

A number of different learning approaches have been applied to the shepherding problem including genetic algorithms acting on symbolic rules [23], expression trees [6], reinforcement learning (RL) systems[4, 11, 12], genetic algorithms acting on artificial neural nets (ANN) [13, 17, 21] and a GA optimized RL [4]. The following section summarizes the learning shepherding literature with the aim of identifying outstanding challenges and opportunities for research within learning shepherding systems.

Although a wide variety of learning approaches have been applied to the shepherding problem, a smaller number of motivations for the work exist. These include investigation into knowledge transfer and generalized learning [6, 23], transparency of learning systems [23] and maximization of learning efficiency and fitness [4, 13, 17, 19, 21]. These motivations will be used to guide the following discussion.

Two papers concerned with generalizable learning and knowledge transfer are Brulé et al. [6] and Schultz et al. [23]. The approach described in both papers utilizes a genetic algorithm (GA) to improve on predefined symbolic rules. Schultz et al. [23] demonstrated a learning shepherding system that was trained in simulation before transferring its learned rules to robotic systems. The authors utilized the rule learning system SAMUEL (Strategy Acquisition Method Using Empirical Learning), which through learning enables the application of rules to a single shepherd system. It employs a competitive approach to rule selection where, at each iteration, rules are applied based on their applicability to the current system state and their "Strength rating" (analogous to fitness). Based on the level of success of the selected rules, a reward signal is generated, which is used to update rule strengths. A genetic algorithm is then used to generate the new population of rules for the next episode. This approach was applied to a shepherding problem with one dog and one sheep. A successful policy for shepherding was learned in simulation and successfully transferred to real robots. Importantly the authors state that in order for this transition to be successful, noise and variation in the simulated environment needed to be greater the variation encountered in the real world environment.

While [23] specifically focused on transferring knowledge from simulation to robotics, Brulé et al's [6] work more broadly focused on the necessary characteristics for a training simulation to develop generalizable and transferable strategies. A genetic algorithm approach was used to evolve expression trees to control the behaviour of shepherding agents. The system was trained on an environment with 20 sheep and successfully learned effective shepherding strategies for both a single shepherd and a 3 shepherd system. Starting positions for each simulation run were selected at random, while sheep behaviour and numbers were kept constant. Brulé et al. [6] showed that this minimal variation during learning resulted in learned strategies that could perform under a number of different variations (i.e., highly general strategies). The authors found that after sufficient training, learned strategies maintained fitness comparable to a human written strategy when encountering several different previously un-encountered variations. These variations included varying sheep numbers from 5 to 100, tripling of sheep speed, reduction of the clustering force of the sheep and forcing all sheep starting positions to be in the same half of the field.

An important contrast between these two learning systems is the transparency of their actions. Transparency is the ease in which a human observer is able to understand the rationale motivating the actions of a system and is considered a strength of symbolic AI [29]. However, despite both systems acting on symbolic knowledge, only the system developed by Schultz et al. [23] produced human understandable rules. The SAMUEL system used in [23] included a human readable language for the expression of rules and a GUI to support transparency of the system. In contrast, the approach taken by Brulé et al. [6] did not include these tools, and the authors found that the learned rules could not be understood by reading them. This suggests that even in symbolic learning systems, transparency cannot be assumed and must be incorporated into the design.

Brulé et al's work also demonstrated the importance of domain knowledge in designing fitness functions. Initially, the fitness function used was related to proximity of the sheep to the pen. However, the sheep needed to enter the pen via a gate. The initial fitness function resulted in the evolution of strategies that herded sheep to positions on the pen with no entrance (i.e., the shepherding agents were trapped in a local minima). Successful strategies only evolved when the fitness function was changed to only reward the number of sheep inside the pen. In other shepherding problems, where there is no pen and sheep are simply herded to goal area, such as in Strömbom et al. [27], the original fitness function would have been adequate. This highlights the importance of choosing fitness functions that match the problem at hand, showing that even in learning systems the application of some domain knowledge can be necessary.

While [6, 23] are concerned with knowledge transfer and transparency, a more common focus in the literature is on learning efficiency and fitness. One of the major challenges for learning efficiency is dimensionality. The more dimensions in a problem, the slower the learning will be. This problem is addressed in a multi-agent shepherding system in Baumann [4]. Here, the author uses reinforcement learning (RL) approach, reducing dimensionality through aggregation of similar system states into abstract state regions. They refer to this process as "State Aggregation". The problem with state aggregation is that it involves a trade-off between learning speed and fitness of the learned solutions. Over-aggregation will occlude details required for learning, while reducing aggregation increases dimensionality, slowing down learning. Unfortunately, optimization of this trade-off has been shown to be NP hard [4]. The usual approach to state aggregation is to rely on domain knowledge to choose appropriate level of quantization giving the desired fidelity. Baumann proposes an alternative technique, Neural Gas Q learning (GNG-Q), that combines Q learning with unsupervised gas vector quantization. This technique updates state aggregation during learning, taking responsibility for aggregation selection away from the designer.

Baumann [4] extends this approach further by applying a smoothing function to the value function of the Q learner. This technique, Interpolating GNG-Q (I-GNG-Q), works by approximating the value function of each state (at run time) by considering not only the abstract state region that the state falls in but also the value function of the k nearest abstract state regions. The author applies GNG-Q, I-GNG-Q and epsilon greedy Q learning to a shepherding problem with one shepherd and one sheep. I-GNG-Q was shown to be the fastest learner; however, both I-GNG-Q and GNG-Q were unable to match the success rate of the baseline epsilon greedy Q learner. This demonstrates that neither approach is able to eliminate the previously discussed trade-off between speed and fitness. Instead each method operates on a different point on the speed vs. fitness Pareto curve.

While in Baumann[4] a single controller is learned and applied to each shepherding agent (homogeneous learning), another approach is for each agent to learn independently (heterogeneous learning). The selection of heterogeneous or homogeneous learning is an important design decision as it directly impacts on learning efficiency. This issue is addressed in the research of Gomes et al. [13] and Potter et al. [21]. Both utilize a GA approach to learn the parameters for an ANN to control shepherding agents. These authors are concerned with heterogeneous co-evolution in co-operative multi-agent systems, i.e., the simultaneous evolution of non-identical agents that interact within a single system to achieve a common goal.

Potter et al. [21] investigated the effect of heterogeneous learning on learning efficiency and fitness. They found that when homogeneous and heterogeneous learning systems were applied to a three shepherd shepherding problem, the homogeneous learning system learned faster. This is not surprising that in the heterogeneous system the dimensionality is increased if the behaviour of peer shepherding agents is considered.

In the standard shepherding problem, the fitness of strategies found by the two learning systems was comparable. The advantage of heterogeneity did not become apparent until a second task was added to the shepherding simulation (protecting the sheep from a fox agent). In the case where a fox was present, although the heterogeneous system was slower to develop a solution, the final solution had a higher fitness rating.

Heterogeneity also has an impact on resilience to agent loss. Smith et al. [26] showed that specialization, a feature of heterogeneous learning [21], introduces vulnerability to agent loss due to the difficulty in recovering from the loss of specialists. These trade-offs between efficiency, fitness and resilience mean that a designer must carefully apply domain knowledge to make a choice between heterogeneous and homogeneous learning to achieve optimal learning outcomes.

The work of Gomes et al. [13] provides an alternative to the designer making this trade-off. The authors are concerned with the scalability of co-operative co-evolution algorithms (CCEA), an approach to heterogeneous multi-agent learning systems. When utilizing CCEA, the number of evaluations (and required computing power) increases rapidly as the number of agents increases. The authors propose Hyb-CCEA to address this. In Hyb-CCEA, the team of heterogeneous agents is divided into homogeneous sub-teams. Each member of a sub-team is controlled by an identical controller. This partial homogeneity can drastically reduce the number of learning operations required. In Hyb-CCEA, both behaviour and team composition are learned. Behaviour is learned by GA. When the behaviour of two teams converges, the teams are merged. Heterogeneity of the teams is maintained by stochastically selecting sub-teams to be split. This approach not only eliminates the design decision of heterogeneity vs. homogeneity, it changes the trade-off from a binary choice to a sliding scale. The Hyb-CCEA approach was applied to teams of 5–10 shepherds controlling one or two sheep and compared to a traditional CCEA algorithm. In all cases, Hyb-CCEA learned faster than CCEA, and as expected, the improvement was more pronounced as the number of shepherds increased. Performance of the learned solutions was not compared to programmed shepherding solution. In the experimental setup, two different initial conditions were explored. In one, the population of agents was initially completely homogeneous, and in the other, it was completely heterogeneous. This was shown to have minimal impact on the fitness score of the evolved teams. However, the authors hypothesize that as the evolved teams are always partially heterogeneous it could prove advantageous to learning efficiency if domain expertise could be applied to select a partially heterogeneous starting point that is close to optimal.

The importance of selecting appropriate starting points for learning efficiency is also seen in Linder et al. [17]. Here, the authors use neuroevolution of augmenting technologies (NEAT) to evolve both the structure and weights of an ANN for controlling a single shepherding agent controlling a single sheep. This approach initially underperformed previous work utilizing GA for control of shepherding agents. Two problems were identified. First, they had included irrelevant sensory input into their learning system. This unnecessarily increased the dimensionality of the problem, slowing down learning. Second, the starting point for the structure of their ANN had zero hidden nodes. The system for modifying the structure of the ANN simply took too long to add an appropriate number of hidden nodes to solve the problem. The authors found the removal of irrelevant sensor data improved learning efficiency. This demonstrates the positive impact that domain knowledge, when applied to the selection of learning inputs, can have on learning performance. The author's finding that an increase in the starting number of hidden nodes improved learning efficiency also highlights the importance of selecting appropriate starting points for learning.

A different approach to learning efficiency is taken in [19]. While the authors in [4, 13, 17, 21] were concerned with finding strategies with the highest fitness function as quickly as possible, Ozdemir et al. [19] instead focused on learning shepherding strategies that utilize minimal computing resources. The goal here was to show that a strategy for multi-agent shepherding could be learned for a system using agents with no computational capability or memory, i.e., optical input is mapped directly to wheel speed. The authors apply covariance matrix adaptation evolution strategy (CMAES) with a RL approach that is optimized with evolution techniques. Using CMAES, a successful control system was learned in simulation for a set of 10 homogeneous shepherding agents that could herd up to 20 sheep. This work highlights an important consideration when designing learning systems, namely, *the fastest learner is not always the best.* Domain knowledge needs to be applied to ensure that the learned strategies are fit for the purposes of the end user.

A recurring theme in much of the previous discussion is the opportunity for improving learning through the application of domain knowledge. An attempt to do this is described in Go [11] and Go et al. [12]. The authors applied reinforcement learning to the Strömbom [27] model for shepherding. In their system, a sub-goal of being in the correct position according to the Strömbom model is used to generate a reward signal. This results in a system that attempts to learn the best method for maintaining the "correct" position according to Strömbom. A disadvantage of the author's method is that the learning system does not have the ability to learn strategies that improve on the behaviour dictated by the Strömbom model for shepherding. This highlights an important consideration when using domain knowledge: *in learning systems exploration vs. exploitation is an important trade-off*; in Go et al. [12], utilizing prior knowledge runs the risk of skewing this balance in favour of exploitation.

While the balance between exploration and exploitation may have been problematic in [11, 12], the discussion above suggests that there are many opportunities for applying domain knowledge to learning shepherding systems. The following sections of this chapter explore those opportunities further.

6.3 Prior Knowledge in Learning Systems

As discussed in Sect. 6.2, the application of domain knowledge has the potential to improve learning by shaping input data [17] and starting points [4, 13, 17], improving reward functions [6], increasing transparency [23] and directing learning [12].

When considering how learning systems should utilize domain knowledge, we may draw parallels to how such knowledge is used by humans. As humans, we are able to apply domain knowledge even when solving previously un-encountered problems; we leverage knowledge gained from previous experiences with related problems [9]. This is an advantage we have over machine learning systems, where generally, each new problem is solved without the benefit of past experiences. There is ongoing research into how humans apply prior knowledge and how this can be applied to learning systems. Dubey et al. [9] has proposed a taxonomy of generic priors that includes "Concept of object, similarity, semantics, affordance, physics and motor control". By modifying a computer game's GUI, Dubey et al's work demonstrated that when humans are unable to take advantage of these generic priors, their learning performance was significantly degraded. Modifications were made to the GUI of an Atari game that systematically eliminated the usefulness of the prior knowledge types defined in the taxonomy. It was found that without the benefit of prior knowledge the human players' performance degraded, approaching that of a RL system. This suggests that the humans' performance advantage for this task over the RL was related to their ability to leverage prior knowledge rather than superiority of their learning mechanism.

While the priors outlined in Dubey et al. [9] were descriptive and specifically related to an agent navigating through an environment, Benigio et al. [5] described a taxonomy that contained more abstract concepts of generic prior knowledge. The authors hypothesize this taxonomy will help learners to "disentangle underlying factors of variation". Benigio et al [5] reviewed recent techniques for unsupervised feature learning and emphasized the importance of data representation. Their categories for generic prior knowledge are "Smoothness, multiple explanatory factors, hierarchical organization of explanatory factors, semi-supervised learning, shared factors across tasks, manifolds, natural clustering, temporal and spatial coherence, sparsity and simplicity of factor dependencies". Benigio's taxonomy, although application agnostic may provide a structure that an application specific ontology of prior knowledge, could be built on.

Recent advances in AI have generally used non-symbolic techniques [14, 15, 18, 25]. In contrast to this, domain knowledge generated by human experts is generally symbolic [24, 29]. Taking advantage of domain knowledge and state of the art learning techniques will therefore require a combined approach, i.e., ***symbolic non-symbolic AI***. Such approaches, combining symbolic knowledge representation and non-symbolic learning, are referred to in the literature as hybrid learning [29]. Section 6.4 reviews hybrid learning techniques.

6.4 Hybrid Learning

Symbolic and non-symbolic artificial intelligence systems have largely complementary strengths and weaknesses [29]. Non-symbolic learning systems such as artificial neural networks (ANN) and reinforcement learning (RL) systems are brittle (i.e., their solutions are specific to a particular situation rather than generalized), are unable to reason on abstract concepts and are difficult to understand [10]. Their advantage is that they can learn to solve problems with little prior knowledge of the domain, that is, they offer a model free approach to solving problems. In contrast, symbolic artificial intelligence systems assume complete domain theory, requiring extensive knowledge of the domain on the part of the designer [29]. Learning in symbolic systems is also generally very limited. The main advantages of symbolic systems are that reasoning on abstract concepts is possible and reasoning processes can be more easily exposed and explained. Because of these complementary strengths, hybrid symbolic/non symbolic learning is an attractive proposition. [2]. An example of such an approach is C-Net, a system proposed by Abbass et al. [1], which utilizes an ANN to generate multivariate decision trees (MVDT) in order to create algorithms that are more accurate than decision trees (DT) and more expressive than ANNs.

The use of prior knowledge is a recurring theme in the hybrid learning systems literature. A common approach is to use a symbolic representation of prior knowledge that is then inserted into a non-symbolic learning system. An often referenced example of this is Towel and Shavlik's Knowledge Based Artificial

Neural Network (KBANN) [24, 29]. Shavlik [24] defined a generic framework for combining symbolic and non-symbolic learners; prior knowledge is represented as symbolic knowledge, inserted into the neural network, refined by the learning process and then extracted into symbolic information. In their paper, the authors posed three open questions for symbolic/non-symbolic learning systems that are arguably relevant for the design of any hybrid learning system. They are:

- "How can we represent symbolic knowledge and learning tasks so that powerful numeric-optimization search methods are applicable?"
- "How can symbolic knowledge about the task at hand guide network refinement?"
- "How can we extract a small and comprehensible symbolic version of a trained network without losing (much) accuracy?"

Towell et al's paper [29] utilized Shavlik's framework [24] and proposed Knowledge Based Artificial Neural Networks (KBANN). In [29], the authors address the first and third of Shavlik's open questions by providing mechanisms for a non-symbolic learning system to act on and generate symbolic knowledge. In KBANN, prior knowledge is represented as "approximately correct" symbolic rules that are then refined by an ANN before being translated back into symbolic knowledge. In the shepherding example, this could be applied by utilizing the "approximately correct" rules of a widely accepted symbolic approach such as Strömbom et al. [27] and then allowing KBANN to optimize.

The use of hybrid approaches that aim to refine predefined domain knowledge is particularly interesting when considering the problems encountered in policy based control systems. Policy based control is a form of symbolic AI that utilizes a prioritized list, i.e., a policy, to exercise control. As the motivation of the AI is clearly explained in human readable language, this policy can be a very useful tool for transparency. However, it is critical that the policy is well designed or unexpected behaviour can occur. For example, consider the policy based controller for resilient positioning systems described by Pearce et al. [20]. This controller utilized the policy based control framework proposed by Rosa et al. [22]. In [22], self-management was achieved with an event-condition-action (ECA) policy, which consisted of a prioritized list of goals, triggers to activate the goals and action sets expected to serve those goals. A ranked eager algorithm was used for action selection [22]. In Pearce et al. [20], the authors used this to control three different GPS devices, minimizing performance degradation from various sources of interference. This was a fairly simple problem to define, and the authors had little difficulty utilizing domain knowledge to describe goals and actions. The challenge was found in the goal prioritization and selection of appropriate trigger thresholds that did not result in unexpected and undesirable behaviour of the controller. This type of policy based design only becomes more difficult as the system under control becomes more complex and can become a barrier to the implementation of such an approach.

Alternatively, consider a system where designers are able to describe policy to the best of their ability and then a learning system such as KBANN improves the policy in order to remove undesired effects. In this case, policy based controllers such as those in [22] and [20] could be used on much more complex systems without relying on a designer with expertise in policy creation. Additionally, the new policy generated by the hybrid system would be a useful tool in providing explainable behaviour of the learning system.

Published in 1994, KBANN can be considered a seminal technique for combining symbolic and non-symbolic knowledge. It has been criticized by, among others, Wang [31] and Teng et al. [28]. Wang states that KBANN is not suitable for deep learning due to its inability to represent dependency structures across high dimensional data. Wang proposes rule-embedded Neural networks (ReNN) as an alternative. ReNN first makes local-based inferences to detect local patterns and then uses rules based on the domain knowledge about the local patterns to generate a rule-modulated map. After that, ReNN makes global-based inferences that "synthesize the local patterns and the rule-modulated map".

Teng et al. [28] also propose an alternative to KBANN for inserting symbolic knowledge into non-symbolic learning systems called Fusion Architecture for Learning and Cognition (FALCON). In FALCON, human defined propositional (if-then) rules are translated into vector patterns before being inserted into a RL system. At the time of [28]'s publication, FALCON was still a system under development and did not have an adequate solution for the explore vs. exploit problem in larger state space applications. This resulted in inserted knowledge being ignored in preference for learned knowledge. The authors proposed a reward vigilance adaptation strategy to address this. At the time of publication, they had instead applied the simple fix of using a greedy exploitation strategy in order to prevent exploration if a satisfactory solution existed.

Approaches such as KBANN, ReNN and FALCON, where symbolic knowledge is applied to data first, are contrasted by the approach taken by Garnelo et al. [10], where a non-symbolic system is utilized first to process the data. In their proposed architecture, an ANN is used to generate a symbolic representation of the state space based on the generic prior knowledge. Reinforcement learning is then applied to generate a Q function. It is interesting to relate this work to Shavlik's three open questions [24], because although symbolic knowledge is used to guide network refinement, and knowledge has been translated between symbolic and non-symbolic representations, this translation has not occurred in the order that Shavlik assumed it would, e.g., symbolic to non-symbolic and back to symbolic. Instead, a non-symbolic learner has been used to categorize data symbolically and that symbolic representation has been used to shape a second non-symbolic learning system. This approach has been shown to generate strategies that are more generalizable than the standard Deep Q networks (DQN). However, the authors provide no commentary on the transparency of the system. To apply this approach to shepherding, the ANN would be used to recognize relevant situations such as when the sheep are scattered and the RL system would be used to learn the behaviour to be followed by the shepherding agents.

Based on their results, Garnelo et al. [10] have described an architectural manifesto for learning systems as a step towards general purpose AI. It has four fundamental principles:

- conceptual abstraction,
- compositional structure,
- common sense priors, and
- causal reasoning.

These principles may prove to be useful guides in the design of learning systems where generalizable, non-brittle strategies are desirable.

Another approach to hybrid learning is proposed by Cai et al. [8], where a symbolic system describing prior knowledge is used to generate training data for a deep neural network (DNN). Cai et al's [8] approach is focused on enabling non-symbolic systems to perform symbolic reasoning. This approach could be a tool to provide Garnelo et al's prescribed conceptual abstraction and causal reasoning. However, as it is focused on categorization rather than control, it could not generate a complete strategy for shepherding agent control. The approach still has applicability to the shepherding problem as a potential classifier to recognize relevant situations; however, a second system would be required for action learning and control.

6.4.1 Guided Learning Systems

Guided learning systems can be considered a sub-class of hybrid learning systems. The term refers to systems where domain knowledge is used to shape and guide a learner. In these systems, the domain knowledge is used to "bootstrap" the learning system so that it is not starting from scratch. The advantages of this were shown by Dubey et al. [9], discussed in Sect. 6.3. Generally, guided learning systems apply symbolic representations of domain knowledge to focus on learning and reduce dimensionality[3]. They can be considered a response to Shavlik's second open question for hybrid learning systems "How can symbolic knowledge about the task at hand guide network refinement?"

The approach described by Ulam et al. [30] is a good example of guided learning. In this system, prior knowledge is used to divide the state space into relevant partitions. If the control system fails to achieve a goal, model based reflection is used to identify the partition or partitions that were the cause of the failure. RL is then applied specifically to the partitions that are identified as problems. This significantly decreases the size of the search space for the RL system. This approach was applied to an agent playing the FreeCiv city defense game and was shown to improve performance compared to both a pure RL approach and a pure model based reflection approach. Such an approach applied to the shepherding task could see the state space divided into partitions such as "flock scattered" and "flock collected". The model would describe the expected behaviour in these two partitions. In this way, herding and collecting behaviour could be learned separately. A limitation of this approach is that there is no mechanism for exploring alternative partitions. This means that learned behaviours will be limited by the initial domain knowledge.

Zhang et al. [33] presented guided learning in the form of a hierarchical ontology used to represent the problem space at multiple levels of abstraction. This structure is then combined with a supervised learning approach to generate decision trees corresponding to the different abstraction levels. This approach was demonstrated by applying it to a categorization problem where a shopper's age was predicted based on their purchases. As this approach only provides categorization, a second learning system would be required to learn behaviour in a shepherding application.

Zhang et al. [33] and Garnello et al. [10] both recommended the use of an ontology of prior knowledge to improve learning; however, they use the ontology in different ways. In Garnello's approach, the ontology is used to reduce "workload" on a RL system by reducing dimensionality. This is achieved by grouping the state/action space based on the groupings described in the ontology. In Garnelo's approach, there is still only one level of learning. In contrast, Zhang et al. utilized the hierarchy described in the ontology to create hierarchical levels of abstraction with separate learning occurring at each level of abstraction.

Guidance through ontology is also utilized by Wang et al. [32], which present a guided classification rather than guided learning system. In Wang et al's approach, concepts defined in an ontology are used to reduce dimensionality when categorizing documents to a particular domain. Their approach requires a large pre-existing domain knowledge ontology. This approach is interesting for guided learning because it is an example of pre-existing human readable domain knowledge being used to reduce dimensionality and generate new knowledge. If a similar approach was applied to shepherding, an ontology could be used to describe relevant situations (e.g., flock scattered, driving occurring), in human readable language, and used for categorization of the state space, reducing dimensionality for a learning system.

6.5 Ontology Guided Shepherding

The potential to utilize domain knowledge to improve learning is a recurring theme in the learning literature. The review conducted for this chapter found opportunities to apply domain knowledge in shepherding to inform the structure of learning systems [13, 21], provide guidance towards potential solutions [11, 12], reduce dimensionality [4, 17], design fitness functions [6], eliminate irrelevant data [17] and provide intelligent starting points for searches [13, 17]. In order to make use of this domain knowledge, it needs to be transferred from a human expert to the learning system. Hierarchical ontologies are a well-established approach to representation of symbolic information in a scalable and re-usable way that is accessible to both humans and machines [7, 10, 16, 32, 33]. Encoding prior knowledge from expert operators in hierarchical ontologies and then using that knowledge to guide a learning system could enable a non-symbolic learning system to take advantage of prior knowledge and reason on abstract concepts.

This presents as a hybrid learning approach where domain knowledge is encoded in an ontology to describe the problem state space symbolically. This allows the ontology to be used to shape the design of the learning system, e.g., “guide network refinement” [24], so as to allow reasoning on abstract concepts. Additionally, the ontology could also be used to partition the state space to reduce the dimensionality of the problem. Once learning is complete, the ontology enables explanation of system actions using human understandable language. This system would address the three questions of hybrid learning systems design proposed by Shavlik [24] and the principles of the “architectural manifesto” for generalizable AI proposed by Garnello et al. [10].

When a non-symbolic learning approach such as reinforcement learning is applied to the shepherding problem, the position of every sheep and every shepherding agent is a dimension in the state space. The action space consists of every possible combination of force vectors that can be applied to the various shepherding agents. As the number of sheep and shepherds increases, dimensionality is likely to prove problematic. In ontology guided learning, prior knowledge from a symbolic approach to shepherding would be encoded into the ontology. If the Strömbom et al. [27] model is used as an example, a learning system controlling a shepherd agent would have two actions to select from, collect and drive. The domain knowledge from the Strömbom model would be codified as an ontology describing the semantics of the relevant states: scattered herd, collected herd and herd within goal area. The ontology would also include the relationships between these states and the actions, collect and drive. If the ontology was used to perform state aggregation (see the discussion of Baumann [4] in Sect. 6.2 for state aggregation), the number of dimensions in the state space would reduce from the position of every agent to only the three relevant states. This significantly improves scalability by removing the link between dimensionality and population size. Additionally, it would allow learning to focus on the parametric values of the Strömbom model, i.e., shepherd speed, driving and collecting positions and the threshold between a collected and scattered flock [27]. This approach takes advantage of domain knowledge while searching for improvements in behaviour. Such an approach also allows the system to reason on the abstract concepts (pertaining to the flock) of scattered and not scattered. Unfortunately, this approach does limit learning to the situations and behaviours described in [27]. If this is undesirable, the semantic domain knowledge could simply be considered a good starting point. A secondary stage of learning could be employed, which utilizes strategies such as the CARTMAP approach suggested by Teng et al. [28] to search for new behaviours. Alternatively, state splitting and merging methods such as I-GNG-Q [4] could be utilized to find new relevant situations.

6.6 Future Work

This review of learning approaches to shepherding identifies the potential benefits that may be afforded by prior knowledge guidance of a learning system. This helps shape our immediate future work that will investigate the performance gains offered to a learning shepherding system by the use of ontologies to guide a RL. Variations of the shepherding problem as described in Strömbom et al. [27] will form the basis of the domain knowledge described in the ontology. The performance of the ontology guided system will be compared to a pure RL system. Metrics for comparison will include learning efficiency (e.g., number of learning cycles to learn a successful shepherding strategy), fitness of the learned strategies (e.g., success rate, distance travelled by the agents and elapsed time of the learned strategies to complete the shepherding task) and generalizability (e.g., ability of the learned strategies to complete the task under previously un-encountered variations). Transparency of the learned strategies will also be explored by investigating techniques to describe the actions of the shepherding system using the ontology.

6.7 Conclusion

In summary, shepherding offers a rich application for the study of learning for the control of robotic systems. Shepherding is scalable in terms of both complexity and dimension and allows for the transition of learning from simulation to real robotic systems. These properties make it suitable for the study of the generalizability of learned strategies and knowledge transfer for simulated to real environments. This chapter has reviewed previous learning strategies for shepherding and highlighted the advantages of applying prior knowledge to the design of learning systems for shepherding. This was followed by a summary of previous hybrid learning approaches that apply prior knowledge to learning. Finally, a potential approach to a learning shepherding system was described. This approach, ontology guided learning, will enable the application of prior knowledge to a learning system so as to enable a non-symbolic system to reason on abstract concepts, an important factor in the generalizabilty of AI [10]. It will also reduce dimensionality by partitioning the state/action space. As the semantics of the problem will have been encoded in the ontology using domain knowledge, this approach will also enable the learning system to focus on the parametric factors of the shepherding problem. Finally, the use of an ontology to describe the state and action space should facilitate the explanation of a non-symbolic learning system's actions using symbolic human readable language, greatly improving the transparency of the system.

References

1. Abbass, H., Sarker, R., Newton, C.: PDE: A pareto-frontier differential evolution approach for multi-objective optimization problems. In: Proceedings of the IEEE Congress on Evolutionary Computation (CEC2001), vol. 2, pp. 971–978. IEEE Press, Piscataway (2001)
2. Arulkumaran, K., Deisenroth, M.P., Brundage, M., Bharath, A.A.: Deep reinforcement learning: a brief survey. IEEE Signal Process. Mag. **34**(6), 26–38 (2017)
3. Barto, A.G., Mahadevan, S.: Recent advances in hierarchical reinforcement learning. Discrete Event Dyn. Syst. **13**(1–2), 41–77 (2003)
4. Baumann, M., Büning, H.K.: Learning shepherding behavior. Ph.D. Thesis, University of Paderborn (2016)
5. Bengio, Y., Courville, A., Vincent, P.: Representation learning: A review and new perspectives. IEEE Trans. Pattern Anal. Mach. Intell. **35**(8), 1798–1828 (2013)
6. Brulé, J., Engel, K., Fung, N., Julien, I.: Evolving shepherding behavior with genetic programming algorithms (2016). Preprint arXiv:1603.06141
7. Bundy, A.: Why ontology evolution is essential in modeling scientific discovery. In: AAAI Fall Symposium: Automated Scientific Discovery, pp. 8–9 (2008)
8. Cai, C.H., Ke, D., Xu, Y., Su, K.: Symbolic manipulation based on deep neural networks and its application to axiom discovery. In: 2017 International Joint Conference on Neural Networks (IJCNN), pp. 2136–2143. IEEE, Piscataway (2017)
9. Dubey, R., Agrawal, P., Pathak, D., Griffiths, T.L., Efros, A.A.: Investigating human priors for playing video games (2018). Preprint arXiv:1802.10217
10. Garnelo, M., Arulkumaran, K., Shanahan, M.: Towards deep symbolic reinforcement learning (2016). Preprint arXiv:1609.05518
11. Go, C.K.C.: A reinforcement learning model of the shepherding task. Masters Thesis (2016)
12. Go, C.K., Lao, B., Yoshimoto, J., Ikeda, K.: A reinforcement learning approach to the shepherding task using SARSA. In: 2016 International Joint Conference on Neural Networks (IJCNN), pp. 3833–3836. IEEE, Piscataway (2016)
13. Gomes, J., Mariano, P., Christensen, A.L.: Cooperative coevolution of partially heterogeneous multiagent systems. In: Proceedings of the 2015 International Conference on Autonomous Agents and Multiagent Systems, pp. 297–305. International Foundation for Autonomous Agents and Multiagent Systems (2015)
14. Guo, X., Singh, S., Lee, H., Lewis, R.L., Wang, X.: Deep learning for real-time atari game play using offline monte-carlo tree search planning. In: Advances in Neural Information Processing Systems, pp. 3338–3346 (2014)
15. Kempka, M., Wydmuch, M., Runc, G., Toczek, J., Jaśkowski, W.: Vizdoom: A doom-based AI research platform for visual reinforcement learning. In: 2016 IEEE Conference on Computational Intelligence and Games (CIG), pp. 1–8. IEEE, Piscataway (2016)
16. Li, X., Bilbao, S., Martín-Wanton, T., Bastos, J., Rodriguez, J.: SWARMs ontology: a common information model for the cooperation of underwater robots. Sensors **17**(3), 569 (2017)
17. Linder, M.H., Nye, B.: Fitness, environment and input: Evolved robotic shepherding, pp. 1–8 (2010)
18. Mnih, V., Badia, A.P., Mirza, M., Graves, A., Lillicrap, T., Harley, T., Silver, D., Kavukcuoglu, K.: Asynchronous methods for deep reinforcement learning. In: International Conference on Machine Learning, pp. 1928–1937 (2016)
19. Özdemir, A., Gauci, M., Groß, R.: Shepherding with robots that do not compute. In: Artificial Life Conference Proceedings 14, pp. 332–339. MIT Press, Cambridge (2017)
20. Pearce, G., Campbell, B., Perry, A., Sims, B., Zamani, M., Newby, L., Nesbitt, D., Bowering, G., Franklin, S., Hunjet, R.: An adaptive policy based control framework for land vehicle systems. In: International Conference on Intelligent Robotics and Applications, pp. 208–222. Springer, Berlin (2018)

21. Potter, M.A., Meeden, L.A., Schultz, A.C.: Heterogeneity in the coevolved behaviors of mobile robots: The emergence of specialists. In: International Joint Conference on Artificial Intelligence, vol. 17, pp. 1337–1343. Citeseer (2001)
22. Rosa, L., Rodrigues, L., Lopes, A., Hiltunen, M., Schlichting, R.: Self-management of adaptable component-based applications. IEEE Trans. Softw. Eng. **39**(3), 403–421 (2012)
23. Schultz, A., Grefenstette, J.J., Adams, W.: Roboshepherd: learning a complex behavior. Rob. Manuf. Recent Trends Res. Appl. **6**, 763–768 (1996)
24. Shavlik, J.W.: Combining symbolic and neural learning. Mach. Learn. **14**(3), 321–331 (1994)
25. Silver, D., Huang, A., Maddison, C.J., Guez, A., Sifre, L., Van Den Driessche, G., Schrittwieser, J., Antonoglou, I., Panneershelvam, V., Lanctot, M., et al.: Mastering the game of go with deep neural networks and tree search. Nature **529**(7587), 484 (2016)
26. Smith, P., Hunjet, R., Aleti, A., Barca, J.C., et al.: Data transfer via UAV swarm behaviours: rule generation, evolution and learning. Aust. J. Telecommun. Digital Econ. **6**(2), 35 (2018)
27. Strömbom, D., Mann, R.P., Wilson, A.M., Hailes, S., Morton, A.J., Sumpter, D.J.T., King, A.J.: Solving the shepherding problem: heuristics for herding autonomous, interacting agents. J. R. Soc. Interf. **11**(100) (2014). https://browzine.com/articles/52614503
28. Teng, T.H., Tan, A.H., Zurada, J.M.: Self-organizing neural networks integrating domain knowledge and reinforcement learning. IEEE Trans. Neur. Netw. Learn. Syst. **26**(5), 889–902 (2015)
29. Towell, G.G., Shavlik, J.W.: Knowledge-based artificial neural networks. Artif. Intell. **70**(1–2), 119–165 (1994)
30. Ulam, P., Goel, A., Jones, J., Murdock, W.: Using model-based reflection to guide reinforcement learning. In: Reasoning, Representation, and Learning in Computer Games, p. 107 (2005)
31. Wang, H.: ReNN: Rule-embedded neural networks (2018). Preprint arXiv:1801.09856
32. Wang, B.B., Mckay, R.I., Abbass, H.A., Barlow, M.: A comparative study for domain ontology guided feature extraction. In: Proceedings of the 26th Australasian Computer Science Conference-Volume 16, pp. 69–78. Australian Computer Society, Darlinghurst (2003)
33. Zhang, J., Silvescu, A., Honavar, V.: Ontology-driven induction of decision trees at multiple levels of abstraction. In: International Symposium on Abstraction, Reformulation, and Approximation, pp. 316–323. Springer, Berlin (2002)

Chapter 7
Activity Recognition for Shepherding

Adam J. Hepworth

Abstract Activity recognition for shepherding is a way for an artificial intelligence system to learn and understand shepherding behaviours. The problem we describe is one of recognising behaviours within a shepherding environment, where a cognitive agent (the shepherd) influences agents within the system (sheep) through a shepherding actuator (sheepdog), to achieve an intent. Shepherding is pervasive in everyday life with AI agents, collections of animals, and humans all partaking in different forms. Activity recognition in this context is the generation of a transformation from sensor stream data to the perceived behaviour of an agent under observation from the perspective of an external observer. We present a method of classifying behaviour through the use of spatial data and codify action, behaviour, and intent states through a multi-level classification mapping process.

Keywords Activity recognition · Behaviour classification · Swarm shepherding · Multi-agent

Activity recognition for shepherding is a way for an artificial intelligence system to learn and understand shepherding behaviours. The problem we describe is one of recognising behaviours within a shepherding environment, where a cognitive agent (the shepherd) influences agents within the system (sheep) through a shepherding actuator (sheepdog), to achieve an intent. Shepherding is pervasive in everyday life with AI agents, collections of animals, and humans all partaking in different forms. Activity recognition in this context is the generation of a transformation from sensor stream data to the perceived behaviour of an agent under observation from the perspective of an external observer. We present a method of classifying behaviour

A. J. Hepworth (✉)
School of Engineering and Information Technology, University of New South Wales, Canberra, ACT, Australia
e-mail: a.j.hepworth@student.adfa.edu.au

H. A. Abbass, R. A. Hunjet (eds.), *Shepherding UxVs for Human-Swarm Teaming*, Unmanned System Technologies,
https://doi.org/10.1007/978-3-030-60898-9_7

through the use of spatial data and codify action, behaviour, and intent states through a multi-level classification mapping process.

7.1 Introduction

Shepherding is a problem that is pervasive in everyday life. Artificial intelligence (AI) agents (robots), collections of animals, and humans partake in different forms of shepherding [77], with practical applications including the control of crowds for safe guidance in an environment [45], detection of suspicious behaviour in crowds [54], herding animals in agriculture [77], limiting birds from flying over a runway [61], controlling uncrewed vehicles [42], and assisting in cleaning the environment after an oil spill [7]. Applications for activity recognition are also systemic in everyday life, from surgeons treating patients with novel methods of drug delivery [52], the detection of covert leaders in a swarm [78], recognition of behaviours for crowd control [47], assisting air traffic controllers to manage swarms of uncrewed vehicles, or merely learning how to herd a flock of sheep [77].

7.1.1 Problem Frame

Activity recognition for shepherding is a way for an AI system to learn and understand shepherding behaviours without reliance on a human teacher, learning autonomously from associations amongst activities. Shepherding is described as a type of inherent multi-agent cooperative task [62]. Our problem here is one of recognising behaviours within this shepherding environment, where a cognitive agent (the shepherd) influences agents within the system (sheep) through a shepherding actuator (sheepdog) to achieve the cognitive agent's intent. Our perspective of the system is that of the external observer: the activity recognition agent. The goal of the external observer is to use observable information on the system and its agents to classify behaviours and infer intent.

This chapter describes why activity recognition is central to solving the shepherding problem. We build a foundation of understanding for the problem and discuss the technical challenges of activity recognition, how activity recognition is viewed from a shepherding context and highlight these complexities, consistently. We present an argument for activity recognition as a central element that enables an agent to learn how to shepherd, codifying this by stating the open problems in activity recognition for shepherding. The main contribution of this chapter is to understand the recognition of behaviour within a shepherding problem. The method proposed describes a model that allows mapping from behaviour and action states through to intent. This is a multi-level classification (mapping) process: a shepherding taxonomy for classifying behaviour through spatial data.

7.1.2 *Motivation*

Shepherding is important as it is an abstraction for group control with many applications across different domains. The first and most intuitive one is agriculture. Within the context of the seminal Strömbom et al. [77] setting, an uncrewed system could be observing what tasks the shepherding agent is undertaking in order to learn the role of the sheepdog and conduct a shepherding task. There exists literature on artificial intelligence systems undertaking shepherding-style tasks [18]; however, these agents do not learn the underlying shepherding task from first principles. Learning about shepherding and how to undertake the task from another agent, be it human, animal, or robotic is an emerging area of research.

Consider a medical patient with a malignant tumour. Treatment through a standard delivery anticancer drug may result in an adverse effect, such as increased toxicity levels. Treatments such as a chemotherapy function by destroying rapidly dividing cells, but also damage healthy cells that divide. If a delivery agent being an intelligent nanorobot can deliver high dosages of anticancer medicine directly to a targeted malignant tumour, without damaging healthy cell division, this could limit the impact of adverse effects on the body.

We envisage a surgeon injecting a group of clustered nanorobots, who begin travelling through a blood vessel. The surgeon intends to deploy these nanorobots to treat the tumour, with each nanorobot releasing a package of medicine at the site of the cancerous cells. Each nanorobot's storage capacity is a premium space; the objective is to maximise the amount of medicine delivered to defeat the tumour. As such, there does not exist sufficient storage capacity to install the extensive, sophisticated communications and control equipment required to direct each nanorobot. Control of this swarm requires influencing the constituent agent nanorobots as a collective.

An AI agent is learning how to control the swarm to deliver the medicine by observing the surgeon. Learning by the AI agent is conducted by observing the interactions and behaviours of the nanorobots (sheep), the surgeon (cognitive shepherding agent), and the local swarm controller (shepherding agent). The goal here is for the AI agent to learn to be the shepherding agent [28, 52].

7.2 Activity Recognition

Activity recognition is an exciting field of research that is receiving ever more attention from various sub-fields of computer science [11, 34], as well as in other disciplines [6]. Activity recognition research is rapidly progressing, due in part, to advances in technology such as increases in mobile computing power and realised efficiencies in miniaturised energy storage solutions, which enables the realisation of complex processes in the real-world.

The dominant discourse of research within the field of activity recognition is that of a single agent conducting sequential, linear activities in a closed environment, with a limited taxonomy of behaviours. Investigation of more complex activities such as interleaved, composite, and concurrent activities [1] remains a further challenge [74]. Many areas of daily life act as application motivations for activity recognition research, such as applications for pervasive and mobile computing, context-aware computing, surveillance-based security, task analytics, and ambient assisted living in a smart-home [46]. A drawback of many approaches is the requirement of large datasets to evaluate a framework, or a comprehensive understanding of the activities exhibited [56]. Classical approaches to activity recognition typically investigate the physical domain, requiring specific combinations of sensors located at predefined orientations and positions. This approach has limited utility within the real-world for scenarios that deal with agent uncertainty and system complexity [40]. Moreover, the generation of datasets used for activity recognition tasks is nominally from ideal, controlled environments. However, real-world endeavours are seldom in like-situations where noise, imperfect data stream observations, and sensor data capture issues detract from accurate and precise classifications [73].

7.2.1 Elements of Activity Recognition

Key to an activity recognition paradigm is understanding the agents and behaviours within the system, as well as defining what activity recognition means within our problem frame. We begin by providing definitions for critical elements being the agent under observation, the action/activity that the agent is conducting, and process of recognising this, activity recognition.

7.2.1.1 Agent

Our first set of definitions regard the entity under observation, the agent. As given in Lettmann et al. [43] (paraphrasing from [82]), an agent may be characterised as:

Definition 7.1 (Agent) Autonomous, computational entities that can be viewed as perceiving their environments through sensors and acting upon their environment through effectors.

More expansive and explicit definitions exist (such as the work of Ferber [23] or Woolridge [85]); however, the selected definition nests well with this application. Common trends appear across all definitions such as self-agency (autonomy), context (environment), and motivation (objectives and goals), although presented from distinct perspectives. The set of generic agents is defined as $\mathscr{A} = \{a_1, a_2, \ldots, a_{\mathscr{A}}\}$. With consideration for generalised shepherding, the physical representation of the set of system agents is equivalent to $\mathscr{A} = \{B \cup \Pi\}$ (defined in the next section).

7.2.1.2 Agent Types

There exist four system agents within our formulation of activity recognition for the shepherding problem. From the perspective of activity recognition, each entity is considered to be a distinct agent, mapped as unique (1:1) between their role and agent type. An understanding of this assumption is that we preclude complex scenarios, for example, where one agent is shepherding a flock, while also being a sheep (follower) from a higher-level shepherding agent's perspective. Future research will need to address multi-level, multi-role agent behaviours and types, in complex environments.

In contrast to the generalised shepherding notation construct [75], we differentiate the role of the cognitive agent from that of the shepherding actuator. The term shepherd here refers to the cognitive agent who is the controlling decision maker. To remain linked with the generalised shepherding notation, however, we refer to the sheepdog as the environmental agent.

Shepherd (Υ) We characterise a shepherd as being environmentally non-active, cognitively a decision maker, and physically a non-actuator. These properties indicate that the shepherd is not a physically engaging agent within the system, although it makes decisions that influence the configuration state and outcomes of the system.

Sheepdog (B) A sheepdog is characterised as being environmentally proactive, cognitively a non-decision maker, and physically an actuator. These properties indicate that the sheepdog is a physically interacting agent within the system. Our underlying model understanding tells us that the decisions and intent of the shepherd motivate the sheepdog. The sheepdog does not set system-level objectives, although it does directly influence other agents within the system to achieve the shepherd's intent.

Sheep (Π) Sheep are characterised as being environmentally reactive, cognitively non-decision makers, and physically as actuators. These properties indicate that the sheep are a physical agent that is reactive. From our understanding of the underlying model, we know that the sheep are reactive only to their immediate, local environmental influence forces. While a transfer of information of sheep within the flock occurs, this is not a designed process.

External Observer (K) An external observer is characterised as being environmentally non-active, cognitively observational, and physically a non-actuator. These properties indicate that the system properties do not bind the external observer and nor are they able to influence agents within the system.

Figure 7.1 depicts our categorisation of each unique combination of environmental, cognitive, and physical properties.

Each agent represents a unique combination of environmental, cognitive, and physical properties; there are multiple properties for which an agent may share with other agent types within the system. The properties codified in Fig. 7.1 are

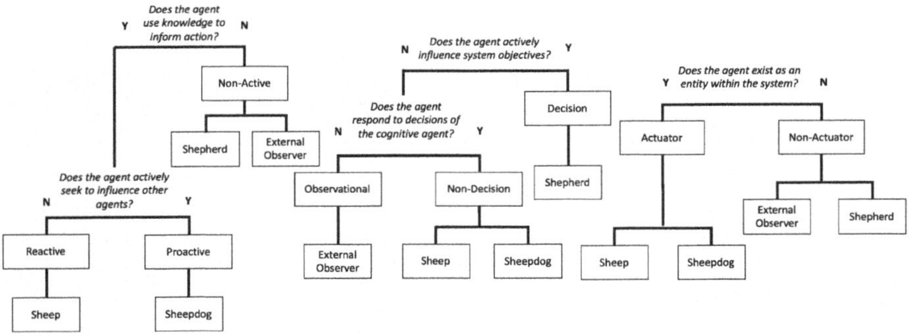

Fig. 7.1 Tree describing the classification of each agent type across environmental, cognitive, and physical properties

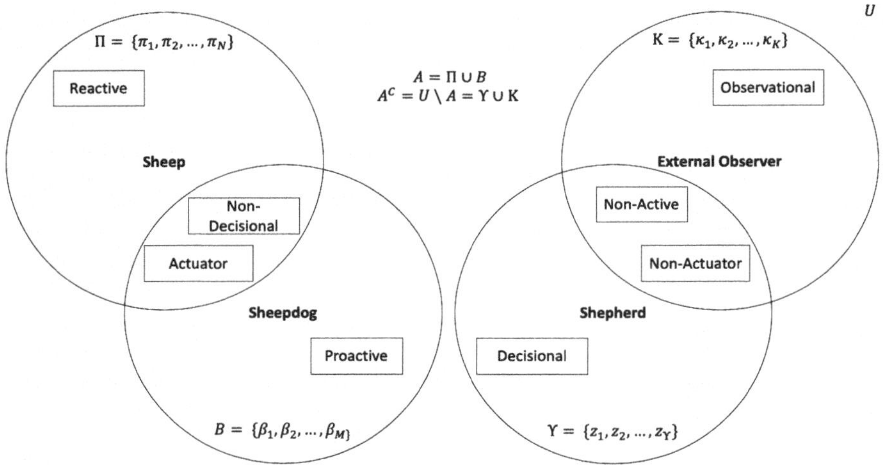

Fig. 7.2 Set of agent type properties

the properties of agents depicted in Fig. 7.2, which describes the set of agent type property similarities, from the perspective of agents.

7.2.1.3 Action and Activity

For an agent to operate within an environment, it must understand its contextual surroundings (sense), determine an appropriate outcome aligned with its objectives (decide), and conduct actions (act). We are interested in defining the final stage of this sense–decide–act architecture [53]. The definitions of action and activity begin to build a taxonomy covering atomic actions and low- and high-level activity. Liu et al. [50] define an action as:

Definition 7.2 (Action) Primitives that fulfil a function or simple purpose, such as walking, jumping, or opening the fridge.

We observe that actions are the direct outputs of actuators, in the simplest form. Continuing from this definition, Liu et al. [50] define an activity as:

Definition 7.3 (Activity) Consist[ing] of a pattern of multiple actions over time. Typical examples include cooking, basketball-playing, and coffee time.

Establishment of a boundary between these definitions is described in the work of Ikizler and Forsyth [31], stating that "we distinguish between short-timescale representation (acts); medium timescale actions, like walking, running, jumping, standing, waving, whose temporal extent can be short (but may be long) and are typically composites of multiple acts; and long-timescale activities, which are complex composites of actions". Keywords such as complex and composite emerge as important in this boundary classification approach, informing the development of action and activity taxonomy levels within. For the generalised shepherding problem, an activity is known as a behaviour and given by $\Sigma = \{\sigma_1, \sigma_2, \ldots, \sigma_l\}$. In our formulation, the set Σ represents the distinct set of lower-level behaviours.

7.2.1.4 Defining Activity Recognition

Prominent definitions for activity recognition are contained within the literature (see for example [16, 33, 56, 63, 65, 80]). Common themes are clear, although divergence remains between application lenses such as the existence of definition gaps within the ontological derivatives of knowledge-driven approaches to activity recognition [64]. Cook et al. [17] provide a succinct definition for activity recognition, amongst others.

Definition 7.4 (Activity Recognition) Activity recognition maps a sequence of sensor data to a corresponding activity label.

More formally, activity recognition within our problem is a type of (supervised) pattern recognition classification task (see: Definition 7.5), and states that for an unknown transformation $\mathscr{S} \rightarrow \Sigma$ which maps sensor stream observation instances $s \in \mathscr{S}$ to behaviour (labels) $\sigma \in \Sigma$, given a set of training sensor stream observation examples $\{s_l, \sigma_l\}$, to find the function f which approximates as closely as possible $f : \mathscr{S} \rightarrow \Sigma$, such that $f(s) = \hat{\sigma}_l$.

7.2.2 Problem Components

The central purpose of this chapter is to understand the elements required to classify the behaviour of an agent, determine if the inference of the intent of a cognitive agent is possible, and what predictors may exist for us to determine when behaviour and intent changes. Classification is a type of inductive learning technique that uses a collection of ground-truth, previously known observations and applies an induced model to new, previously unobserved data. Classification is formally defined in de Carvalho and Freitas [21] as:

Definition 7.5 (Classification Task) Given a set of training examples composed of pairs $\{x_i, y_i\}$, find a function $\hat{f}(x)$ that maps each attribute vector x_i to its associated class y_i, $i = 1, 2, \ldots, n$, where n is the total number of training examples.

This definition partially falls short in activity recognition for shepherding as it does not capture the generalisation inherent in our problem space. Specifically, not embodied is the requirement for validation after training of the classification model, such that $\hat{f}(x)$ needs to generalise beyond the training set and as closely as possible approximate $f(x)$. de Carvalho and Freitas [21] discuss this nuance in the paragraphs following their definition, although it remains mathematically missing from it. Understanding what features enable classification and how and when an observed activity is recognised is key. The value proposition of activity recognition here is as a vehicle to build a task analytics framework, from the perspective of an external observer.

7.2.2.1 Agent Design

Singh et al. [75] introduce a generalised form of the sequential process *Sense, Decide, Act*, as described in McVicar [53]. The behaviour lifecycle here contains the logic for each constituent element, from raw sensor data through calculation and decision, and ultimately action. Figure 7.3 depicts the generalised form of each behaviour considered for our problem.

This type of problem presents challenges in identifying activity patterns at both the collective and individual agent level and in certain situations, decoupling an agent's activity from that of the collective may be near impossible. Simpler systems display little to no heterogeneity or adaptivity, where more complex adaptive systems contain a diverse range of agents and are highly adaptable. For complex systems, higher-order statistics are the measures to understand the nature of the system and its invariant properties, such as determining decision change points. We

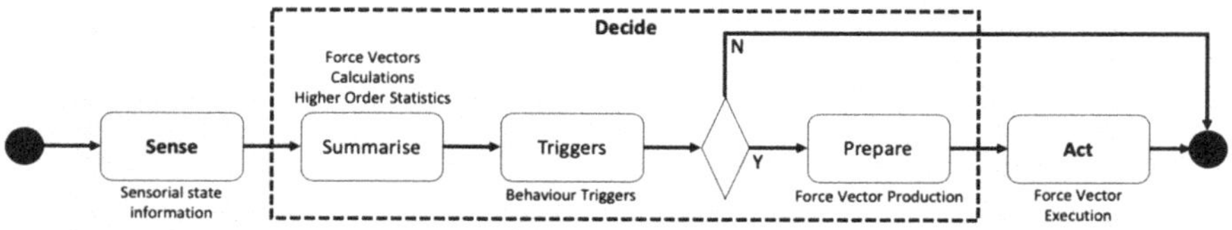

Fig. 7.3 Behavioural logic using a sense–decide–act agent architecture, originally published in Singh et al. [75]

broadly define a decision change point based on the work of McVicar [53] and Singh et al. [75] as

Definition 7.6 (Decision Change Point) The moment in time, which marks the completion of a sense, decide, act (decision) cycle.

What makes our challenge complex is the change in ambiguity associated with a change in decision, and subsequently the associated change in uncertainty (decaying over a window $t_0 \rightarrow \Delta t$) for the duration of an activity, where the time window is an invariant property associated with every sense, decide, act cycle. Agents with competing goals in the system induce a complexity known as deep uncertainty into the system. For each activity, there is an associated time window granularity which must be achieved and recognising the activity as early as possible within this time window is essential in approximating as closely as possible the correct behaviour exhibited. The lack of necessary information in the initial stages of the time window is when activity uncertainty is at a maximum. In contrast, uncertainty is at a minimum at the end of each time window period: knowledge of the activity increases through the window as we sense and collect more information. The new time window is the decision change point for an activity, and where the process commences again. If we know when an agent switches between decisions (a discovery requirement), then we can learn the chain of actions leading to this behaviour and allowing for the inference of intent. The intent is a way in which to recognise the activity of a collective. The distinguishability, or indistinguishability, of a single agent from within the collective of agents, leads us to consider the role of entropy, providing insight into the randomness of change within our system [2]. A classification mapping for this problem presents an opportunity to contribute to this problem space, and activity recognition for shepherding is a vehicle to such an approach.

7.2.3 Approaches

7.2.3.1 Data-Driven Approaches

The primary approaches to activity recognition include data-driven, knowledge-driven, and hybrid approaches [5, 30]. Data-driven approaches infer activities based on a probabilistic or statistical classification of information (i.e. a bottom-up approach). Sensor-based activity recognition arose in the early 1990s and formed a broad base of the available literature [13]. Both knowledge- and data-driven approaches focus on heterogeneous-sensor, single-agent, single-environment problem spaces, and these approaches have been extensively studied [4, 15]. Examples of broader data-driven approaches are in [24, 66]. Most common methods of human activity recognition utilise supervised learning techniques, with less research focusing on unsupervised or semi-supervised methods [39]. Supervised learning generally works well in the classification of more straightforward activities, whereas more complex recognition tasks that are data-driven do not perform well in terms of portability, extensibility, and support for common-sense knowledge [30]. Supervised learning excels when employed to learn a model of the data-instance for situations in which annotated (labelled) data exists.

Activity discovery is the use of unsupervised learning techniques for the discovery of activity labels that lack the presence of a known data-instance [6]. While data-driven approaches offer many advantages, there are severe shortfalls in their implementation to address, such as the cold-start problem. To address the cold-start problem requires enormous amounts of data and computational power, presenting challenges due to model inflexibility, which are central to scientific critiques for these approaches [13]. While robust in dealing with various levels of data uncertainty, these type of data-driven approaches offer only a single-solution path, which must be trained sufficiently for each agent under observation, activity to be classified, and environment deployed. Implementation can be computationally expensive, as well as inflexible to accommodate the stochastic and dynamic nature of the behaviour, and are therefore potentially unsuitable in the real-world [56]. Many attempts have been made to overcome these shortfalls, and ensembles of these methods have gained interest in research communities [30].

7.2.3.2 Knowledge-Driven Approaches

The most popular forms of knowledge-driven activity recognition are based on ontological approaches [60, 67]. The performance of a knowledge-driven approach declines with increased uncertainty in sensor data. Thus the majority of earlier research within the activity recognition field focused on single-agent paradigms. This focus is not practical within the majority of complex scenarios, such as the inclusion of activity interleaving spatiotemporally across multiple agents [88]. Knowledge-driven approaches require a rich domain knowledge for successful implementation, where data-driven approaches create activity recognition models that can include public, open-source data [13]. As discussed, advanced data-driven techniques suffer from high computational complexity and large training data requirements which are unsuitable to real-world applications. Reasoning approaches offer an alternative solution in dealing with sensor data to infer activities, although they may require significant state-spaces to enact [37, 51, 89]. Multi-user concurrent, complex activities represent challenges tackled by utilising knowledge-driven approaches [86].

Ontological approaches have recently gained popularity within the literature for dealing with context-awareness in the activity recognition problem [8, 9, 12, 13]. The key concept here is one of a knowledge-transfer from low-level activity information to high-level activity information, in order to infer a context. A drawback of this approach is the rigid nature of rules required to support the underlying information structure assumptions [22]. Logical formalism approaches have been successful in overcoming the data-driven problems [25, 57], although this approach is yet to support robust probabilistic reasoning, dramatically impairing the inference mechanism [91]. The main advantage identified of this method is the requirement for significantly fewer data observations for training processing. However, these methods are not robust in situations with low signal to high noise, where uncertainty in sensor data exists. Fuzzy logic represents a research effort in overcoming this limitation [6].

Knowledge-driven approaches are poor at capturing intrinsic knowledge of the real-world and are as yet unsuitable for full deployment. Time-slicing is a common approach to this problem within a knowledge-driven paradigm, although it suffers from other ontological flaws [14, 58, 68]. Dynamic temporal windows may overcome this issue but are immature in their deployment currently. Necessary for this approach will be to deliver accurate activity classifications under uncertainty, within spatiotemporal environmental constraints. Similar semantic relations can be used to derive new knowledge across various dimensions of information [87], where generic context descriptors for activity recognition may lead to novel methods in handling information uncertainty in a multi-sensor, multi-user environment, beyond the rigidity of pattern and background knowledge requirements for data-driven approaches [13]. A limitation of knowledge-driven approaches that use ontologies is that of static modelling, precisely an inability to accept unforeseen activities without an explicit, manual intervention and redevelopment of the underlying model [56]. Also contributing is a lack of uniformity and formality within guidelines for knowledge heterogeneity; however, computer science sub-fields recognise that ontological-based approaches represent a promising means for the sharing and reuse of knowledge [87].

7.2.3.3 Hybrid Approaches

Hybrid approaches are often successful methodologies as they offer the ability to achieve temporal sensor alignment, as well as handle uncertainty within data. The handling of temporal alignment for different modalities, combined with diverse and complex sources of uncertainty is a relatively recent addition to activity recognition literature. Data-driven solutions nominally prescribed through the proposal of rigid and specific problem representations, which rapidly degrade when underlying assumptions as to their operation or parameters change [20]. Through the application of multiple methods, models, and techniques, hybrid approaches are employed to combine distinct elements into a single, holistic approach and as such, have been successfully implemented for human activity recognition, as well as other classification tasks [36]. While normal to consider hybrid approaches as the intersection between knowledge- and data-driven approaches, there exist examples of hybrid models which combine disparate elements of a single domain. Applications of single-approach hybrid methods within the data-driven hierarchy fuse supervised and unsupervised methods, advancing the understanding of fundamental label-generation issues, finding unnoticed, novel activity, and identifying local attribute associations [44]. For example, one approach combines techniques such as principal component analysis and back-propagating artificial neural networks, for use in human activity recognition endeavours (with varying levels of success) [36]. Further discussion of hybrid-based approaches is described in Helaoui et al. [29] and Steinhauer et al. [76].

7.3 Shepherding

Shepherding is the interaction between one or many agents influencing a collection of hetero- or homogeneous agents by exerting a force to achieve the intent of the cognitive shepherding agent. A key component of this is the exertion of attraction or repulsive force by an agent, leading to the aggregate flocking behaviour of the agents herded: a function of self-avoidance from the predator [26]. A concise definition is provided in Lien et al. [49], stating that:

Definition 7.7 Shepherding behaviours are a flocking behaviour in which outside agents guide or control members of a flock.

The shepherding problem presents a computationally interesting, complex challenge as a result of the large resultant state-space [48]. As a collection of problems, shepherding has received attention centred around group behaviours and more recently, attempts to optimise path planning methodologies. The problem considered throughout this research is a form of the shepherding problem, where a shepherding agent (sheepdog) collects and drives potentially resistant individual agents (sheep, agents) towards a common goal location [77]. The base-model employed throughout this work considers a plausible framework of interactions for which to experiment and is described earlier in the generalised shepherding chapter.

7.3.1 Open Challenges

Activity recognition for shepherding research must advance our understanding of handling multiple agents with various sensor types and performance simultaneously, multi-agent environments, and sensor data fusion. Accuracy for data fusion research within this frame remains an open area of research [64]. Activity recognition in the shepherding problem has many interesting, analogous applications to the real-world which includes the recognition of behaviours for crowd control[47], situational awareness for a group of vehicles to recognise what another vehicle is doing, identification of specific behaviours within an intelligence collection process, and path planning [3]. Research gaps presented here are a non-exhaustive list.

7.3.1.1 Activity Verification

Activity verification extends the field of activity recognition to include the notion of correctness for a sequence of activities, particularly for composite, complex environments and contexts. In expressing the correct execution sequence of activities, verification is necessary to understand the complex nature of activities involving task switching [35]. Activity verification research is not conventional within the literature, nor the associated complex analysis framework required to address this gap [59, 71]. Shepherding applications include recognising specific behaviour types from multiple agents [54], learning the correct sequence of activities to enable the discovery of underlying processes, to assess present activity for conformance to previously observed processes, and to assist in activity prediction and intent inference.

7.3.1.2 Adversarial Activity Recognition

Adversarial activity recognition is a concept where the recognition agent seeks to maximise the classification of an observed agent's activity, and where the observed agent attempts to minimise the ability of the activity recognition agent to do so. Adversarial image classification has recently begun emerging in computer vision algorithms [32], although research through the lens of activity recognition remains in its infancy. From a shepherding perspective, the observed shepherding agent may seek to minimise the ability of an activity recognition agent to classify actions and activities, temporally disrupting the observer's ability to not only classify activity but also to limit their ability to infer intent. An application of research into adversarial activity recognition could be within the national security and intelligence realm, such as recognising certain behaviours within a crowd at a protest. In this scenario, the activity recognition agent may seek to classify the behaviour of a leader within a crowd (shepherding agent) that is directing a specific activity. The behaviour they seek to mask may be that of inciting violence or trying to turn the protest into a riot. The leader does not want to rise above the detection threshold as this would stop the achievement of their goals, as the dynamics of the crowd will not become more violent, failing to achieve their intended outcome.

7.3.1.3 Context-Aware Activity Recognition

Social interactions provide the essential context in learning and understanding conditional nuances of activities [72]; further, the context itself has been suggested to be an activity [64]. Activity recognition is a crucial enabler to understand an agent's intent and the environmental factors influencing the activity [19, 70]. Spatiotemporal, online context-based activity recognition exists for applications within adaptive learning and irregular behaviour classification situations. Potential solutions to the context-recognition problem may investigate the fusion of sub-symbolic and symbolic-based approaches, together with abductive reasoning, to create meaningful, real-time guidance [6]. Solutions to this problem may be employed to solve gaps within both activity verification, and adversarial activity recognition scenarios where minimisation of activity classification occurs, or identified activity sequences are incorrect. For significant state-space problems with a comprehensive taxonomy of activities, context-aware activity recognition has the potential to decrease classification times, through prediction of plausible activity-sequences, temporally.

Approaches to solving the context-aware problem have applied a combination of ontological and statistical reasoning to recognise an activity given a context, understanding spatiotemporal correlation and complex dynamic agent coupling [27]. Intent recognition is a motivational use-case for context-aware activity recognition, where recognising an activity rather than the observed actions is the goal. We could consider intent recognition here as meta-activity recognition to recognise the higher-level activities and the intent of the observed lower-level activities and actions.

7.3.1.4 Cross-Domain (Multi-Modality) Activity Recognition

Approaches to activity recognition generally employ either object-usage information or require explicit action modelling, where few exploit additional source domains for activity recognition [92]. Approaches such as transfer learning from one domain to another offer potential solutions to this problem [69]. Solutions have been proposed that look at cross-domain inferences for context-awareness as a way of solving this problem, although this research remains in its infancy [90]. Understanding of the source and target domain for information representation is vital to ensure that the assumptions from the information of one domain hold within another and more importantly, where underlying assumptions may be no longer valid within the target domain. Temporal synchronisation of multiple modality sensors too remains an open challenge [92]. With a large number of agents available for observation within the shepherding problem, it is possible to measure the transfer of information between one agent and another. With an assumed level of sensor prevalence within a system, the classification of activity for adjacent agents to the shepherding agent of interest may minimise system-level uncertainty and provide a solution path to also deal with the adversarial activity recognition problem.

7.3.1.5 Dynamic Activity Recognition

A primary challenge in advancing activity recognition research is developing robust frameworks that combine changing sensor ensembles with different capabilities to provide reliable inference [40]. A future requirement may be to enable autonomous systems to interpret and understand multi-modal behaviour [72]; such a paradigm is inherently dynamic, online and must allow for incremental assessment of streaming sensor data. Shepherding offers us a motivational use-case for the development of online systems to deal with (1) changing sensor configurations, (2) evolving activities over time both in terms of activity conduct and the agent under observation, and (3) the identification and classification of new activities.

7.3.1.6 Inter- and Intra-Activity Delay and Task Selection

Interleaving task delays and re-prioritisation change the efficiency of task completion, potentially limiting the effectiveness of a task, or ceasing its achievement altogether [41]. Recognising factors which may impact these temporal and contextual task delays, as well as how these impact future decision cycles are essential in many applications. Dynamic activity recognition may offer solutions to deal with this problem, although this may not deal with intent changes. Shepherding offers us a way to study these changes in a contained system and develop methods to deal with recognising when these cases occur, and when the intent of a shepherding agent has changed. Determining the observed agent's decision change point through the transfer of information throughout the system offers a path to solve this challenge [81].

7.3.2 Solving the Activity Recognition for Shepherding Problem

Solving the activity recognition for shepherding problem spans not only the range of open challenges within the activity recognition literature as discussed, but too invokes further challenges directly from shepherding research. Combining the fields of activity recognition and shepherding leads us to formulate a central, open question to answer, *what is the sheepdog's behaviour and how does it contribute to understanding the shepherd's intent?* This section addresses this expansive question by formulating the core problem space and designing systems models that each elicit facets of the challenge. We will employ a standardised nomenclature used to describe and model the shepherding problem [75]. The generalised shepherding notation system expands on the notation described within the seminal paper [77], framing each agent through the lens of the Sense–Decide–Act (SDA) architecture control cycle [53].

7.3.2.1 Shepherding Taxonomy

The developed shepherding taxonomy is a multi-layered template-matching classifier that transforms raw sensor data into atomic actions, sequences atomic actions to classify lower-level behaviour, and sequences lower-level behaviours to classify higher-level behaviour. The key idea with this template-matching classifier is to generate the mapping from raw positional sensor data to intent, through a series of sequential feature transformations, expressed as $\mathscr{S} \rightarrow \Sigma$, $\Sigma \rightarrow \Psi$, and $\Psi \rightarrow \mathfrak{I}$. The taxonomy of shepherding atomic actions, lower- and higher-level behaviours, and intents is given in Fig. 7.4, and has been developed from the generalised shepherding nomenclature [75]. Note the use of P to indicate individual agent positions (such as $P^t_{\pi_i}$), collective agent positions (such as P_q), and position locations for agents (such as P_A and P_B). Note the mathematical logic for each action represents a threshold, for each behaviour represents a trigger, and for each intent represents the formulation of the verbiage.

The higher-level intent, $\mathfrak{I}$, is the aim of the cognitive agent, translated through unobserved commands to the sheepdog (the shepherding actuator). Higher-level behaviour, $\Psi = \{\psi_1, \psi_2, \ldots, \psi_\Psi\}$ is a function of lower-level behaviours and is the physical enactment of the shepherd's intent. Lower-level behaviours, explicitly defined to replace general behaviours at all levels, given as Σ, are a function of atomic actions, $A = \{a_1, a_2, \ldots, a_A\}$ to achieve a near-term effect. Atomic actions are the underlying set of individual actions that are generated through raw sensor stream data observations, $S = \{s_1, s_2, \ldots, s_S\}$.

The taxonomy does not explicitly describe the observed hierarchy of interaction rules which all agents within the system follow. The interaction rules are invariant of agent type, which we describe in terms of individual, local, and global interaction rules. A special case of this scenario is described by Perry (Chap. 4), known as the stalling distance. This occurs as $\beta_j \rightarrow P_d$ approaches $\Lambda^t_{\beta_j}$ for any π_i, a β_j slows their movement until stopping, short of their target location. This special case has been generalised such that the interaction and subsequent reaction occurs with priority firstly on the individual level, secondly on the local level, and lastly on the global level.

We assume that intent may be demonstrated through activity, which to an observer, may be used to infer the original intent and reinforce understanding of a system. Within the construct of the shepherding taxonomy, the observation of higher-level behaviour over a sufficiently long time window informs an observers understanding of intent; higher-level intent is the cognitive aim of the shepherd, translated through commands to the sheepdog. Lower-level behaviour represents the lowest level of intent execution, through a behaviour, of which behaviour-chain sequences inform an understanding of higher-level behaviour, and a function of lower-level behaviours forms a higher-level behaviour, subsequently informing an intent (Fig. 7.5).

Atomic actions are observed intent invariant without context, and afford little to no insight into the shepherd's intent; they are the underlying set of individual actions, raw sensor stream data observations. Atomic actions do not form action chains but critically are the building blocks that enable behaviour and intent inference, enabling the achievement of a near-term effect. Higher-level behaviours are observed at a system-level and require the observation of a critical mass of agents to infer this behaviour state. Lower-level behaviours are observed and classified at the agent level and do not require the observation of other agents within the system.

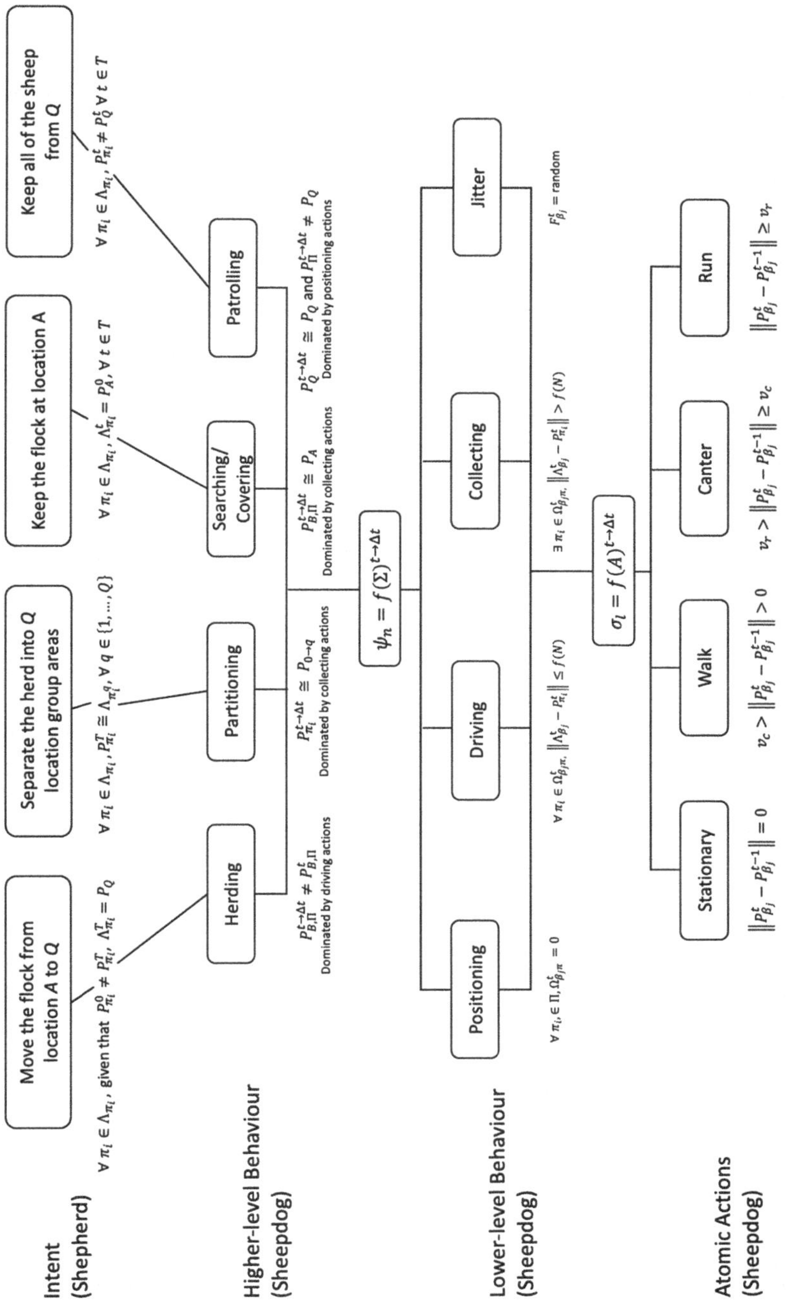

Fig. 7.4 Shepherding action, behaviour, and intent taxonomy

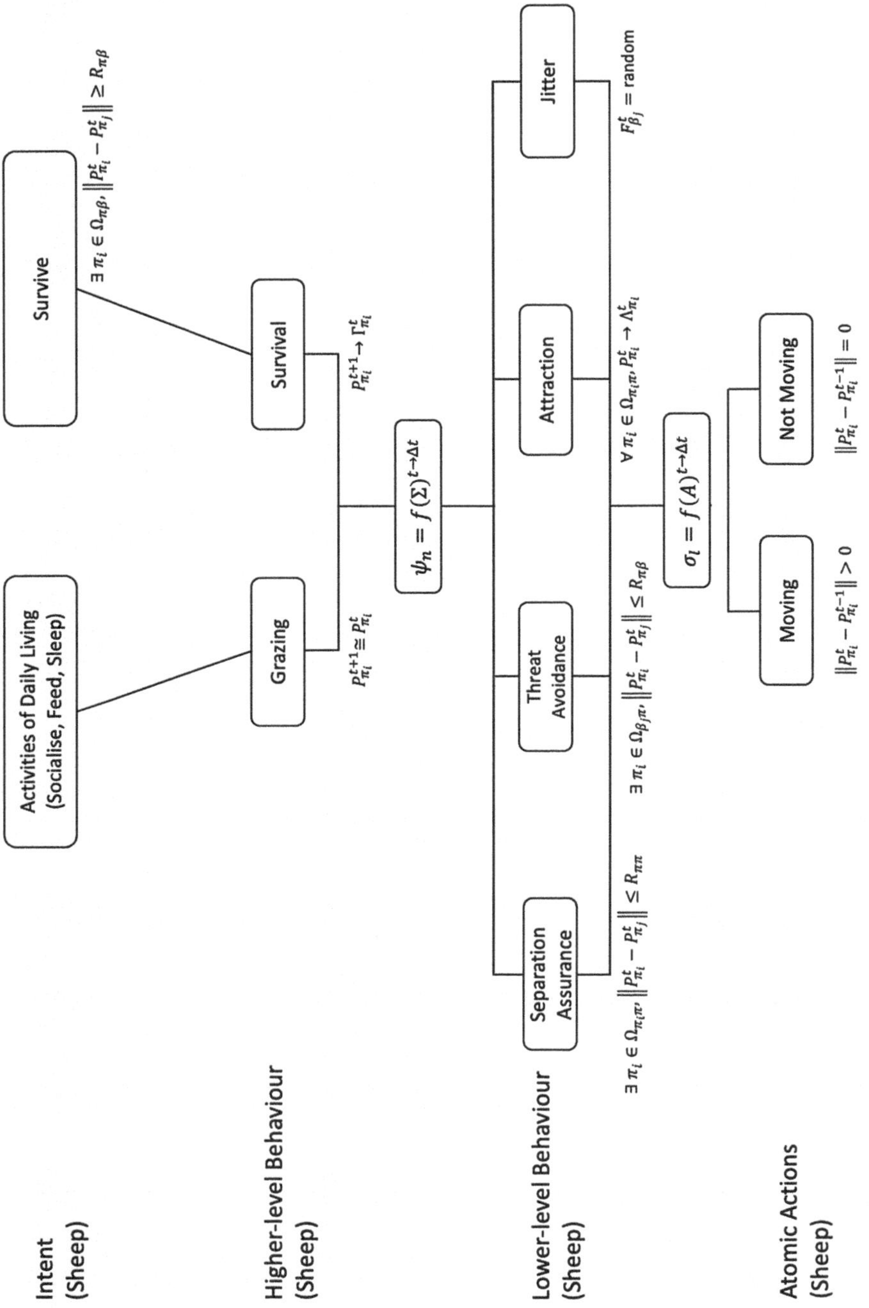

Fig. 7.5 Sheep action, behaviour, and intent taxonomy

Table 7.1 Comparison of the different agent and collective systems

	Shepherding agent	Reactive agent	Collective herd of reactive agents
Intent	Given to the shepherding agent by the shepherd as a mission (complex task) to be completed	Viscerally reactive to the situation; unbounded in intensity or priority and likely to change based on lower-level behaviour triggers	Changing: situational and environmental, founded on the constituents in the herd
High-level behaviour	The execution of an intent	Natural behaviour that is possible to intuitively observe and classify for a human when considering an individual sheep	Electoral: agents of the herd consider and "vote" for an action, with locally "winning" decisions competing globally
Low-level behaviour	Anticipatory: achieving a near term effect as the first level of intent execution	Reactive: trigger-based behaviour addressing a visceral response	Reactive: trigger-based behaviour that cascades through the collective herd
Atomic action	Positional-based sensor stream data observations at different thresholds, transformed into rich feature sets		

The sheep action and behaviour taxonomy are not overly dissimilar to that of the shepherding taxonomy; however, it is contextualised through the lens of a single sheep as opposed to the shepherding agent. While positional atomic actions are the same as those of the shepherd, we observe divergence between lower-level behaviours and higher-level behaviours. Whereas in the shepherding taxonomy lower-level behaviours are anticipatory, those of the sheep are reactive to their situation. Table 7.1 describes the primary differences of behaviours between each action-agent type, as well as the collective flock. The key perspective of each agent type is that the shepherding agent is anticipatory, whereas the reactive agent is only reactive.

Our central claim is that there exists sufficient information from the spatial position of an agent within the system to infer an agent's behaviour, interaction (influence) with other agents, and the intent of a cognitive agent. The critical assumption here is that the rich features required can be generated from positional data, alone. Strömbom et al. [77] allude to and recognise this in their seminal paper, in which they visualise trajectories of agents within their simulation and specifically highlight the trajectory change for both driving and collecting phases. We see that driving phases are iterative and visually predictable, whereas collecting phases are volatile and visual pattern prediction is challenging. We address this directly.

The baseline taxonomy is proposed to be simulated such that higher-level behaviours will have access to only a specific subset of the total lower-level behaviours. The taxonomy model here is not a result of unsupervised learning from underlying simulation data; the data is only representative with a theory instantiated into the taxonomy model, and capturing this is what truly matters. The taxonomy model here is about defining an appropriate type-system. Determining how lower-level behaviours are composed into higher-level behaviours, according to the composition rules contained within the taxonomy, is focused on the underlying system processes as opposed to the function transformation described. Assumptions at each level are not yet clearly defined, and the understanding of how these assumptions, and behaviours, nest and/or couple together is yet to be understood. Furthermore, the system-level assumptions may be different from the taxonomy generated ones, compounding the complexity of the collective adaptive agents established. Research efforts that contribute to answering this overarching question will need to address the following conjectures:

1. There exists sufficient positional information within sensor data to generate the mapping $f : \mathscr{S} \rightarrow \Sigma$, such that $f(s) = \hat{\sigma_l}$ with probability, p.
2. There exists a sufficient sequence of lower-level behaviours (Σ) to generate the mapping $g : \Sigma \rightarrow \Psi$, such that $g(\{\sigma_1^{t=0}, \sigma_2^{t=1}, \ldots, \sigma_l^{t=T}\}) = \hat{\psi_n}$.
3. There exists a sufficient sequence of higher-level behaviours to generate the mapping $h : \Psi \rightarrow \mathfrak{I}$, such that $h(\{\psi_1^{t=0}, \psi_2^{t=1}, \ldots, \psi_k^{t=T}\}) = \hat{\mathfrak{I}_k}$.
4. For any sequence of positional sensor time-series senor stream data observations, $\{s_1, s_2, \ldots, s_S\}$, the demonstrated behaviour σ_l is the physical representation of (1) $\mathfrak{I}_k$ and (2) the observed agent's decision cycle.

Note that the approximation term is not carried through in each subsequent conjecture, for example that $g(\sigma_l^t)$ with real-world data is represented as $g(\hat{\sigma_l^t})$. This difference in nomenclature use is not a mistake. The purpose of separating each of the individual mapping transformations is to present the ground truth; there is a requirement to approximate this ground truth as closely as possible, within each conjecture. While discovering these transformations with real-world data that is noisy, the estimates and associated error must be carried through in each subsequent classification task.

7.3.2.2 Framework

A research framework of assumptions must support determining the validity of the conjectures. The assumptions presented are preliminary and require refinement as future research explores the intersection of activity recognition and shepherding. They constrain the problem space as a means of addressing central issues and are intended to be relaxed as solutions address the conjectures. Relaxation of assumptions is important to ensure that the proposals solve real-world problems, and are not artificial. Problem scenario space assumptions include:

1. There exists a necessary and sufficient (1:1) mapping between the cognitive agent's intent, $\mathfrak{I}_k$, and an exhibited higher-level behaviour, ψ_k, that is invariant for the time window of classification. $\mathfrak{I}_k. \Longleftrightarrow \psi_n k$
2. The intent of the cognitive agent is demonstrated through the behaviour of the actuator agent. $\Psi \rightarrow \mathfrak{I}$.
3. The shepherd is the only cognitive agent within the problem space.

4. The goal of the sheep is survival, and therefore, the sheep are reactive agents.
5. The shepherding taxonomy contains the complete set of atomic actions, lower- and higher-level behaviours; any novel, unobserved behaviours do not exist within the system.
6. The shepherd's messages to the sheepdog (shepherding actuator) are not observable, such that they are encoded or hidden from the activity recognition agent.
7. The sheepdog does not have global free will or intent, although it does exhibit a local intent to achieve the shepherd's global intent.
8. The local shepherding actuator behaviour-decision change point does not occur until the previous sense–decide–act cycle is complete. $\sigma_1 \prec \sigma_2 \prec \cdots \prec \sigma_l$.
9. The observed time-series data stream does not experience stationarity degeneracy while within a single sense–decide–act decision cycle. This assumption is explicit only when considered within a single decision cycle, and not across multiple decision points, which invokes a situation of deep uncertainty.
10. Internal decision-making processes are not visible, and only limited, partial information exists about the system (communication, perception, and cognition).
11. If an agent does not have sensor stream observations, then we cannot determine their exact, present behaviour (although we can estimate and infer this from assuming their role within the system, which must be unique).
12. The cognitive agent's "intensity of intent" does not change over time.
13. Adaptations to the system environment are as a result of a combination of reactions.
14. An individual or collection of herded agents will not undertake an adversarial response to the shepherding agent (there is no effect of the repulsion from the shepherd encountered).

The development of assumptions naturally leads us to consider the requirement of research challenges as objectives. These objectives are the milestone indicators that both assess the validity of the framework of the assumptions, as well as break down multiple facets of the conjectures into distinct, measurable areas of research. The collection of objectives below are open problems of research that are at the core of activity recognition for shepherding. These specific objectives are linked and nested within the open challenges previously discussed, bounding specific areas of research within the broader challenge domains.

1. Determine what the algebraic structures that describe how viable candidate component behaviours fit together to make a viable candidate higher-level (shepherding) behaviours.
2. Understand system dynamic properties, such as the cascade of information throughout the flock, and the impact of sensor stream observation noise, signal loss, and adversarial behaviour.
3. Determine which features contain the greatest classification and anticipatory value within the system (invariant properties).

4. In online, streaming settings, determine which system classification features best predict future behaviours (at all levels), and how these features contribute to an understanding of the cognitive agent's intent.
5. Quantify the impact of time-constrained behaviour discrimination on classifier performance, for a single decision cycle only (such that there is no time-series stationarity degeneracy or violation of assumptions that underpin ergodic time-series behaviour).
6. Understand the readability of classification outputs as a methodological issue, discovering how we can identify the structural features of classification fit together, concerning any particular classification problem instance.
7. Determine the invariant system properties which categorise and describe the behaviour of both individual agents within the flock, as well as the aggregate system (entropy, designed measures, and metrics).
8. Develop candidate, higher-order statistics as measures for the nature of the system and its invariant properties (such as determination of decision change points).
9. Quantify decision change points within the system and describe the effect of these on the configuration of the system, as well as the (reactive) agents within (structure of information transfer and associated thresholds).
10. Describe the nature (properties and effect) of deep uncertainty observed within the system, including the effects of agent dynamic coupling, long action-chain feedback, and observed reward states.
11. Learn the underlying decision strategies through the observation of spatial position data.
12. Determine how classification systems perform through time with changing and adapting behaviours resulting from system feedback and reward structures (note that this is learning an optimal behaviour pattern sequence as well as evolving individual behaviours).
13. Identify and explain novel, individual agent behaviours as measured by the invariant system properties.
14. For low-observability systems (such as those with limited bit-rate throughput, changing sensor stream observations and patterns, and low signal-to-noise ratio) identify the classification trade-off between high-level observation (such as the system or flock as a collective entity) and low-level observation (such as an individual agent or local cluster of agents).
15. Identify the constituent fields and relevant theories that bound the problem space and support novel contributions.
16. Determine the impact of violating the assumptions of the system (such as A_1, where intra- or inter-activity delay occurs).
17. Determine the mechanism(s) of external influence demonstrated by the shepherding actuator (such as physical, cognitive, environmental, or communicative).
18. Understand the reaction chain of an agent (for further discussion of this, see Tinbergen [79, p. 47]).

19. Quantify the intensity of an agent's response and the effect this has on the conduct of observed behaviour.
20. Quantify the threshold of external stimuli required (if an action is complete) and the threshold at which behaviour changes (linked to the intensity of response).
21. Identify the impact of external stimuli on both an individual agent and the collective herd, and how this changes the threshold for an individual's response and their willingness to be within closer proximity to other members of the herd.

7.3.2.3 Central Challenge

The central challenge of activity recognition for shepherding is, in part, due to the dynamic coupling of agents; long action-chain feedback and rewards; violation of stationarity time-series data assumptions; the evolution of behaviour patterns over time; unknown decision change points, resulting in maximum uncertainty at the beginning of each decision time window; and varying degrees of resistance between repulsive and attractive forces changing through time, where deep uncertainty may come into the model (if the resistance is a function of, for instance, just the current state, then there may be a logical cyclic dependency influencing agents). Contained within the preliminary analysis section is a discussion of initial results that have sought to define and understand these problem complexities, and explain them well.

The shepherding taxonomy provides an initial analysis of a herding system. Similar classification approaches have developed ontologies that require specific domain knowledge, and are potentially fragile in their application. Datasets generated using on-agent sensors have been employed to classify behaviours of herding animals in predominantly offline settings [38, 55]. These studies have allowed the placement of specific sensor types on and around agents to enable more accurate classification, such as the use of accelerometers and magnetometers [84]. Our hybrid approach fuses both data- and ontology-driven approaches to the activity recognition problem. An advantage of this taxonomy is the ability to identify multi-level indicators to determine what higher-level behaviour is occurring, and in this, we address the cold-start problem faced by many simple data-driven approaches to activity recognition. Limited by this approach is our ability to understand the ergodicity and stationarity of the system. The taxonomy allows us to make a high-level assessment for shepherding systems where deep uncertainty is not present. Systems, where deep uncertainty is present, are covered in the following section.

7.4 Formulating Activity Recognition for Shepherding

7.4.1 Describing Shepherding Behaviours

To describe shepherding behaviours, we first conceive of a scenario for which a sheepdog is tasked to move a flock of sheep from point A to point B. The cognitive shepherding agent has directed the sheepdog to herd the flock from point A to point B. Once the sheepdog has successfully moved into their initial herding position, the sheepdog commences their behaviour sequences to achieve this intent. In this case, the higher-level activity of herding.

We see in Fig. 7.6 that there is a sequence of low-level behaviours for the shepherding agent. Initially, the shepherding agent is *positioning* from the starting location (star), followed by a sequence of *collecting* behaviours, and subsequently a sequence of *driving* behaviours. The final core behaviour codified within the simulation environment (no movement, an atomic action) is not represented here. What we observe is the decision change point between each positioning, collecting and driving sequence, visually identified though different dashed-lines (in this figure, representing a behaviour classification change). For this specific example, the execution of the positioning behaviour represents the commencement of a new decision change point. When observing the entire behaviour sequence, we can identify some patterns which emerge, such as directional changes between behaviours, as well as the length of a behaviour. The longer we can observe the system and its agents, the greater our certainty of intent and behaviours should be.

Visual inspection of positional information intuitively leads us to conclude that our understanding of the certainty of behaviour increases across a time window, as we obtain additional observations. Certainty is a monotonically increasing function.

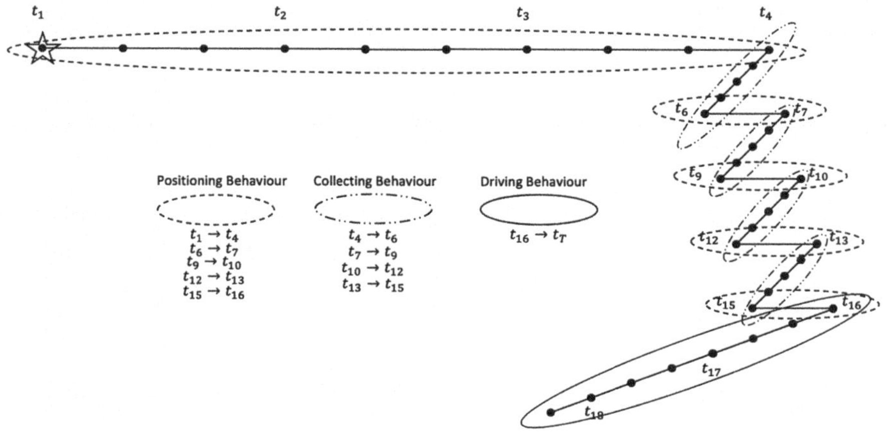

Fig. 7.6 Shepherding positioning (dash), driving (solid), and collecting behaviours (mixed dot/dash)

In contrast, we cannot make the same assertion regarding intent or higher-level behaviour. If we take the example in Fig. 7.6, the limited portion of observation allows for the inference of atomic actions and lower-level behaviour, although no clear understanding of higher-level behaviour or global intent. We can think about this problem in similar terms to a global versus local optima, in which a solution may become fixated around an incorrect solution (such as is observed with gradient descent methods). Throughout the short period of observation here, we infer visually as to potential global properties of the agents, as well as some of the local agent properties. The key insight here is that relative changes give us an understanding of the agent (local environment), whereas absolute changes give us an understanding of the global environment.

Figure 7.7 are the codified behaviours of Fig. 7.6, in which key time periods are listed. We allocated atomic actions to unique groups, then aggregate these through a transformation to indicate a lower-level behaviour. The lower-level behaviours correspond to the higher-level behaviour discussed above, and as represented in Fig. 7.6. From observing this scenario with full information, we can see aggregation at each level towards a single higher-level activity and ultimately, the achievement of the intent. For a situation with only partial or limited information, it becomes possible to infer behaviour through the classification of patterns from the sensor streams observations. When used in conjunction with assumptions detailed in the shepherding section, we can infer lower-level and higher-level behaviours, and ultimately the shepherd's intent. To allow this process to occur, we must design suitable features to classify the spatial behaviour of agents within the system.

7.4.2 Classifying Behaviour Through Spatial Data

Positional-based spatial features used to classify agent atomic actions into more complex behaviour within the system have been designed to allow for the inference of behaviours otherwise limited in situations where only partial information is available. Four observable sensor streams were determined to be the core, irreducible set of data that allowed for the classification of agent and system behaviour. The four pieces of information in each period collected are

1. The goal location of the cognitive agent, P^t_G.
2. The shepherding actuator (sheepdog's) location, $P^t_{\beta_j}$.
3. The centre of mass of the flock of sheep agents, Γ^t_{Π}.
4. The furthest sheep from the cognitive agent's goal location, $\max_{\pi \in \Pi} ||P_G - P^t_{\pi_i}||$.

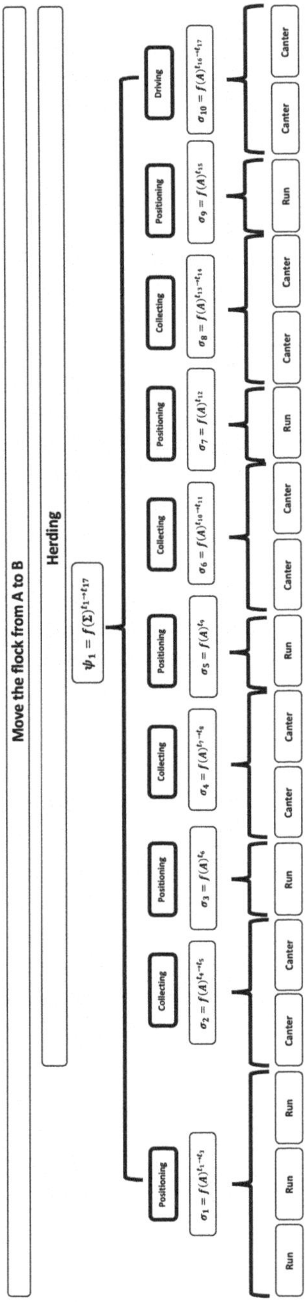

Fig. 7.7 Classified behaviours from Fig. 7.6 represented through the shepherding taxonomy

7.4.2.1 Methodology

Positional data contains a large amount of information about the agents within the system. The design of features that best use a transformation of the raw sensor data that are both meaningful and interpretable is an essential element of classification tasks. An angular velocity-based measure is presented as a transformation of the raw sensor data, which iteratively calculates an angle, ϕ, between the previous coordinate position of the sheepdog $P_{\beta_j}^{t-1}$ and the present coordinate position of the sheepdog $P_{\beta_j}^{t}$, relative to that of a static goal location (selected by the cognitive agent). The formulation of ϕ in each period is a modified implement ion of the Law of Cosines [83], defined as

$$\phi = \cos^{-1}\left(\frac{||P_{\beta_j}^{t} - P_{\beta_j}^{t-1}||^2 + ||P_G - P_{\beta_j}^{t-1}||^2 - ||P_G - P_{\beta_j}^{t}||^2}{2 \cdot ||P_G - P_{\beta_j}^{t-1}|| \cdot ||P_{\beta_j}^{t} - P_{\beta_j}^{t-1}||}\right). \tag{7.1}$$

The general idea here is that ϕ is a measure of variability of the sheepdog's behaviour within the system, relative to the goal location. Two boundary cases exist being for

- Small ϕ: indicating that the sheepdog is moving along a shortest-path in the expected direction towards to the goal location. We interpret this as likely being the driving behaviour (σ_1).
- Large ϕ: indicating that the sheepdog is moving off from the shortest-path, away from the goal location. We interpret this as likely being a collecting behaviour (σ_2).

The second feature, δ, calculates the rate of change for the direction of the angle of the sheepdog's motion. We do not use origin information within the measure; however, it does provide an understanding of the sheepdog's angular movement direction. We say that $P_{\beta_j}^{t}$ is the present coordinate position of the sheepdog, that $P_{\beta_j}^{t-1}$ is the previously sampled position, and that $P_{\beta_j}^{t+1}$ is the next sampled position. Assuming some level of smoothness in each period of change, then we state that:

$$\delta = \frac{\operatorname{atan2}\left(||P_{\beta_j}^{t+1} - P_{\beta_j}^{t}||\right) - \operatorname{atan2}\left(||P_{\beta_j}^{t} - P_{\beta_j}^{t-1}||\right)}{\Delta t}. \tag{7.2}$$

The central insight here is that these features rely on different levels of information certainty. The magnitude of angular position change (ϕ) requires global information and absolute position data, whereas the angular velocity rate (δ) requires only local, single-agent information. The second feature can be implemented in nearly all circumstances of observation in this problem, although it offers little information as to the global behaviour of the agent (or system). The first feature requires a global perspective of information; however, it can provide information on higher-level behaviour and intent. Forward work to evolve these measures will need

Table 7.2 Exploratory classification data analysis

	Overall accuracy (percentage)		Avg class accuracy (percentage)		Out-of-bag error (percentage)
Algorithm	Validation data	Test data	Validation data	Test data	–
Linear model	90.20	89.10	60.95	65.78	–
Decision tree	94.30	93.50	70.13	69.43	–
Support vector machine	90.70	89.80	61.98	63.25	–
Random forest	95.90	94.40	85.15	79.45	4.96

to employ common information sources better and understand limitation application constraints, including how assumptions such as the goal location may change the higher-level behaviour and classification outcomes. Understanding where it is appropriate to relax information constraints such as specifying or updating goal locations is essential to ensure measures are both mathematically feasible, as well as produce interpretable outcomes.

7.4.2.2 Analysis

Preliminary analysis employing these features for behaviour classification is in Table 7.2. Algorithms selected for this analysis include standard classifiers used for offline datasets. Exploration of ground-truth data reveals a high imbalance between the number of observations for each behaviour, confounding the ability to discriminate between each behaviour exhibited. We describe the performance of in terms of overall accuracy, defined as $\frac{\text{True Positive} + \text{True Negative}}{\text{Total Population}}$ and averaged classification accuracy, defined as the arithmetic means of all classes. A further measure for the random forest classifier, being the out-of-bag error, is a method for measuring prediction error that uses bootstrap aggregating sub-sample samples of data used for training. We note that within Table 7.2 that the average class accuracy is substantially less than the overall accuracy, due to the number of instances in each class (i.e. the high-class imbalance).

Previous studies have investigated different classifier techniques on disparate datasets, agent types, and the number of behaviours to classify. Generated simulation data does not subject classifiers to the issues of real-world data, such as signal-to-noise complexity; however, artificial complexities can be increased to simulate classification response problems. High performing offline approaches often utilise similar techniques to those employed above, such as the decision tree [38]. The highest performing offline approach in Table 7.2 employs a collection of decision trees known as a random forest [10], indicating the feasability of this approach in offline cases.

7.5 Conclusion

Activity recognition is an exciting field of research that is receiving increased attention from both within academic and practitioner communities [11]. Activity recognition remains a complex undertaking and while still relatively immature in academic literature presents many opportunities to advance complementary fields of research in computer science and beyond [6, 13]. Both activity recognition research and implementation is progressing rapidly, in part, due to advances in technology such as increases om mobile computing power and the efficiency of mobile energy storage solutions, enabling the deployment of complex processes in the real-world. When combined with decreasing technology costs that lower the barriers to entry, activity recognition is prime to harness these advancements [13]. For shepherding, activity recognition offers an avenue to understand the complexity of an agent's behaviour through understanding its spatiotemporal distribution, motivating the need for new methodologies to meet this challenge.

Advancing activity recognition for shepherding requires a more in-depth investigation of the intersection between activity recognition and shepherding, addressing complexities from within each field. We have addressed a selection of known activity recognition challenges through the shepherding lens, including briefly the promising area of adversarial activity recognition which elicits interest for intelligence and national security applications. The problem necessitates the generation of methods which are suitable for a wide range of applications and robust to high-noise sensor stream data. Future research presented throughout offers a range of plausible areas for exploration; our research focus is centred around advancing classification techniques in both offline and online settings through the use of observable spatial position data, enabling an artificial intelligence system to learn and understand exhibited behaviour without cognitive guidance from a human.

References

1. Allen, J.F.: Maintaining knowledge about temporal intervals. Commun. ACM **26**(11), 832–843 (1983). https://doi.org/10.1145/182.358434
2. Anderson, S., Bredeche, N., Eiben, A., Kampis, G., van Steen, M.: Adaptive Collective Systems: Herding Black Sheep. Fundamentals of Collective Adaptive Systems (2013)
3. Aroor, A., Epstein, S.L., Korpan, R.: Online learning for crowd-sensitive path planning. In: Proceedings of the 17th International Conference on Autonomous Agents and MultiAgent Systems (2018)
4. Artikis, A., Sergot, M., Paliouras, G.: An event calculus for event recognition. IEEE Trans. Knowl. Data Eng. **27**(4), 895–908 (2015). https://doi.org/10.1109/TKDE.2014.2356476
5. Azkune, G.: Learning for dynamic and personalised knowledge-based activity models. Ph.D. Thesis, Universidad de Deusto (2015)
6. Bakar, U.A.B.U.A., Ghayvat, H., Hasanm, S.F., Mukhopadhyay, S.C.: Activity and Anomaly Detection in Smart Home: A Survey, pp. 191–220. Springer International Publishing, Cham (2016). https://doi.org/10.1007/978-3-319-21671-3_9

7. Baumann, M., Büning, H.K.: Learning shepherding behavior. Ph.D. Thesis, University of Paderborn (2016)
8. Bettini, C., Brdiczka, O., Henricksen, K., Indulska, J., Nicklas, D., Ranganathan, A., Riboni, D.: A survey of context modelling and reasoning techniques. Pervasive Mob. Comput. **6**(2), 161–180 (2010). https://doi.org/10.1016/j.pmcj.2009.06.002
9. Bikakis, A., Antoniou, G., Hasapis, P.: Strategies for contextual reasoning with conflicts in ambient intelligence. Knowl. Inf. Syst. **27**(1), 45–84 (2011). https://doi.org/10.1007/s10115-010-0293-0
10. Breiman, L.: Random forests. Mach. Learn.**45**(1), 5–32 (2001). https://doi.org/10.1023/A:1010933404324
11. Bruno, B., Mastrogiovanni, F., Saffiotti, A., Sgorbissa, A.: Using fuzzy logic to enhance classification of human motion primitives. In: Laurent, A., Strauss, O., Bouchon-Meunier, B., Yager, R.R. (eds.) Information Processing and Management of Uncertainty in Knowledge-Based Systems, pp. 596–605. Springer International Publishing, Cham (2014)
12. Chen, L., Khalil, I.: Activity Recognition: Approaches, Practices and Trends, pp. 1–31. Atlantis Press, Paris (2011). https://doi.org/10.2991/978-94-91216-05-3_1
13. Chen, L., Hoey, J., Nugent, C.D., Cook, D.J., Yu, Z.: Sensor-based activity recognition. IEEE Trans. Syst. Man Cybern. Part C (Appl. Rev.) **42**(6), 790–808 (2012). https://doi.org/10.1109/TSMCC.2012.2198883
14. Chen, L., Nugent, C.D., Wang, H.: A knowledge-driven approach to activity recognition in smart homes. IEEE Trans. Knowl. Data Eng. **24**(6), 961–974 (2012). https://doi.org/10.1109/TKDE.2011.51
15. Chen, L., Nugent, C., Okeyo, G.: An ontology-based hybrid approach to activity modeling for smart homes. IEEE Trans. Human-Machine Syst. **44**(1), 92–105 (2014). https://doi.org/10.1109/THMS.2013.2293714
16. Cicirelli, F., Fortino, G., Giordano, A., Guerrieri, A., Spezzano, G., Vinci, A.: On the design of smart homes: a framework for activity recognition in home environment. J. Medical Syst. **40**(9), 200 (2016). https://doi.org/10.1007/s10916-016-0549-7
17. Cook, D.J., Crandall, A.S., Thomas, B.L., Krishnan, N.C.: Casas: A smart home in a box. Computer **46**(7), 62–69 (2013). https://doi.org/10.1109/MC.2012.328
18. Cowling, P., Gmeinwieser, C.: Ai for herding sheep. In: Proceedings of the Sixth AAAI Conference on Artificial Intelligence and Interactive Digital Entertainment, AIIDE'10, pp. 2–7. AAAI Press, California (2010). http://dl.acm.org/citation.cfm?id=3014666.3014668
19. Craig I. Schlenoff Sebti Foufou, S.B.B.: An approach to ontology-based intention recognition using state representations. In: 4th International Conference on Knowledge Engineering and Ontology Development (KEOD 2012) (2014)
20. Crispim-Junior, C.F., Buso, V., Avgerinakis, K., Meditskos, G., Briassouli, A., Benois-Pineau, J., Kompatsiaris, I.Y., Bremond, F.: Semantic event fusion of different visual modality concepts for activity recognition. IEEE Trans. Pattern Anal. Mach. Intell. **38**(8), 1598–1611 (2016). https://doi.org/10.1109/TPAMI.2016.2537323
21. de Carvalho, A.C.P.L.F., Freitas, A.A.: A Tutorial on Multi-label Classification Techniques, pp. 177–195. Springer, Berlin (2009). https://doi.org/10.1007/978-3-642-01536-6_8
22. Eiter, T., Ianni, G., Krennwallner, T., Polleres, A.: Reasoning Web. Springer, Berlin (2008). https://doi.org/10.1007/978-3-540-85658-0_1
23. Ferber, J.: Multi-Agent Systems: An Introduction to Distributed Artificial Intelligence, 1st edn. Addison-Wesley Longman, Boston (1999)
24. Fleury, A., Vacher, M., Noury, N.: SVM-based multimodal classification of activities of daily living in health smart homes: sensors, algorithms, and first experimental results. IEEE Trans. Inf. Technol. Biomedicine **14**(2), 274–283 (2010). https://doi.org/10.1109/TITB.2009.2037317
25. Guo, K., Li, Y., Lu, Y.: An alternative-service recommending algorithm based on semantic similarity. China Commun. **14**(8), 124–136 (2017). https://doi.org/10.1109/CC.2017.8014353
26. Hamilton, W.D.: Geometry for the selfish herd. J. Theoret. Biol. **31**(2), 295–311 (1971)
27. Hasan, M., Roy-Chowdhury, A.K.: Context aware active learning of activity recognition models. In: 2015 IEEE International Conference on Computer Vision (ICCV), pp. 4543–4551 (2015). https://doi.org/10.1109/ICCV.2015.516

28. He, X.: Swarm robotics: The future of medicine? (2015). https://medtechboston.medstro.com/blog/2015/10/06/swarm-robotics-what-you-need-to-know-about-the-future-of-medicine/
29. Helaoui, R., Niepert, M., Stuckenschmidt, H.: Recognizing interleaved and concurrent activities using qualitative and quantitative temporal relationships. Pervasive Mob. Comput. **7**(6), 660–670 (2011). https://doi.org/10.1016/j.pmcj.2011.08.004
30. Helaoui, R., Riboni, D., Stuckenschmidt, H.: A probabilistic ontological framework for the recognition of multilevel human activities. In: Proceedings of the 2013 ACM International Joint Conference on Pervasive and Ubiquitous Computing, UbiComp '13, pp. 345–354. ACM, New York (2013). https://doi.org/10.1145/2493432.2493501
31. Ikizler, N., Forsyth, D.: Searching video for complex activities with finite state models. In: 2007 IEEE Conference on Computer Vision and Pattern Recognition, pp. 1–8 (2007). https://doi.org/10.1109/CVPR.2007.383168
32. Ilyas, A., Engstrom, L., Athalye, A., Lin, J.: Proceedings of the 35th International Conference on Machine Learning, PMLR **80**, 2137–2146, (2018)
33. Incel, O., Kose, M., Ersoy, C.: A review and taxonomy of activity recognition on mobile phones. BioNanoScience **3**(2), 145–171 (2013). https://doi.org/10.1007/s12668-013-0088-3
34. Iqbal, M., Pao, H.K.: Activity recognition from minimal distinguishing subsequence mining. AIP Confer. Proc. **1867**(1), 020,046 (2017). https://doi.org/10.1063/1.4994449. https://aip.scitation.org/doi/abs/10.1063/1.4994449
35. Iwamoto, S., Ohmura, R.: Towards concurrent task verification in context-aware applications. In: Adjunct Proceedings of the 2015 ACM International Joint Conference on Pervasive and Ubiquitous Computing and Proceedings of the 2015 ACM International Symposium on Wearable Computers, UbiComp/ISWC'15 Adjunct, pp. 1473–1477. ACM, New York (2015). https://doi.org/10.1145/2800835.2801618
36. Kishore, S., Bhattacharjee, S., Swetapadma, A.: A hybrid method for activity monitoring using principal component analysis and back-propagation neural network. In: 2017 International Conference On Smart Technologies For Smart Nation (SmartTechCon), pp. 885–889 (2017). https://doi.org/10.1109/SmartTechCon.2017.8358499
37. Kleiminger, W., Mattern, F., Santini, S.: Predicting household occupancy for smart heating control: a comparative performance analysis of state-of-the-art approaches. Energy Build. **85**, 493–505 (2014). https://doi.org/https://doi.org/10.1016/j.enbuild.2014.09.046. http://www.sciencedirect.com/science/article/pii/S037877881400783X
38. Kuankid, S., Rattanawong, T., Aurasopon, A.: Classification of the cattle's behaviors by using accelerometer data with simple behavioral technique. In: Signal and Information Processing Association Annual Summit and Conference (APSIPA), 2014 Asia-Pacific, pp. 1–4 (2014)
39. Kumar, R.C., Bharadwaj, S.S., Sumukha, B.N., George, K.: Human activity recognition in cognitive environments using sequential ELM. In: 2016 Second International Conference on Cognitive Computing and Information Processing (CCIP), pp. 1–6 (2016). https://doi.org/10.1109/CCIP.2016.7802880
40. Kunze, K.: Real-life activity recognition – focus on recognizing reading activities. In: Iwamura, M., Shafait, F. (eds.) Camera-Based Document Analysis and Recognition, pp. 179–185. Springer International Publishing, Cham (2014)
41. Lasecki, W.S., Marcus, A., Rzeszotarski, J.M., Bigham, J.P.: Using microtask continuity to improve crowdsourcing. Technical Report (2014)
42. Lee, W., Kim, D.: Autonomous shepherding behaviors of multiple target steering robots. Sensors **17**(12), 2729 (2017)
43. Lettmann, T., Baumann, M., Eberling, M., Kemmerich, T.: Modeling Agents and Agent Systems, pp. 157–181. Springer, Berlin (2011). https://doi.org/10.1007/978-3-642-24016-4_9
44. Li, F., Dustdar, S.: Incorporating unsupervised learning in activity recognition. In: Proceedings of the 4th AAAI Conference on Activity Context Representation: Techniques and Languages, AAAIWS'11-04, pp. 38–41. AAAI Press, Cambridge (2011). http://dl.acm.org/citation.cfm?id=2908613.2908620

45. Li, M., Hu, Z., Liang, J., Li, S.: Shepherding behaviors with single shepherd in crowd management. In: Xiao, T., Zhang, L., Ma, S. (eds.) System Simulation and Scientific Computing, pp. 415–423. Springer, Berlin (2012)
46. Liao, L., Fox, D., Kautz, H.: Hierarchical conditional random fields for GPS-based activity recognition. In: Proceedings of the International Symposium of Robotis Research (ISRR 2005). Springer, Berlin (2005)
47. Licitra, R.A., Hutcheson, Z.D., Doucette, E.A., Dixon, W.E.: Single agent herding of n-agents: a switched systems approach. IFAC-PapersOnLine **50**(1), 14374–14379 (2017). 20th IFAC World Congress. https://doi.org/10.1016/j.ifacol.2017.08.2020. http://www.sciencedirect.com/science/article/pii/S2405896317326587
48. Lien, J.M., Pratt, E.: Interactive planning for shepherd motion. In: AAAI Spring Symposium: Agents that Learn from Human Teachers (2009)
49. Lien, J.M., Bayazit, O.B., Sowell, R.T., Rodriguez, S., Amato, N.M.: Shepherding behaviors. In: IEEE International Conference on Robotics and Automation, vol. 4, pp. 4159–4164. Citeseer (2004)
50. Liu, Y., Nie, L., Liu, L., Rosenblum, D.S.: From action to activity. Neurocomputing **181**(C), 108–115 (2016). https://doi.org/10.1016/j.neucom.2015.08.096
51. Loke, S.W.: Representing and reasoning with situations for context-aware pervasive computing: a logic programming perspective. Knowl. Eng. Rev. **19**(3), 213–233 (2004). https://doi.org/10.1017/S0269888905000263
52. Luz, G., Barros, K., AraÃžjo, F.V., Barbosa da Silva, G., Augusto Ferreira da Silva, P., Condori, R., Brasil, L.: Nanorobotics in drug delivery systems for treatment of cancer: a review. J. Materials Sci. Eng. A **6** (2016). https://doi.org/10.17265/2161-6213/2016.5-6.005
53. McVicar, K.E.: C3: The challenge of change. IEEE Trans. Aerosp. Electron. Syst. **AES-20**(4), 401–413 (1984). https://doi.org/10.1109/TAES.1984.4502061
54. Mould, N., Regens, J.L., III, C.J.J., Edger, D.N.: Video surveillance and counterterrorism: the application of suspicious activity recognition in visual surveillance systems to counterterrorism. J. Polic. Intell. Count. Terror. **9**(2), 151–175 (2014). https://doi.org/10.1080/18335330.2014.940819.
55. Nadimi, E.S., Jørgensen, R.N., Blanes-Vidal, V., Christensen, S.: Monitoring and classifying animal behavior using ZigBee-based mobile ad hoc wireless sensor networks and artificial neural networks. Comput. Electron. Agric. **82**, 44–54 (2012). https://doi.org/10.1016/j.compag.2011.12.008
56. Okeyo, G., Chen, L., Wang, H., Sterritt, R.: Ontology-enabled activity learning and model evolution in smart homes. In: Yu, Z., Liscano, R., Chen, G., Zhang, D., Zhou, X. (eds.) Ubiquitous Intelligence and Computing, pp. 67–82. Springer, Berlin (2010)
57. Okeyo, G., Chen, L., Wang, H., Sterritt, R.: Ontology-Based Learning Framework for Activity Assistance in an Adaptive Smart Home, pp. 237–263. Atlantis Press, Paris (2011). https://doi.org/10.2991/978-94-91216-05-3_11
58. Okeyo, G., Chen, L., Wang, H., Sterritt, R.: A hybrid ontological and temporal approach for composite activity modelling. In: 2012 IEEE 11th International Conference on Trust, Security and Privacy in Computing and Communications, pp. 1763–1770 (2012). https://doi.org/10.1109/TrustCom.2012.34
59. Okeyo, G.O., Chen, L., Wang, H.: An agent-mediated ontology-based approach for composite activity recognition in smart homes. J. UCS **19**, 2577–2597 (2013)
60. Okeyo, G., Chen, L., Wang, H., Sterritt, R.: Dynamic sensor data segmentation for real-time knowledge-driven activity recognition. Pervasive Mob. Comput. **10**, 155–172 (2014). https://doi.org/10.1016/j.pmcj.2012.11.004
61. Paranjape, A.A., Chung, S.J., Kim, K., Shim, D.H.: Robotic herding of a flock of birds using an unmanned aerial vehicle. IEEE Trans. Rob. **34**(4), 901–915 (2018)
62. Parker, L.: Multiple Mobile Robot Systems, pp. 921–941. Springer, Berlin (2008). https://doi.org/10.1007/978-3-540-30301-5_41
63. Quero, J., Orr, C., Zang, S., Nugent, C., Salguero, A., Espinilla, M.: Real-time recognition of interleaved activities based on ensemble classifier of long short-term memory with fuzzy temporal windows. Proceedings **2**(19), 1225 (2018). https://doi.org/10.3390/proceedings2191225

64. Ranasinghe, S., Machot, F.A., Mayr, H.C.: A review on applications of activity recognition systems with regard to performance and evaluation. Int. J. Distrib. Sens. Netw. **12**(8), 1550147716665520 (2016). https://doi.org/10.1177/1550147716665520
65. Rashidi, P., Cook, D.J.: COM: A method for mining and monitoring human activity patterns in home-based health monitoring systems. ACM Trans. Intell. Syst. Technol. **4**(4), 64:1–64:20 (2013). https://doi.org/10.1145/2508037.2508045
66. Rashidi, P., Cook, D.J., Holder, L.B., Schmitter-Edgecombe, M.: Discovering activities to recognize and track in a smart environment. IEEE Trans. Knowl. Data Eng. **23**(4), 527–539 (2011). https://doi.org/10.1109/TKDE.2010.148
67. Riboni, D., Bettini, C.: Context-aware activity recognition through a combination of ontological and statistical reasoning. In: Zhang, D., Portmann, M., Tan, A.H., Indulska, J. (eds.) Ubiquitous Intelligence and Computing, pp. 39–53. Springer, Berlin (2009)
68. Riboni, D., Pareschi, L., Radaelli, L., Bettini, C.: Is ontology-based activity recognition really effective? In: 2011 IEEE International Conference on Pervasive Computing and Communications Workshops (PERCOM Workshops), pp. 427–431 (2011). https://doi.org/10.1109/PERCOMW.2011.5766927
69. Roggen, D., Tröster, G., Lukowicz, P., Ferscha, A., del R. Millán, J., Chavarriaga, R.: Opportunistic human activity and context recognition. Computer **46**(2), 36–45 (2013). https://doi.org/10.1109/MC.2012.393
70. Sadri, F.: Logic-based approaches to intention recognition. In: Handbook of Research on Ambient Intelligence and Smart Environments: Trends and Perspectives. Citseer (2009)
71. Saguna, S., Zaslavsky, A., Chakraborty, D.: Complex activity recognition using context-driven activity theory and activity signatures. ACM Trans. Comput.-Hum. Interact. **20**(6), 32:1–32:34 (2013). https://doi.org/10.1145/2490832
72. Salah, A., Oudeyer, P., c li, C.M., del Solar, J.R.: Guest editorial behavior understanding and developmental robotics. IEEE Trans. Auton. Ment. Dev. **6**(2), 77–79 (2014). https://doi.org/10.1109/TAMD.2014.2328731
73. Sato, K., Fujinami, K.: Active learning-based classifier personalization: A case of on-body device localization. In: 2017 IEEE 6th Global Conference on Consumer Electronics (GCCE), pp. 1–2 (2017). https://doi.org/10.1109/GCCE.2017.8229317
74. Satyanarayanan, M.: Pervasive computing: vision and challenges. IEEE Pers. Commun. **8**(4), 10–17 (2001). https://doi.org/10.1109/98.943998
75. Singh, H., Campbell, B., Elsayed, S., Perry, A., Hunjet, R., Abbass, H.: Modulation of force vectors for effective shepherding of a swarm: A bi-objective approach. In: 2019 IEEE Congress on Evolutionary Computation (CEC), pp. 2941–2948. IEEE, Piscataway (2019). https://doi.org/10.1109/CEC.2019.8790228
76. Steinhauer, H.J., Chua, S.L., Guesgen, H.W., Marsland, S.R.: Utilising temporal information in behaviour recognition. In: AAAI Spring Symposium: It's All in the Timing (2010)
77. Strömbom, D., Mann, R.P., Wilson, A.M., Hailes, S., Morton, A.J., Sumpter, D.J.T., King, A.J.: Solving the shepherding problem: heuristics for herding autonomous, interacting agents. J. R. Soc. Interf. **11**(100) (2014). https://browzine.com/articles/52614503
78. Sun, Y., Rossi, L., Luan, H., Shen, C.C.: Modeling and analyzing large swarms with covert leaders. In: 2013 IEEE 7th International Conference on Self-Adaptive and Self-Organizing Systems, pp. 169–178 (2013). https://doi.org/10.1109/SASO.2013.32
79. Tinbergen, N.: The Study of Instinct. Clarendon Press, Oxford (1951)
80. Vail, D.L., Veloso, M.M., Lafferty, J.D.: Conditional random fields for activity recognition. In: Proceedings of the 6th International Joint Conference on Autonomous Agents and Multiagent Systems, AAMAS '07, pp. 235:1–235:8. ACM, New York (2007). https://doi.org/10.1145/1329125.1329409
81. Wang, X.R., Miller, J.M., Lizier, J.T., Prokopenko, M., Rossi, L.F.: Measuring information storage and transfer in swarms. In: European Conference on Artificial Life, Paris (2011)
82. Weiss, G. (ed.): Multiagent Systems: A Modern Approach to Distributed Artificial Intelligence. MIT Press, Cambridge (1999)

83. Weisstein, E.W.: Law of cosines. MathWorld–A Wolfram Web Resource (2019). http://mathworld.wolfram.com/LawofCosines.html
84. Williams, H.J., Holton, M.D., Shepard, E.L.C., Largey, N., Norman, B., Ryan, P.G., Duriez, O., Scantlebury, M., Quintana, F., Magowan, E.A., Marks, N.J., Alagaili, A.N., Bennett, N.C., Wilson, R.P.: Identification of animal movement patterns using tri-axial magnetometry. Mov. Ecol. **5**(1), 6 (2017)
85. Wooldridge, M.: An Introduction to MultiAgent Systems. Wiley, New York (2009). https://books.google.com.au/books?id=X3ZQ7yeDn2IC
86. Ye, J., Stevenson, G.: Semantics-driven multi-user concurrent activity recognition. In: Augusto, J.C., Wichert, R., Collier, R., Keyson, D., Salah, A.A., Tan, A.H. (eds.) Ambient Intelligence, pp. 204–219. Springer International Publishing, Cham (2013)
87. Ye, J., Stevenson, G., Dobson, S.: A top-level ontology for smart environments. Pervasive Mob. Comput. **7**(3), 359–378 (2011). Knowledge-Driven Activity Recognition in Intelligent Environments. https://doi.org/10.1016/j.pmcj.2011.02.002. http://www.sciencedirect.com/science/article/pii/S1574119211000277.
88. Ye, J., Dobson, S., McKeever, S.: Situation identification techniques in pervasive computing: a review. Pervasive Mob. Comput. **8**(1), 36 – 66 (2012). https://doi.org/10.1016/j.pmcj.2011.01.004. http://www.sciencedirect.com/science/article/pii/S1574119211000253
89. Ye, J., Dasiopoulou, S., Stevenson, G., Meditskos, G., Kontopoulos, E., Kompatsiaris, I., Dobson, S.: Semantic web technologies in pervasive computing. Pervasive Mob. Comput. **23**(C), 1–25 (2015). https://doi.org/10.1016/j.pmcj.2014.12.009
90. Ye, J., Fang, L., Dobson, S.: Discovery and recognition of unknown activities. In: Proceedings of the 2016 ACM International Joint Conference on Pervasive and Ubiquitous Computing: Adjunct, UbiComp '16, pp. 783–792. ACM, New York (2016). https://doi.org/10.1145/2968219.2968288
91. Yordanova, K., Krüger, F., Kirste, T.: Context aware approach for activity recognition based on precondition-effect rules. In: 2012 IEEE International Conference on Pervasive Computing and Communications Workshops, pp. 602–607 (2012). https://doi.org/10.1109/PerComW.2012.6197586
92. Zheng, V.W., Hu, D.H., Yang, Q.: Cross-domain activity recognition. In: Proceedings of the 11th International Conference on Ubiquitous Computing, UbiComp '09, pp. 61–70. ACM, New York (2009). https://doi.org/10.1145/1620545.1620554

Chapter 8
Stable Belief Estimation in Shepherd-Assisted Swarm Collective Decision Making

Aya Hussein and Hussein A. Abbass

Swarm collective decision making refers to the case where a swarm needs to make a decision based on different pieces of evidence collected by its individuals. This problem has been investigated by several recent studies which proposed strategies to enable the swarm to perform fast and accurate collective decision making. However, the performance of these strategies (in terms of its accuracy, speed and level of consensus) suffers significantly in complex environments. The aim of our work is to propose a collective decision-making strategy that promises a consistent performance across different levels of scenario complexity and achieves superiority over the existing strategies in highly complex scenarios. To achieve this aim, our proposed algorithm employs a shepherding agent to boost the performance of the swarm. The swarm members are only responsible for sensing the state of a feature distributed in the environment. Only the shepherd needs to be able to process position and navigation abilities to collect the swarm. The algorithm consists of two phases: exploration and belief sharing. In the exploration phase, swarm members navigate through an environment and sense its features. Then, in the belief sharing phase, a shepherding agent collects the swarm members together so that they can share their estimates and calculate their decisions. The results demonstrate that the proposed shepherding algorithm succeeds across different levels of scenario complexity. Additionally, the approach achieves high levels of accuracy and consensus in complex non-homogeneous environments where the baseline state-of-the-art algorithm fails.

A. Hussein (✉) · H. A. Abbass
School of Engineering and Information Technology, University of New South Wales, Canberra, ACT, Australia
e-mail: z5139608@ad.unsw.edu.au; h.abbass@adfa.edu.au

H. A. Abbass, R. A. Hunjet (eds.), *Shepherding UxVs for Human-Swarm Teaming*, Unmanned System Technologies,
https://doi.org/10.1007/978-3-030-60898-9_8

8.1 Introduction

Collective decision making has been the focus of numerous research studies on swarm and multi-agent systems [2, 5, 7, 9, 12, 13] due to its widespread real-life applications. These include robot selection within human interaction with multi-robots [3], the selection of the shortest path to be traversed by the swarm [4], swarm leader election [1], best site selection [12] and abnormal behaviour detection [11]. The collective decision-making problem has two aspects that are discussed in the following paragraphs: the evaluation criteria and factors of complexity.

Similar to single-agent decision making, accuracy and speed are key performance measures to evaluate collective decision-making strategies. Decision accuracy, in terms of the proportion of runs that ends with a consensus on the correct decision, is the main measure of the effectiveness of the decision-making strategy. Meanwhile, speed, as measured by the amount of time needed for all the swarm members to lock in a final decision, is generally an indicator of the efficiency of the decision-making strategy. Speed can also affect mission success for time-critical tasks, for instance, in emergencies. The speed versus accuracy trade-off in collective decision-making algorithms has been acknowledged by past researchers [2, 12].

Besides accuracy and speed, high levels of consensus, in terms of the proportion of runs where all the swarm members reach the same final decision, can be desirable or sometimes necessary for swarm operation. This can be the case when the swarm subsequent action is conditioned upon the decision reached by the swarm. For instance, a swarm making a decision on the shortest path might need to achieve high levels of consensus if network connectivity among its members is demanded, as in [6]. Similarly, high levels of consensus while selecting a goal position might be required for successful object collective transport [8]. Failure to achieve consensus in this scenario could lead the swarm members to push the object in opposite directions. A strategy for improving consensus among swarm members has been proposed in [7]. Together, the three indicators of accuracy, speed and consensus can be helpful in evaluating different aspects of the performance.

Valentini et al. [13] formulated an abstraction of the collective decision-making problem in which a swarm of agents explores a grid-based environment to evaluate the abundance of an environmental feature that is scattered across the environment. The swarm is required to collectively determine whether the feature is frequent in the environment, i.e., the feature exists in at least half of the cells. The problem serves as an abstraction which is applicable to a swarm search for precious metals, pollutants or cancer cells. The level of difficulty (or complexity) in this problem formulation is impacted by a number of factors, including feature frequency and the homogeneity of feature distribution.

Feature frequency is the ratio of an environment's cells where the feature is present to the total number of cells in the environment. Feature frequency is an important factor that affects the level of complexity of a collective decision-making problem. When feature frequency is very low, it is likely that most of the cells encountered by an agent will not contain the feature and that within a short time window, most of the agents will maintain that the feature is not frequent in the environment. An analogous scenario is expected when feature ratio is very high. On the contrary, as feature ratio approaches 0.5, it is likely that an agent's opinion will swing frequently and that agents will have conflicting opinions, which increases the difficulty of the decision-making process. Previous studies showed that the performance of collective decision making (in terms of its speed and accuracy) suffers more in environments with feature ratios approaching 0.5 [5, 13].

In addition, the lack of homogeneity in the feature distribution was also shown to considerably increase the task complexity by increasing decision time while decreasing its accuracy [5]. A non-homogeneous environment implies that the estimates obtained by different agents can vary widely depending on the locations explored by each agent. This increases the likelihood that agents will have contradicting opinions, which impedes achieving consensus.

Our work addresses these challenges by proposing a shepherd-assisted algorithm that performs consistently well across environments with various levels of complexity. We propose a shepherd-assisted decision-making (SDM) algorithm with the aim of solving the swarm decision-making problem while maintaining the success of the algorithm, in terms of its accuracy, speed and consensus, in highly complex environments (i.e., environments with feature ratio close to 0.5 and non-homogeneous distribution) where the state-of-the-art algorithms fail.

Moreover, our algorithm is designed to eliminate the burden of heavy processing a swarm member needs to make to localise, orient or navigate in the environment. Swarm members only need to sense the state of a feature distributed in the environment. The shepherd takes the responsibility for finding the members of the swarm to collect and navigate them to a destination. During collection, belief update and revision take place to enable reaching consensus.

8.2 Related Work

Variations of the collective decision-making problem have been considered by recent studies. For instance, Valentini et al. [12] considered the problem of site selection in which swarm members are required to explore two candidate sites to select the one with better quality. The quality of each site is broadcast by beacons which can only be heard from within their corresponding sites. Being initialised with a random opinion on the preferred site, a swarm member visits the site associated with its current opinion to estimate its quality. Then, the swarm member returns back to the nest and broadcasts its opinion for a duration proportional to the sensed quality. Swarm members update their beliefs by applying a majority vote or copying the opinion of a random neighbour (voter model). Valentini et al. [12] reported that the majority rule resulted in faster but less accurate decision making than the voter model.

A more difficult version of the problem was proposed by Valentini et al. [13] in which a swarm of robots performs collective perception of a binary environmental feature. The feature is scattered across the environment, and the robots are required to explore the environment and collectively evaluate the abundance of this feature. The feature is chosen to be the colour of the environment, such that each cell is either black or white, and the swarm is required to collectively decide whether the environment as a whole is mostly black. In the absence of mapping and sharing route/trajectory information, this collective decision-making problem is inherently more difficult than the collective site selection. This is because the distributed feature problem implies that different swarm members form their opinions based on different but possibly overlapping pieces of evidence (as they can visit different but non-disjoint sets of cells within the environment). Meanwhile, in the site selection problem, a swarm member can directly evaluate the quality of a certain site by visiting it.

Valentini et al. [13] considered using three swarm decision-making strategies to solve the feature detection problem: majority based, voter model and direct comparison. The direct comparison strategy is similar to the voter model with the exception that a robot copies the opinion of its neighbour only if the quality of neighbour's opinion is better than its own opinion quality. Each of these strategies consists of two states: exploration and dissemination. In the exploration state, a robot performs a random walk while sensing the colour of the cells it visits. The robot estimates its opinion as the most common feature encountered (black or white) and calculates the opinion quality as the percentage of time the colour associated with its opinion was encountered. Then, the robot moves to the dissemination state in which it broadcasts its opinion for a duration proportional to its opinion quality.

Valentini et al. compared the three strategies under different levels of complexity, by manipulating the frequency of the feature in the environment. Valentini et al. found that the direct comparison strategy had the fastest operation in low-complexity settings but suffered the most as the problem complexity increased. The majority-based strategy was faster but less accurate than the neighbour-based one, and they both were less sensitive to problem complexity than the direct comparison.

Recently, Ebert et al. [5] proposed another algorithm for combining swarm members' opinions to collectively decide on the most common feature in the environment. Each swarm member maintains an estimate, a belief and a belief concentration. The estimate contains the agent's estimation of the environment colour based only on its most recent exploration. The belief is formed by applying a majority voting on the estimates received within the last exploration–dissemination cycle. Finally, the belief concentration represents the agent's integration of the received beliefs over time. When the belief concentration crosses a predefined threshold and stays there for 30 s, the agent makes its own final decision (non-revertible belief) but also resumes its exploration–dissemination operation.

Ebert et al. [5] tested their algorithm under two levels of homogeneity of feature distribution: homogeneous and non-homogeneous distributions. In a homogeneous environment with a feature ratio of $r \in [0, 1]$, each cell contains the feature with a probability r independently of other cells. On the other hand, a non-homogeneous environment with a feature ratio of r has the feature present only within a continuous region containing r of the environment cells [5, Figure 2]. That is, $100r\%$ of the cells clustered together contains the feature. Ebert et al. found that the performance of their algorithm (in terms of its speed and accuracy) dropped considerably in non-homogeneous as compared to homogeneous settings. The drop in performance was most profound as r approached 0.5.

In fact, the results reported in [5, 13] are understandable in the light of the discussion, in Sect. 8.1, on the factors increasing the complexity of the collective decision making. Nonetheless, the design of the algorithm itself might also exacerbate the problem. For instance, in the algorithms presented in [5, 13], an agent disseminates its opinion for a duration proportional to the quality/confidence of its opinion, while the quality/confidence is higher when most cells explored by the agent have the same colour. This means that as the feature ratio approaches 0.5 in a homogeneous environment, the quality/confidence decreases, which in turn decreases the communication between the swarm members, leading to slower decision making.

8.3 Problem Definition and Assumptions

In this work, we use the same version of the collective decision-making problem as [5, 9, 13]. The problem can be formulated as follows: consider an $L_{env} \times L_{env}$ grid-based environment, E, where each cell c_i is characterised by a binary value representing a feature f, such that $f : c_i \rightarrow \{0, 1\}, \forall c_i \in E$. The feature f represents the colour of the cell and can be either black or white.

At mission commencement, a swarm of N agents $\Pi = \{\pi_1, \pi_2, \dots \pi_N\}$ is deployed with random positions and orientations within an $L_{nest} \times L_{nest}$ nest at the top left corner of the environment. The swarm is required to explore the environment and perform collective decision making to decide which feature value is the most frequent in the environment. The environment is bounded by four walls that are detectable by the swarm members.

The swarm members have some limitations on their actions, similar to the limitations described in [5]. First, an agent π_i can sense the colour of a cell c_j only if the position of π_i lies within the boundaries of c_j. In addition, at any time step, an agent π_i can either sense the colour of the environment or disseminate its belief, but not both. Meanwhile, π_i can listen to other agents at each time step, regardless of whether it is sensing or disseminating. Additionally, the swarm members have no means for calculating their absolute positions within the environment. Thus, they cannot determine which cells have or have not been visited by the swarm as a whole. An agent π_i can locally communicate with another agent π_j, if the distance between π_i and π_j is less than the communication range, R_{comm}:

$$\|P^t_{\pi_i} - P^t_{\pi_j}\| < R_{comm}, \tag{8.1}$$

such that $P^t_{\pi_i}$ and $P^t_{\pi_j}$ are the positions of agents π_i and π_j at time t, respectively. Using these local communications, estimates calculated by individual swarm members can be shared to form an overall estimate of the feature in question.

A shepherding agent β is used to assist the swarm with the collective decision making. The shepherding agent is not able to sense the colour of the cells. However, β is assumed to be spatially aware of its position as well as the positions of all the swarm members. An agent π_i can sense the shepherd β if the distance between the two is less than $R_{\pi\beta}$. Thus, if appropriately configured, β can be used to collect the swarm individuals and/or drive them to a pre-specified location within the environment. The maximum velocity is $V_{\pi max}$ for a swarm member and is $V_{\beta max}$ for the shepherding agent. The swarm members and the shepherding agent are assumed to be time synchronised. Besides, all the agents are assumed to have physical bodies such that no two agents can be in the same position at the same time, and the environment is appropriately scaled down such that the precision in measuring the position of the agent is sufficiently adequate to eliminate the effect of the body size for the swarm while maintaining enough resolution to detect the objects of interest.

8.4 Shepherd-Assisted Algorithm

In this section, we propose a shepherd-assisted algorithm that is aimed to solve the swarm decision-making problem while maintaining consistent performance across different environment configurations. As discussed earlier, existing algorithms suffer when the complexity of the problem increases due to feature ratios approaching 0.5 or a lack of homogeneity in feature distribution. We argue that making the length of the dissemination period proportional to the level of confidence in the local estimate can harm the decision-making problem by reducing the communication between the agents in environments with r close to 0.5.

In non-homogeneous environments, an agent's estimate of feature prevalence within the environment is strongly biased by the area which the agent explores. Thus, estimates made in the beginning of the mission are expected to have a large variance, which prevents belief convergence. However, as time progresses, agents moving randomly in the environment will visit a monotonically increasing number of cells, and hence their estimates are expected to slowly approach the actual feature ratio. Therefore, communicating estimates becomes more likely to result in belief convergence.

We conclude that early belief dissemination might not be useful in accelerating the decision-making process. As agent exploration is not affected by its current belief, belief dissemination can be delayed after completely finishing the exploration phase. However, an important benefit of alternating periods of exploration and belief dissemination, as in [5, 13], is that it increases agent's exposure to a large number of spatially diffused agents as each agent's position changes over time. That is, we need to design an algorithm with the following features:

- The duration of opinion dissemination should not be proportional to the level of confidence in the local estimate.
- Early opinion dissemination should be avoided.
- Agent's exposure to other agents should be increased, so that each agent can receive the beliefs of a sufficient number of agents.

Based on these desirable properties, we propose a shepherd-assisted decision-making (SDM) algorithm in which swarm members perform exploration followed by belief sharing. In the exploration phase, the swarm members navigate through the environment to sense its features. Then, after a predefined amount of time, the exploration phase comes to an end, and a shepherding agent is introduced to collect the swarm members together to facilitate belief sharing. The details of the algorithm employed by both the swarm members and the shepherding agent are provided in the following subsections.

8.4.1 Swarm Members' Behaviour

As mentioned earlier, the algorithm employed by a swarm member consists of two phases: exploration and belief sharing. Each phase implies different behaviours for the swarm members in terms of their motion, feature sensing, communication and belief update. The exploration phase lasts for T_{exp} seconds during which a swarm member behaves as follows:

- Motion: in this phase, each swarm member π_i navigates through the environment by performing a random walk. The random walk method is selected due to its simplicity and because it does not assume that agents can obtain their positions within the environment. Random walk is achieved by a repeated sequence of straight line movement, pause and on-the-spot rotation. Agent π_i moves in a straight line for a duration t_s, which is a random variable with an exponential distribution $t_s \sim EXP(\lambda)$. Then, it pauses for zero time steps, similar to the algorithm in [5]. This is followed by an on-the-spot rotation with an angle drawn from a uniform distribution $U[\frac{-\pi}{2}, \frac{\pi}{2}]$. This sequence of straight line movement and rotation is repeated till the end of the exploration phase. Collision is resolved by a random rotation followed by a new cycle of random walk.
- Feature sensing: during the exploration phase, each swarm member keeps a record of the numbers of black n_{black} and white n_{white} cells it encounters within the environment. These numbers are set to zero in the beginning of the mission and are then updated after the colour of each cell is sensed. At the end of the exploration phase, a swarm member π_i calculates its initial belief $b_i^{T_{exp}}$ based on its observations, such that $b_i^{T_{exp}}$ is set to 1 if more white cells are observed than black cells and is otherwise set to 0, according to Eq. (8.2). No further belief updates or communication with other agents takes place in the exploration phase.

$$b_i^{T_{exp}} = round\left\{\frac{n_{white-i}}{n_{black-i} + n_{white-i}}\right\}. \tag{8.2}$$

When the exploration phase comes to an end, the belief sharing phase is activated by imposing a different set of rules on the behaviour of the swarm members, as follows:

- Motion: the motion of the swarm members in this phase follows the motion specified by the sheep model in [10], according to the equation:

$$F_{\pi_i}^t = W_{\pi_\upsilon} F_{\pi_i}^{t-1} + W_{\pi\Lambda} F_{\pi_i \Lambda_{\pi_i}^t}^t + W_{\pi\beta} F_{\pi_i\beta}^t + W_{\pi\pi} F_{\pi_i\pi_{-i}}^t + W_{\pi obst} F_{\pi_i obst}^t$$
$$+ W_{e\pi} F_{\pi_i \epsilon}^t, \tag{8.3}$$

such that $F_{\pi_i}^t$ is the total force applied on π_i at time t, $F_{\pi_i \Lambda_{\pi_i}^t}^t$ is the attraction force towards the center of mass of π_i's neighbours, $F_{\pi_i\beta}^t$ is the repulsion force from the shepherding agent, $F_{\pi_i\pi_{-i}}^t$ is the sheep collision avoidance force, $F_{\pi_i obst}^t$ is the obstacle avoidance force, $F_{\pi_i \epsilon}^t$ is a random force and W_{π_υ}, $W_{\pi\Lambda}$, $W_{\pi\beta}$, $W_{\pi\pi}$, $W_{\pi obst}$ and $W_{e\pi}$ are the weights determining the strength of each force.

The inertia of the swarm member π_i is considered by including its previous force, $F_{\pi_i}^{t-1}$, in the calculation of its current force. The swarm member stays close to its neighbour swarm members without colliding with them due to appropriately weighted swarm attraction and repulsion forces, $F_{\pi_i \Lambda_{\pi_i}^t}^t$ and $F_{\pi_i\pi_{-i}}^t$. The shepherding agent repulsion force, $F_{\pi_i\beta}^t$, is what allows the swarm member to be driven by the shepherding agent towards a target position.
- Communication and belief update: in each time step in the belief dissemination phase, a swarm member π_i disseminates a message containing its ID and current belief, (i, b_i^t). This message is successfully received by all π_js for which Eq. (8.1) is satisfied. Each swarm member π_i keeps a list of the other swarm members' beliefs B_i. The list is empty in the beginning of the dissemination phase. When a belief b_j^t is received by π_i, an entry of the form $[j, b_j^t, t]$ is added to B_i. If B_i already contains an entry for the agent π_j, the entry is updated rather than adding new entry. The entry is valid if the belief has been received within the last T_{valid} time steps. Agent π_i maintains a belief concentration γ_i, which is a moving average that fuses the beliefs received from the other agents over time. The concentration is set to 0.5 in the beginning of the belief dissemination phase and can range from 0 to 1. When π_i receives a belief b_j^t, π_i updates its concentration if B_i does not contain an entry for π_j or if the value of b_j stored in B_i is different from b_j^t in the newly received message. In these cases, the belief concentration is updated as follows:

$$\gamma_i^t = 0.9\gamma_i^t + 0.1 b_j^t. \tag{8.4}$$

The weight values used in the above equation are the same as those used in [5]. At the end of each time step, π_i updates the value of its belief and belief concentration, based on the equations:

$$b_i^{t+1} = round\{\gamma_i^t\} \tag{8.5}$$

$$\gamma_i^{t+1} = \gamma_i^t. \tag{8.6}$$

Agent π_i turns its belief into a final decision if at least one of the following two conditions is satisfied. First, if the belief concentration crosses a predefined threshold and stays there for a certain amount of time; that is, if $\gamma_i^t < \theta_{decision}$ or $\gamma_i^t > 1 - \theta_{decision}$ for $T_{decision}$ seconds; such that $\theta_{decision}$ is the decision threshold and $T_{decision}$ is the decision threshold time. Second, if the number of valid beliefs in B_i exceeds $0.75N$. After making its decision, π_i continues disseminating its belief, which is the same as its decision, but it does not listen to its neighbours' beliefs anymore as it does not need to update its own belief. Algorithms 6 and 7 show the pseudo-code of a swarm member's behaviours and its procedure for belief update.

8.4.2 Shepherd's Behaviour

The shepherding agent β can be in one of the three states: idle, collecting and driving. The behaviour of β in these states is as follows:

- Idle: the initial state of β is set to idle in which the shepherding agent does not perform any action or observe any data. The β agent remains in this state throughout the exploration phase as the swarm members' operation in this phase does not rely on it. When time reaches T_{exp}, β transitions to the next state: collecting.
- Collecting: during this phase, β collects swarm members into clusters so that they can be easily driven to a collection point in the next phase. At the start of the collecting phase, β finds the swarm member π_i with the lowest number of neighbours, i.e., $Neighbours(\pi_i) \leq Neighbours(\pi_j) \forall j \neq i$. Then, it selects another swarm member π_k from the remaining agents such that π_k has the lowest number of neighbours and is at least D_{min} away from π_i, i.e., $Neighbours(\pi_k) \leq Neighbours(\pi_j), ||P_{\pi_i}^t - P_{\pi_k}^t|| \geq D_{min}, k \neq i, \forall j \neq k$. After selecting π_i and π_k, β starts driving π_i towards π_k by setting its own heading position, $P_{\beta\sigma_2}^t$, according to the following equation:

$$P_{\beta\sigma_2}^t = P_{\pi_i}^t + R_{\pi\beta}.\frac{P_{\pi_i}^t - P_{\pi_k}^t}{||P_{\pi_i}^t - P_{\pi_k}^t||} \tag{8.7}$$

Algorithm 6 A swarm member's behaviour

```
n_black = 0, n_white = 0, γ = 0.5, decided = false, t = 0
INITIALIZE B = list of N entries of the form (b, age)
for i = 1, ..., N do
    B[i].age = −1
end for
while t < T_exp do
    if current cell is black then
        n_black = n_black + 1
    else
        n_white = n_white + 1
    end if
    random_walk( )
end while
b = ROUND(n_white / (n_white + n_black))
timer_decision = T_decision
while t < T_max do
    message = (ID, b)
    broadcast(message)
    if decided = false then
        incoming ← receive_neighbour_messages( )
        update_belief_and_concentration(b, γ, B, incoming)
        if γ < θ_decision or 1 − γ < θ_decision then
            timer_decision = timer_decision − 1
        else
            timer_decision = T_decision
        end if
        if timer_decision = 0 or valid = 0.75N then
            decided = true
        end if
    end if
    sheep_movement( )
end while
```

Algorithm 7 update_belief_and_concentration(b, γ, B, $incoming$)

```
for i = 1 ... incoming.length do
    j = incoming[i].ID
    if B[j].age = −1 OR B[j].b ≠ incoming[i].b then
        B[j].age = 0
        B[j].b = incoming.b
        γ = 0.9 γ + 0.1 b
    end if
end for
b = ROUND(γ)
valid = 0
for i = 1 ... N do
    if B[i].age < T_valid then
        valid = valid + 1
    end if
    B[i].age = B[i].age +1
end for
```

such that, P_x^t is the position of agent x at time t. The shepherd then calculates its normalised force vector, $F_{\beta_{cd}}^t$, towards the heading position as follows:

$$F_{\beta_{cd}}^t = \frac{P_{\beta\sigma_2}^t - P_\beta^t}{\|P_{\beta\sigma_2}^t - P_\beta^t\|}. \tag{8.8}$$

The total force applied to β is calculated using the following equation:

$$F_\beta^t = W_{\beta\upsilon} F_\beta^{t-1} + W_{\beta_{cd}} F_{\beta_{cd}}^t + W_{\beta\pi} F_{\beta\pi}^t + W_{e\beta} F_{\beta\epsilon}^t, \tag{8.9}$$

such that $F_{\beta\pi}^t$ is the force of collision avoidance from nearby swarm members whose positions are within $R_{\beta\pi}$, and $F_{\beta\epsilon}^t$ is a random component. It should be noted that the parameters of the swarm motion can be chosen such that if π_i already has some neighbours, all the swarm members in its neighbourhood will also be driven towards π_k. The shepherding agent keeps driving π_i towards the current position of π_k till the distance between these agents becomes less than D_{min}.

At that point, β starts the same procedure anew to select another two agents to cluster. If the number of neighbours of these two agents is greater than a predefined threshold, θ_{neigh}, this means that the swarm members are adequately clustered, so β transitions to the next state, driving.

- Driving: in the beginning of the driving state, the swarm members are expected to be in clusters being scattered around the environment. In the driving phase, β drives each of these clusters towards the center of the environment P_{center}. Starting with the swarm member π_i with the furthest distance to P_{center}, β sets its heading position $P_{\beta\sigma_2}^t$ as follows:

$$P_{\beta\sigma_2}^t = P_{\pi_i}^t + R_{\pi\beta} . \frac{P_{\pi_i}^t - P_{center}}{\|P_{\pi_i}^t - P_{center}\|}. \tag{8.10}$$

The force applied by β is then calculated according to Eqs. (8.8) and (8.9). The shepherd continues driving π_i towards P_{center} till the distance between the two is less than D_{min}. Then, β repeats this process again to select another cluster to drive. When the distance between the furthest swarm member and P_{center} is less than D_{min}, the shepherd goes back to the idle state and its algorithm stops. The pseudo-code of the algorithm employed by the shepherd is given in Algorithm 8.

Algorithm 8 The shepherd's behaviour in the SDM algorithm

t =0 $state$ = "idle"
while $t < T_{exp}$ **do**
  $t = t + 1$
end while
$state$ = "collecting"
π_i = null
while $state \neq$ "idle" **and** $t < T_{max}$ **do**
  if $state$ = "collecting" **then**
    if π_i = null **or** $\| P_{\pi_i} - P_{\pi_j} \| < D_{min}$ **then**
      $\pi_i \leftarrow$ find_swarm_member_with_least_neighbours( )
      $\pi_k \leftarrow$ find_swarm_member_with_second_least_neighbours( )
      if $\pi_i.neighbours < \theta_{neigh}$ **then**
        $state$ = "driving"
        π_i = null
      end if
    end if
    $P_{\beta\sigma_2} = P_{\pi_i} + R_{\pi\beta}.\frac{P_{\pi_i} - P_{\pi_k}}{\| P_{\pi_i} - t_{\pi_k} \|}$
  else if $state$="driving" **then**
    if π_i = null **or** $\| P_{\pi_i} - P_{center} \| < D_{min}$ **then**
      $\pi_i \leftarrow$ find_swarm_member_furthest_to_center( )
      if $\| P_{\pi_i} - P_{center} \| < D_{min}$ **then**
        $state$ = "idle"
      end if
    end if
    $P_{\beta\sigma_2} = P_{\pi_i} + R_{\pi\beta}.\frac{P_{\pi_i} - P_{center}}{\| P_{\pi_i} - P_{\pi center} \|}$
  end if
  $F_{\beta_{cd}} = \frac{P_{\beta\sigma_2} - P_\beta}{\| P_{\beta\sigma_2} - P_\beta \|}$
  $t = t + 1$
end while

8.5 Experimental Results

To evaluate the performance of the proposed shepherd-assisted algorithm, we set up several simulation experiments that represent different levels of scenario complexity. The complexity of each scenario is manipulated by changing the ratio of the black feature ($r = 0.8$, $r = 0.65$ and $r = 0.55$) and the homogeneity of feature distribution (homogeneous and non-homogeneous). Without loss of generality, we assume that the black tiles are always the most frequent features. The parameter setting used for the simulation is listed in Table 8.1.

A critical parameter of the SDM algorithm is the exploration time T_{exp}, as it determines the lower limit of the time of a decision-making scenario. Before running the main experiments to evaluate the SDM algorithm performance, we ran some experiments to study the effect of T_{exp} on agents' beliefs. We set $T_{exp} = 125$ min and ran each scenario 20 times. We examined the change in the swarm members' belief over time along the exploration phase, as shown in Fig. 8.1. The left side of Fig. 8.1 shows that in homogeneous environments, the mean of the swarm

Table 8.1 The parameter settings used in the experiments. The dimensions of a cell are 1×1 unit^2

Global parameters								
L_{env}	L_{nest}	N	T_{max}	T_{exp}				
150 unit	26 unit	100	500 min	50 min				
Swarm members' movement parameters								
$V_{\pi max}$	λ	$W_{\pi \upsilon}$	$W_{\pi \Lambda}$	$W_{\pi \pi}$	$W_{\pi \beta}$	$W_{e \pi}$	$W_{\pi obst}$	$R_{\pi \beta}$
1 unit/s	$\frac{1}{240}$ s^{-1}	0.1	0.5	0.5	1	0.1	0.5	13.5 unit
Swarm members' decision-making parameters								
$\theta_{decision}$	$T_{decision}$	T_{valid}	R_{comm}					
0.005	30 s	180 s	13.3 unit					
Shepherd parameters								
D_{min}	$V_{\beta max}$	θ_{neigh}	$W_{\upsilon \beta}$	$W_{e \beta}$	$W_{\beta cd}$	$W_{\beta \pi}$	$R_{\beta \pi}$	
9.3 unit	1.5 unit/s	5	1.5	0.3	0.5	1	6.5 unit	

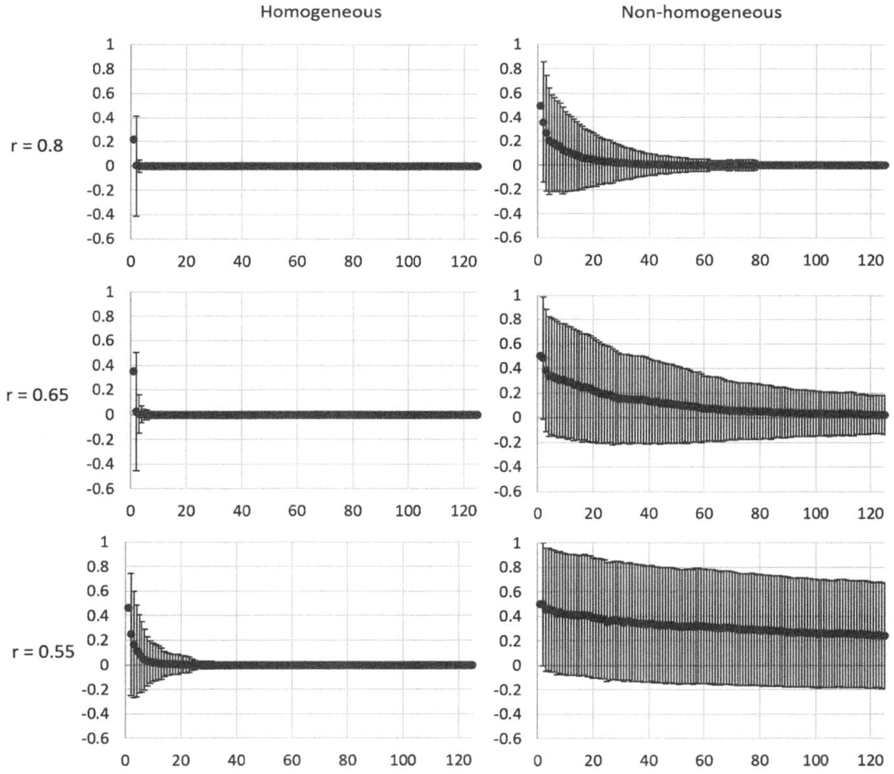

Fig. 8.1 The change of swarm members' belief over time under different settings of scenario complexity. The dots represent the mean value while the error bars represent the standard deviation

members' belief drops quickly over time and reaches 0 within 5 min at maximum. Additionally, the variance of the belief exhibits a sharp decrease over time and vanishes within about 25 min at maximum. Thus, in homogeneous environments, setting T_{exp} to as a small value as 5 min can result in very fast and highly accurate performance.

On the other hand, the right side of Fig. 8.1 shows a much slower decrease in the mean value of the agents' belief over time. Furthermore, the change in the standard deviation is much slower compared to the corresponding homogeneous scenarios. In very complex scenarios with $r = 0.55$, no significant change in the standard deviation can be seen over time. Hence, a large value for T_{exp} will benefit the SDM algorithm performance in non-homogeneous environments. While the belief stabilises after 90 min in non-homogeneous environments, such a value is excessive for homogeneous environments. The value of T_{exp} is set to 50 min to achieve a balance between the speed and accuracy of the SDM algorithm across different environments. This value is used for T_{exp} in the following evaluation experiments.

The SDM algorithm is compared to the algorithm presented in [5] by testing their performance in scenarios with different levels of complexity. A total of 2 (algorithms) $\times$ 3 (feature ratios) $\times$ 2 (homogeneity conditions) $\times$ 30 (random initialisations) $=$ 360 simulation runs have been used to perform the comparison. The maximum time allowed for each scenario was set to $T_{max} = 500$ min. By the end of the scenario, any swarming agent that has not reached a decision is obliged to make a decision based on its current belief. The same parameter setting is used for the two algorithms to ensure a fair comparison.

Figure 8.2 shows a graphical comparison between the time taken by each algorithm to perform the task in different scenarios. In homogeneous environments, Ebert's algorithm was able to finish the task within a considerably less time than the SDM algorithm. Manipulating the feature ratio in homogeneous environments had a significant effect on the time taken by Ebert's algorithm, as it took more time for scenarios with $r = 0.55$ than the other scenarios ($p < 0.001$). In contrast, feature ratio did not have a statistically significant effect on the time taken by our algorithm to finish the task ($p = 0.47$).

Moving to non-homogeneous environments, it can be seen in Fig. 8.2 that Ebert's algorithm outperformed the SDM algorithm only when the feature ratio was 0.8. As the feature ratio decreases, the time taken by Ebert's algorithm exhibited a statistically significant, drastic increase to T_{max}, ($p < 0.001$). In addition, a statistically significant effect of environment homogeneity is found on the time taken by Ebert's algorithm ($p < 0.001$). On the contrary, the SDM algorithm maintained similar run-time values as in homogeneous environments. The SDM algorithm considerably outperforms Ebert's algorithm, in terms of its run-time, in non-homogeneous environments with feature ratios approaching 0.5. Neither feature ratio nor environment homogeneity had a statistically significant effect on the run-time of the SDM algorithm.

Figure 8.3 shows the percentage of swarm members that have made a determination (decided) over time in different scenarios. The upper part of the figure shows that for the SDM algorithm, the percentage of agents decided increases

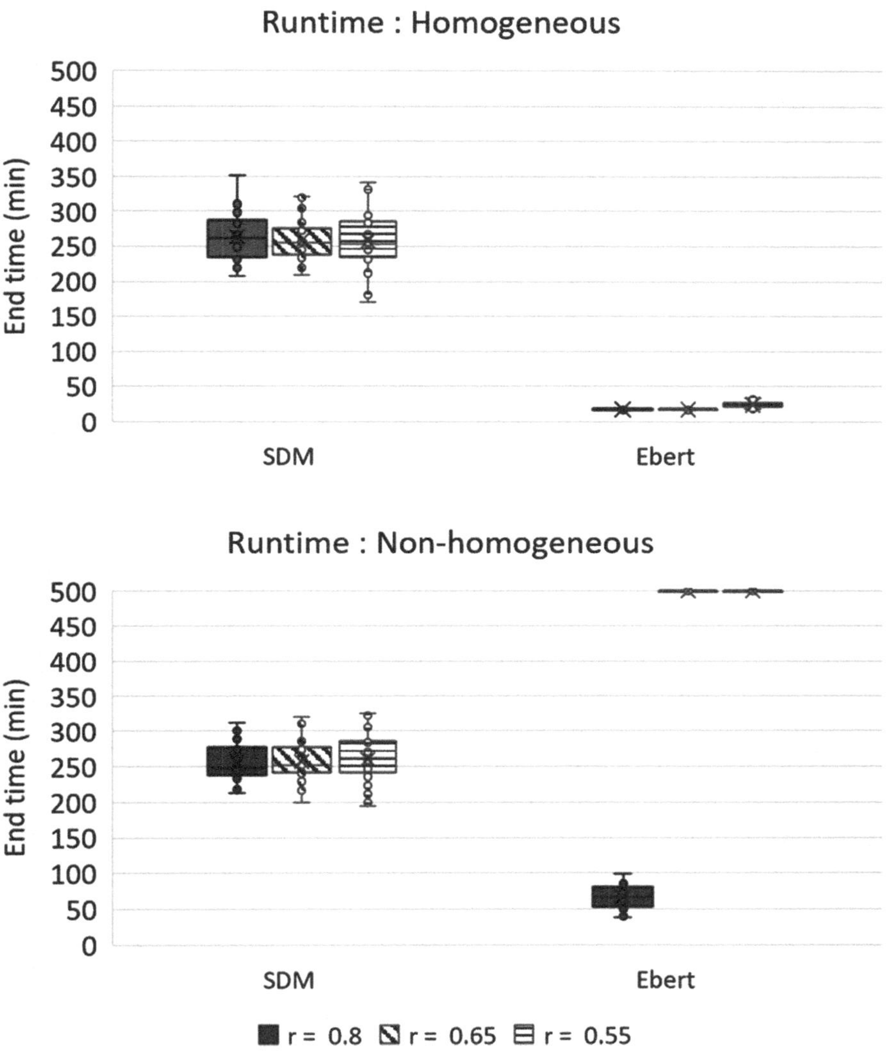

Fig. 8.2 A comparison between the average time taken by each algorithm to perform the collective decision making under different scenarios

gradually over time in homogeneous environments. Similar trends can be seen across scenarios with different feature ratios in homogeneous environments. By inspecting the lower part of the figure, it can be seen that for Ebert's algorithm, the percentage of agents decided increases sharply over time, in homogeneous environments. The curves for $r = 0.8$ and $r = 0.65$ are completely identical, while the curve for $r = 0.55$ has a slightly lower slope.

Moving to non-homogeneous environments, Fig. 8.4 shows that the proposed algorithm maintains similar patterns to those exhibited in homogeneous envi-

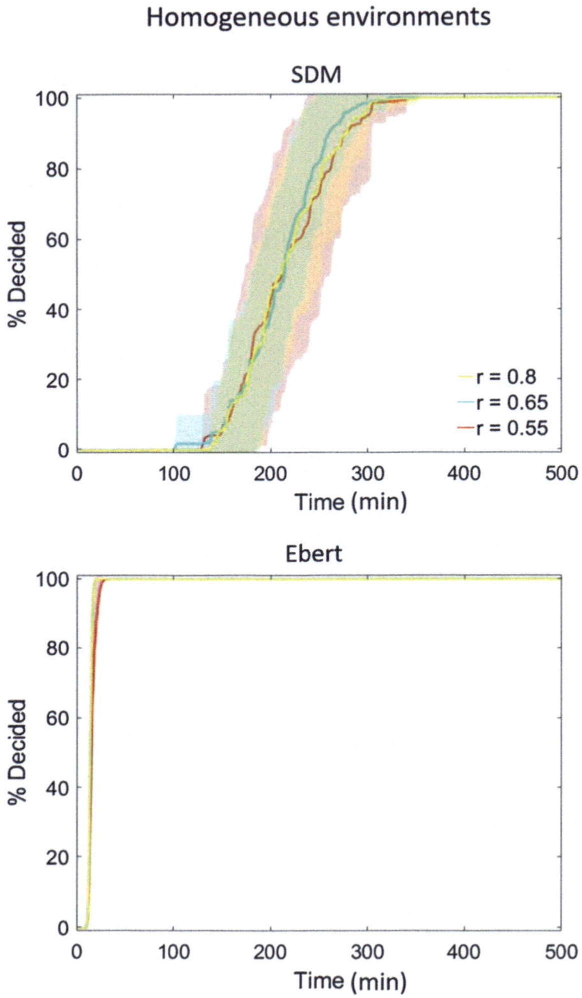

Fig. 8.3 The percentage of swarm members that reached a decision over the simulation time, in homogeneous environments. The thick lines show the mean, while the shades show the standard deviation

ronments. In contrast, Ebert's algorithm exhibits a considerably less steep curve for scenarios with $r = 0.8$ in non-homogeneous environments as compared to homogeneous ones. Furthermore, as the feature ratio approaches 0.5, the percentage of agents decided remains zero along time. That is, none of the agents decide before they are obliged to do so, by end of the scenarios.

The level of consensus of each algorithm is studied based on the number of simulation runs in which all the swarm members reached the same decision, regardless of its accuracy. In homogeneous environments, the two algorithms achieved 100% consensus across scenarios with different feature ratios. The levels of consensus achieved in non-homogeneous environments are shown in the upper

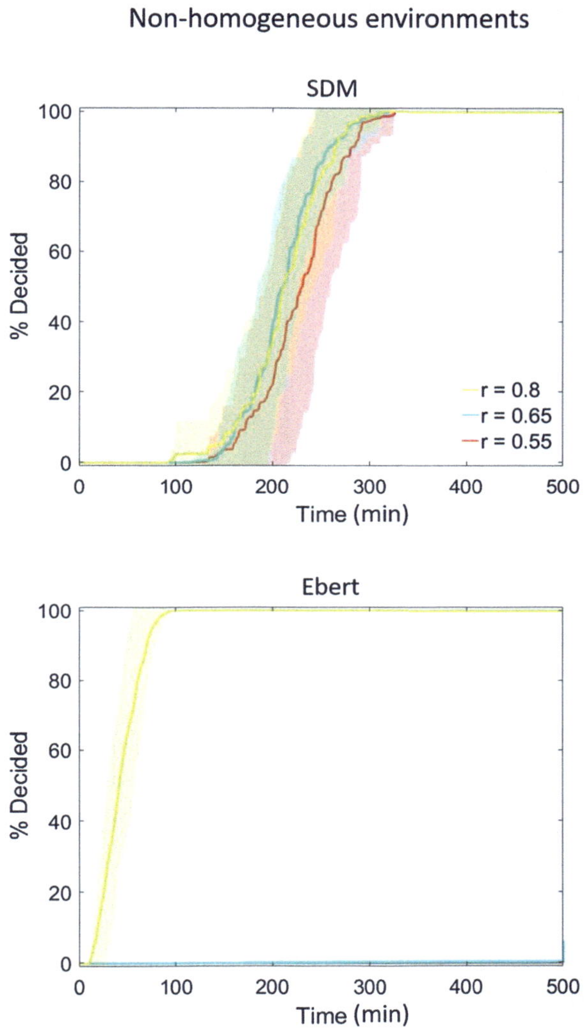

Fig. 8.4 The percentage of swarm members that reached a decision over the simulation time, in non-homogeneous environments. The thick lines show the mean, while the shades show the standard deviation

part of Fig. 8.5. The proposed algorithm achieved consensus in all the scenarios with $r = 0.8$ and $r = 0.65$ and in 90% of the scenarios with $r = 0.55$. In contrast, Ebert's algorithm achieved consensus in all the scenarios with $r = 0.8$, in only 3.33% of the scenarios with $r = 0.65$ and in none of the scenarios with $r = 0.55$.

Algorithm accuracy was calculated as the ratio of runs the swarm reaches consensus on the right decision. The two algorithms achieved 100% accuracy in all the homogeneous environments regardless of the feature ratio. Different levels of accuracy, however, were achieved in non-homogeneous environments, as shown in the lower part of Fig. 8.5. The proposed algorithm maintained 100% accuracy

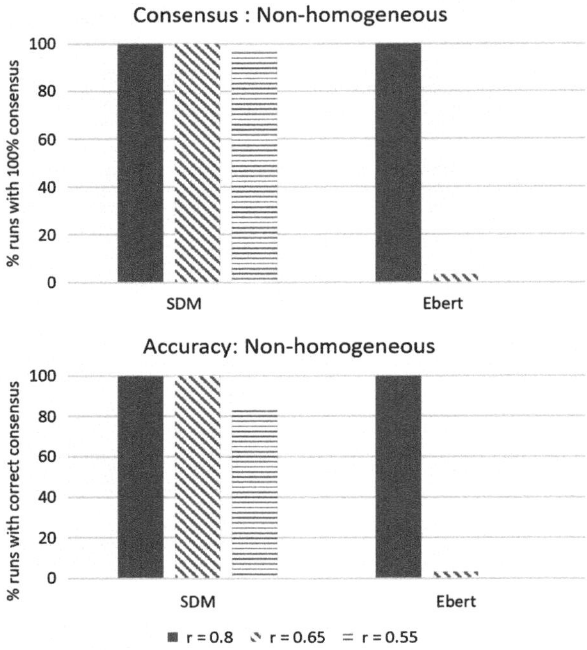

Fig. 8.5 A comparison between the consensus and accuracy of the two algorithms under different feature ratios in non-homogeneous distribution

in scenarios with feature ratios $r = 0.8$ and $r = 0.65$. The accuracy dropped considerably to about 83.3% in the scenario with $r = 0.55$. For Ebert's algorithm, accuracy levels were identical to the consensus levels.

8.6 Discussion

The complexity of the decision-making problem considered in this work is affected by both the feature ratio and the homogeneity of the feature distribution. Nonetheless, the impact of the homogeneity on scenario complexity is much more pronounced than the impact of feature ratio. Based on the results, it is evident that the proposed shepherd-assisted algorithm significantly outperforms Ebert's algorithm in the more complex environments of those tested.

The main premise behind the proposed shepherd-assisted algorithm is to increase exploration time sufficiently before the swarm members are able to share their estimates using the guidance of a shepherd. The practical significance of this premise is twofold. First, members of the swarm do not need to occupy themselves with localisation, mapping or any spatial processing that could consume their batteries. Second, the decision on how much time is sufficient for exploration and the collection of the swarm can be left to the shepherd agent.

An adequate increase in exploration time results in reducing the error in individual estimates and lowering their variance. In effect, when the dissemination phase starts, higher accuracy estimates are more likely to be maintained by the swarm members. In addition, a biased estimate that is based on observing a single-coloured region in a non-homogeneous environment is not given a higher chance of being disseminated than a less-biased estimate. This is because the dissemination period is independent of the value of the estimate. As a result, estimates of relatively high quality (in terms of their representation of the environment) drive the swarm decision making, which enabled the proposed shepherd-assisted algorithm to maintain its performance across scenarios with differing complexity.

On the other hand, Ebert's algorithm performed extremely well in the low complexity scenarios. The performance difference between Ebert's and our shepherd-assisted algorithm in homogeneous environments is not related to accuracy, but speed with Ebert's algorithm having a considerably lower run-time than the SDM. The complexity caused by the lack of homogeneity has a significant impact on the speed of Ebert's algorithm particularly when the feature ratio gets close to 0.5. In these scenarios, the decision-making strategy becomes unable to converge, and none of the agents are able to make a decision before they are forced to, by the end of the scenario. Within the complex settings, the accuracy and consensus of Ebert's strategy also struggled, as they reached 0% in the most complex scenarios. This does not mean that all the agents' decisions were wrong. Rather, it reflects the fact that swarm members applying Ebert's algorithm never achieved consensus in scenarios with $r = 0.55$ and non-homogeneous feature distribution.

8.7 Conclusion and Future Directions

In this work, we propose a shepherd-assisted collective decision-making algorithm that consists of two stages: exploration and belief dissemination. In the first stage, the swarm members independently navigate the environment and sense its features. Then, in the second phase, the swarm members disseminate their belief to obtain aggregated opinions about the feature in question. A shepherding agent is used to gather the swarm members during the belief dissemination stage so that each member can share its opinion with a large number of the swarm members. The results show that the proposed algorithm could maintain similar performance across environments with differing levels of complexity.

There is a room for future work to further improve the performance of the proposed algorithm, as follows: first, the lack of sensitivity of the shepherding algorithm to the task structure is what helped it to maintain its success in highly complex scenarios. However, this has the downside that the shepherding algorithm could not benefit from the ease of the task to improve its performance in lower complexity scenarios. In fact, if the shepherd could autonomously adapt its parameters to the state of the environment, the speed of the proposed algorithm would improve considerably in low-complexity settings.

Second, while the time of the exploration phase was set to 50 min, the average run-time of the proposed algorithm was about 255 min. That is, the dissemination phase was about five times longer than the exploration phase. Extending the algorithm to allow for multiple shepherding agents to collaboratively gather the swarm members could result in much shorter running time. Thus, using multiple shepherds to assist the swarm decision making can yield promising results.

Finally, the values used for the algorithm parameters, T_{exp}, θ_{neigh}, $\theta_{threshold}$ and $T_{threshold}$, can affect its performance. The values of these parameters have been selected manually by comparing the algorithm performance for several parameter settings. However, better algorithm performance could be reached by optimising the values of these parameters, for instance, by using evolutionary algorithms.

References

1. Bounceur, A., Bezoui, M., Noreen, U., Euler, R., Lalem, F., Hammoudeh, M., Jabbar, S.: LOGO: A new distributed leader election algorithm in WSNs with low energy consumption. In: International Conference on Future Internet Technologies and Trends, pp. 1–16. Springer, Berlin (2017)
2. Cai, G., Sofge, D.: An urgency-dependent quorum sensing algorithm for n-site selection in autonomous swarms. In: Proceedings of the 18th International Conference on Autonomous Agents and MultiAgent Systems. International Foundation for Autonomous Agents and Multiagent Systems, pp. 1853–1855 (2019)
3. Couture-Beil, A., Vaughan, R.T., Mori, G.: Selecting and commanding individual robots in a multi-robot system. In: 2010 Canadian Conference on Computer and Robot Vision, pp. 159–166. IEEE, Piscataway (2010)
4. de Oca, M.A.M., Ferrante, E., Scheidler, A., Pinciroli, C., Birattari, M., Dorigo, M.: Majority-rule opinion dynamics with differential latency: a mechanism for self-organized collective decision-making. Swarm Intell. **5**(3–4), 305–327 (2011)
5. Ebert, J.T., Gauci, M., Nagpal, R.: Multi-feature collective decision making in robot swarms. In: Proceedings of the 17th International Conference on Autonomous Agents and MultiAgent Systems, pp. 1711–1719. International Foundation for Autonomous Agents and Multiagent Systems (2018)
6. Kolling, A., Nunnally, S., Lewis, M.: Towards human control of robot swarms. In: Proceedings of the Seventh Annual ACM/IEEE International Conference on Human-Robot Interaction, pp. 89–96. ACM, New York (2012)
7. Lee, C., Lawry, J., Winfield, A.: Combining opinion pooling and evidential updating for multi-agent consensus. In: International Joint Conferences on Artificial Intelligence (2018)
8. Rubenstein, M., Cabrera, A., Werfel, J., Habibi, G., McLurkin, J., Nagpal, R.: Collective transport of complex objects by simple robots: Theory and experiments. In: Proceedings of the 2013 International Conference on Autonomous Agents and Multi-Agent Systems, pp. 47–54. International Foundation for Autonomous Agents and Multiagent Systems (2013)
9. Strobel, V., Castelló Ferrer, E., Dorigo, M.: Managing byzantine robots via blockchain technology in a swarm robotics collective decision making scenario. In: Proceedings of the 17th International Conference on Autonomous Agents and MultiAgent Systems, pp. 541–549. International Foundation for Autonomous Agents and Multiagent Systems (2018)
10. Strömbom, D., Mann, R.P., Wilson, A.M., Hailes, S., Morton, A.J., Sumpter, D.J.T., King, A.J.: Solving the shepherding problem: heuristics for herding autonomous, interacting agents. J. R. Soc. Interf. **11**(100) (2014). https://browzine.com/articles/52614503

11. Tarapore, D., Christensen, A.L., Lima, P.U., Carneiro, J.: Abnormality detection in multiagent systems inspired by the adaptive immune system. In: Proceedings of the 2013 International Conference on Autonomous Agents and Multi-Agent Systems, pp. 23–30. International Foundation for Autonomous Agents and Multiagent Systems (2013)
12. Valentini, G., Hamann, H., Dorigo, M.: Efficient decision-making in a self-organizing robot swarm: On the speed versus accuracy trade-off. In: Proceedings of the 2015 International Conference on Autonomous Agents and Multiagent Systems, pp. 1305–1314. International Foundation for Autonomous Agents and Multiagent Systems (2015)
13. Valentini, G., Brambilla, D., Hamann, H., Dorigo, M.: Collective perception of environmental features in a robot swarm. In: International Conference on Swarm Intelligence, pp. 65–76. Springer, Berlin (2016)

Part III
Sky Shepherding

Chapter 9
Sky Shepherds: A Tale of a UAV and Sheep

Kate J. Yaxley, Nathan McIntyre, Jayden Park, and Jack Healey

Abstract This chapter considers the evolution of a shepherding agent from the traditional sheep dog to a UAV capable of fostering the welfare of sheep throughout the shepherding task. Dorper sheep have been exposed to a single Sky Shepherd, with results of such screening tests discussed. Behaviours identified include an alert response, whereby flock members display curiosity toward the UAV. As testing progressed, flocks displayed synchronous alert and drive behaviour responses, with some sheep, hypothesised to be leader sheep, pausing periodically and displaying curiosity towards the UAV.

Keywords Shepherding · Sheep · Aural cues

This chapter considers the evolution of a shepherding agent from the traditional sheep dog to a UAV capable of fostering the welfare of sheep throughout the shepherding task. Dorper sheep have been exposed to a single Sky Shepherd, with results of such screening tests discussed. Behaviours identified include an alert response, whereby flock members display curiosity toward the UAV. As testing progressed, flocks displayed synchronous alert and drive behaviour responses, with some sheep, hypothesised to be leader sheep, pausing periodically and displaying curiosity towards the UAV.

Electronic Supplementary Material: The online version of this chapter (https://doi.org/10.1007/978-3-030-60898-9_9) contains supplementary material, which is available to authorised users.

K. J. Yaxley (✉) · N. McIntyre · J. Park · J. Healey
School of Engineering and Information Technology, University of New South Wales, Canberra, ACT, Australia
e-mail: k.yaxley@unsw.edu.au; nathan.mcintyre2@defence.gov.au; jayden.park@student.unsw.edu.au; jack.healey@defence.gov.au

H. A. Abbass, R. A. Hunjet (eds.), *Shepherding UxVs for Human-Swarm Teaming*, Unmanned System Technologies,
https://doi.org/10.1007/978-3-030-60898-9_9

9.1 Introduction

With the evolution of technology, we have also seen an evolution of understanding and ability to influence our environment, processes and practices. Many technologists are revisiting the industrial revolutions to gain an understanding of the changing dynamics and how we will interact with new technologies.

As with all industries, the agricultural industry has undergone significant changes during the evolution of technology. In Australia, the sheep industry evolved from pastoral, required to support the establishment of a new British colony [26] in 1788, to an economy stimulant supported by the establishment of a wool industry in 1805. The Australian agriculture industry remains an economic stimulus, with sheep pastoral and wool production contributing to 13% of the Nation's value of production [17].

To support the development and well-being of sheep livestock, the animals are often herded to different areas around a farm. To assist the natural flocking behaviour of the animal is often harnessed. When exposed to an external agent viewed as a potential threat to the herd, prey animals, such as sheep, will flock together and move in order to avoid this external agent and is defined as shepherding [33]. While this action results in stress induced over the animal, prenatal stress in ewes has also been shown to improve birth weight of lamb [30], improving pastoral yield.

In aid of the shepherding task, dogs are often used. Experienced and trained working dog, Glencairn Seven, was bought for $22,200 at the 2018 Australian Kelpie Muster [4]. While it is unknown how many farms operate with similarly experienced and trained working dogs, dogs have bitten sheep during herding activities [18], which induces significantly higher stress during necessary herding activities. In some cases an all-terrain vehicle (ATV) is used by farmers to support shepherding tasks. Since 2001, ATVs have contributed to over 220 farmer deaths across Australia [22], as a result of accidents and roll-overs when using the potentially unstable vehicles.

To enable farmers to eliminate such a high-risk enabler from the workplace, incentives are available to assist with the purchase of unmanned aerial vehicles (UAVs) [31]. As such, there are farmers who are controlling UAVs to shepherd sheep [3], with potential to improve this capability to enable autonomous operation and introduce meaningful human-autonomy teaming (M-HAT), that can be applied beyond the agricultural industry. With legislation [6] and training [35] surrounding the use of UAV aircraft, operation may be optimised with autonomous augmentation of such systems, enabling optimised use for the animal benefits, improvement of agricultural processes and augmentation of the agricultural industry.

UAV shepherding is currently being conducted by some farmers, predominantly in New Zealand [3]. In order to perform this task, the farmer must be a qualified UAV pilot, or hire a suitably qualified pilot to fly the UAV [6]. In most cases, the pilot will require the UAV to fly beyond line of sight, which requires further qualifications and equipment, such as radio. Introducing such additional cognitive load onto the farmer/pilot introduces the potential for mistakes.

Aerial mustering is not new to Australian farms, with the first successful aerial muster occurring in 1968 by Stuart Skoglund on Ivanhoe Station using a Bell 47 D1 helicopter [29]. Helicopters are the predominant choice for aerial mustering, due to their manoeuvrability and relative suitability for low level flying. Some fixed-wing craft are used, however, are prone to defects due to the low level manoeuvring required, increased turbulence and greater flight time with flaps partially deflected.

The industry professionals who conduct aerial mustering conduct a high risk task, due to the low level flying, high manoeuvrability required and remote location of such endeavours [5]. Aerial mustering is predominantly conducted in northern Australia, in support of a vast beef industry. Over the period 2002–2012, one accident occurred every 53.3 million hours, 9% of which were fatalities. By harnessing Sky Shepherd technology, it would be possible to remove this statistic.

Aerial mustering poses significant hazards to both the human operator and the animal (predominantly cattle). The risk to the animal occurs when they are exhausted beyond their physical limits and often occurs when pilots do not realise the stress induced on the animal [29]. By introducing a lower cost option of a Sky Shepherd, risk to operators will be reduced, while also reducing risk to livestock from over-stress or animal bites.

Overall, there are many options to assist a farmer in performing shepherding tasks, however, there is no standardisation for those choosing to introduce UAV technology. In order to develop a suitable framework, this chapter discusses the evolution of shepherding models developed by researchers, inspired by the natural flocking and swarming nature exhibited by collective sheep movement. By parametrising the models to more accurately reflect sheep movement, in response to a herding UAV agent, it will be possible to develop educational models and provide advice to industry, while also furthering research understanding of shepherding, herding and flocking. This chapter discusses the initial screening results for research into developing a future Farmer and Sky Shepherd Team (FaSST) system for herding sheep with UAVs.

9.2 Shepherding Models

Foundation for the development of the shepherding models can be considered as beginning in 1987 when Craig Reynolds developed a computer simulation of a swarm of birds [28]. These birds named *boids* exhibited behaviour with three hierarchical behaviours: repel, align, attract. The behaviour of repel, or collision avoidance, consists of two elements: when stationary and when in motion. Repel is based on relative distance between each pair of boids. Flock cohesion is instantiated by the align behaviour [12]. The attract behaviour, or flock centring, ensures each boid stays close to nearby flockmates [28].

While Reynolds identified three behaviours required for successful flocking simulations, Harvey et al. [12] investigated the impact of using modified behaviour rules in models, confirming models which implemented *repel* and *attract* only, exhibited flocking behaviour, complexity and sensitivity to initial conditions. Consequently, it is possible to efficiently model flocking behaviour using a simplified Reynold's model and has been demonstrated by both Vaughan et al. [36] and Strömbom et al. [33].

In 1998 Vaughan et al. [36] developed a behaviour model for the shepherding of ducks, choosing ducks due to their slower movement than sheep and ability to be herded indoors. This behaviour model utilised the behaviours of repel and attract and was developed for an indoor, circular arena. The duck behaviours are characterised by attraction to each other (forming of a flock), repelled from each other to prevent collisions and maintain spacing, repelled from the arena wall and repelled from the robot. The linear response of the duck movement and robot movement are proportional to the inverse square of distance between the robot and flock of ducks, allowing Vaughan to synthesis robot navigation techniques and particle dynamics between a robot, goal and obstacles. The circular nature of the arena and circular orbit around the flock centre results in a radial movement model, with robot motion enabled by individual wheel movement.

Another behaviour model was by Strömbom et al. [33] in 2014. In this model, again the behaviours of repel and attract are used to illustrate the inter-flock dynamics of sheep. This model was validated using a sheepdog and sheep where the behaviours of *collect* and *drive* were identified. The collect and drive behaviours are exhibited by the sheepdog when exerting influence over the sheep. These simulated behaviours are similar to the trained sheepdog behaviours of *Drive* (move the sheep from behind, regardless of position of the handler, towards the goal destination) and *Head* (arc in front of a breakaway sheep to turn it back) [38]. Given the Strömbom model is a particle model, the dynamics between sheepdog and sheep are not fully modelled, so the term *collect* is accurate.

The *collect* and *drive* behaviours are also used by Fujioka and Hayashi [8] in developing a multi-agent shepherding model that considers the head position of the shepherd, relative to the flock. This evolution of the shepherd was also considered by Lien et al. [21], who introduced multiple shepherding agents to improve efficiency in herding large flocks, or flocks that are difficult to control.

The commonalities between the various shepherding models are many, with both Vaughan [36] and Strömbom [33] using linear expressions to describe the repel and attract force vectors. Both are circular based and are pictorially presented in Fig. 9.1, where β is the shepherd agent and π is the sheep agent. The repel sensing range between sheep agents is illustrated by $R_{\pi\pi}$, while the repel sensing range between sheep and shepherd is illustrated by $R_{\pi\beta}$. The force vector $F_{\pi\Lambda}$ describes the attraction force between a sheep agent and the local centre of mass Λ, while the force vectors $F_{\pi\pi}$ and $F_{\pi\beta}$ describe the repulse forces between sheep agents and sheep and shepherd agents respectively.

Often, the shepherding models consider only a single shepherd and assume the shepherd has knowledge of the whole flock, such as in Strömbom [33] and

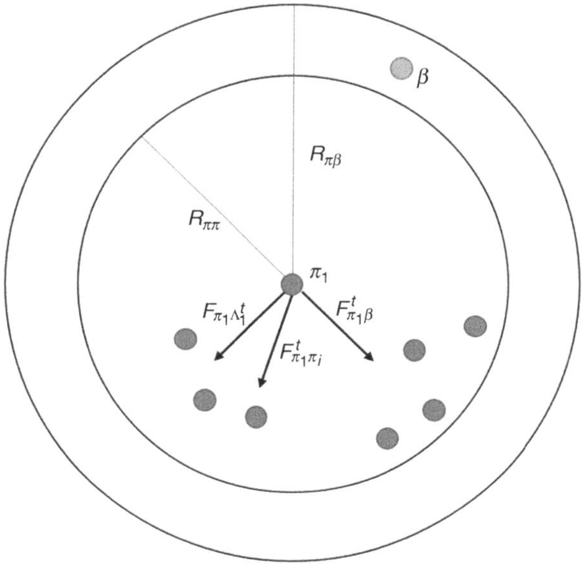

Fig. 9.1 Flock and influence vectors between flock and shepherd agents

Fujioka [8]. However, most shepherds do not have this full knowledge, or are working with flocks of various sizes. It would therefore be more efficient to consider sub-flocks, such as Lien et al. [21].

The behaviours developed by Reynolds are often used to describe the influence of a shepherd agent over a flock of social animals. Together, the flock moves towards a defined goal, known to the shepherding agent, which when reached, the shepherding agent ceases influence over the flock. The ability of the shepherding agent to align the flock towards the goal determines success. At the individual level, the relationship between the agents is described in terms of repel and attract. This generalisation allows for the modelling of the *collect* and *drive* behaviours, yet assumes all flocking agents are of equal value; that is, the force required by the shepherd agent to influence each animal is the same.

In both Vaughan [36] and Strömbom [33], a single shepherding agent is used to influence a flock towards a goal. To introduce efficiency in shepherding, Lee and Kim [20] introduced multiple shepherds, while also introducing flocking agents with orientation and hence higher fidelity relative to observed behaviours.

Building upon Strömbom's [33] model of flocking agents, Gade et al. [9] wavefront analysis of the effect of using a UAV to herd birds away from an airport was presented, where the herding of the birds relied upon influencing a fear response, as opposed to a welfare-fostering collect and drive response. This model was later translated into real-world testing by Paranjape [25], demonstrating the ability to herding flocking agents using a UAV. By influencing a fear response in the flocking agents, the single shepherding agent was successful in achieving the goal of restricting bird access to an airport. This is a further example where the use of technology enables the fostering of welfare of flocking agents (inducing fear as opposed to injury by air vehicles around an airport), which could be adapted to enabling influence of collect and drive response of sheep flocking agents.

To ensure the successful application of Gade's model [9], Paranjape et al. [25] ensured the predator response for individual species of birds was both known and induced by the shepherding UAV. This is the first instance of explicit acknowledgment of individual flock dynamics in a shepherding model, albeit generally applied using wavefront analysis and used to induce a fear response, which are further discussed in Sect. 9.3.

By investigating sheep behaviour, it is apparent that individuals will respond to forces in a different manner. Consequently, while Vaughan and Strömbom have enabled the understanding of basic shepherding and flocking behaviours, their work was insufficient to train an autonomous system. The evolution of shepherding models to include multiple shepherds [21] and the head position of the shepherd [8] has resulted in more efficiency in herding flocking agents, however, there is a need to model flock dynamics to fully capture the dynamics of shepherding.

9.3 Flock Dynamics

As discussed in Sect. 9.2, the investigation of flock dynamics has been studied by Gade [9] and Paranjape et al. [25], identifying force dynamics that would enable a pursuer (unmanned aerial vehicle (UAV)) to scare a flock of birds away from an airport, as opposed to shepherding them. The key difference is by understanding which influence force induces fear to the flocking agents and which do not. This requires an understanding of the species, how they naturally respond to an external agent and how to influence the flock behaviour to deliver an outcome.

Consequently, while Vaughan [36] and Strömbom [33] were seeking to influence natural response of social flocking, Paranjape et al. [25] sought to induce a predatory fear response, inducing the flock of birds to be repelled, while also avoiding fragmentation of the flock. In both cases, an influence vector is used, yet to induce fear, the influence vector is considered larger. As discussed in the Introduction, events of fear response have occurred during aerial mustering activities, resulting in the loss of an animal. In the agricultural industry, such influence vectors are undesired. It is therefore necessary to gain an understanding of what influence vectors exist and how this shapes the shepherding action. By understanding this phenomenon, shepherding to foster welfare may be promoted, using UAVs as opposed to dogs, ATVs or helicopters.

While Paranjape et al. [25] consider the effect of increasing the repel vector, the flock dynamic remains simplified. The model requires flock specific quantities for the fear response (R_{fear}) and aggressor response (R_{agg}). Referring to the $R_{\pi\beta}$ radial measure in Fig. 9.1, which was present in both Vaughan and Strömbom models,

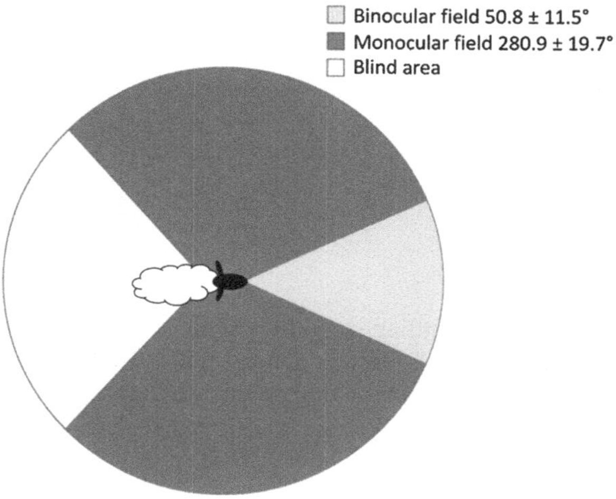

Fig. 9.2 Visual acuity of sheep, as measured by Hutson [14]

then R_{fear} represents a similar relationship, with the bird accelerating radially away from the UAV when exceeds R_{fear}, yet not R_{agg}. However, should the UAV exceed R_{agg}, the birds would attempt to out-manoeuvre the UAV and continue towards their intended flight path that likely encroaches on the airport. The R_{agg} property has not been modelled previously, and allows for the definition of when system dynamics would be made unstable due to flock fracture due to fear. However, it must be noted that the Strömbom model [33] does attempt to account for flock instability by defining a distance of $3R_{\pi\pi}$ at which the shepherd agent does not move for a time-step. While it is part of sheepdog training to ensure the sheepdog does not cause a flock to fracture, the dynamic of *giving the eye* is not a trained behaviour and is often unwanted [38], as it is associated with lack of confidence. The actual behaviour exhibited by the sheepdog and sheep that lead to this arbitrary amount is unclear and therefore unhelpful for developing a model with the intent of capturing flock dynamics and fostering sheep welfare during the shepherding task.

When a new agent approaches a flock, the flock will respond based on both the environment, perceived threat and natural responses [11]. Sheep have a very good visual acuity, as depicted in Fig. 9.2, which is influenced by the position of the eyes [14]. Consequently, this influences how they respond to agents, with a stronger fear response when approached from behind.

Given sheep are highly social animals, when isolated, sheep exhibit signs of stress, including higher heart rate [2], vocalisation [34] and increased flight zone [15]. The flight zone, or distance from an agent when the animal takes flight (runs), is dependent on the perceived threat, environmental conditions and internal flock dynamics.

As prey animals, sheep are often considered to exhibit centre-seeking herding behaviour when exposed to a predator, which was first explored by Hamilton [11] using ‘Guppy’ fish. The conclusions have been widely accepted as being trans-

Table 9.1 Expected individual sheep responses to a Sky Shepherd

Behaviour	Measure	Expected occurrence
Flock leader	Sheep leads the flock	Stronger response to Sky Shepherd
Flock follower	Sheep follows leader	Respond to cues from the leader
Uncooperative follower	Sheep will display mix of leader and follower tendencies	Difficult to initially identify, however, will influence the stability

ferable to all herding animals and have been used for modelling, as discussed in Sect. 9.2. Evolving this general understanding, consideration to internal flock dynamics have been modelled, such as those generally modelled by Yang and Schmickl [39] as coward, explorer and dodger. Coward is synonymous with the selfish-herd behaviour, while explorer is more akin to effective predator avoidance strategies identified by Morrel et al. [23]. The dodger behaviour emerged when individuals were crowded in a large flock, resulting in flock instability. Focusing on the instance of sheep flocking and considering the findings of Syme and Elphick [34], indicating individual sheep respond to stressors dependent on their own natural responses, it is likely that flock stability will be influenced by the individuals, leading to the individual behaviours presented in Table 9.1.

The behaviours presented in Table 9.1 are based on the findings of Syme and Elphik [34], which investigated how sheep responded to isolation. A sheep that responded to isolation with little vocalisation and lower heart rate was later identified to lead a mob of sheep towards handling practices, while sheep that responded to isolation with vocalisation or vocalisation and movement often had higher heart rates and were later identified to position themselves within the middle or back of the mob. Of note, the uncooperative follower sheep were always positioned at the back, were difficult to handle and had higher heart rates during isolation.

To test the hypothesis *flock stability is influenced by individuals*, a phased methodology to understand flock dynamics is presented in Sect. 9.4. Further, if this hypothesis is accepted, it would be possible to develop an updated model to more accurately represent the behaviours between flock individuals and flock stability, which could also be transferred into a model to support shepherding understanding.

9.4 Autonomous Sky Shepherd Methodology

As detailed in Yaxley et al. [40], to aid in the ethical development of an autonomous system, a systematic evaluation of the proposed system is recommended, while also considering the impact on the key stakeholders of the system. Given a Sky Shepherd would be used in a predominately agricultural setting, a key stakeholder is a farmer. To imply a Sky Shepherd would replace a farmer is naive. Instead, the Sky Shepherd and farmer would form a team, not unlike a farmer and sheepdog. Consequently, meaningful interactions between the Sky Shepherd, farmer and sheep would be required. The initial stage to achieve this is by using current UAV technology to establish a baseline interaction between the farmer and sheep.

There are a number of examples of farmers interacting with UAVs to assist with farming [3], however, the impact on animal welfare has not been scientifically considered. To ensure the welfare of sheep is fostered, the response of sheep to presence of a UAV needs to be measured to enable the development of an autonomous system that fosters holistic well-being. By measuring the response of sheep to the presence of a UAV, it will then be possible to develop shepherding models that reflect the dynamics of shepherding, when the shepherd agent is a UAV. Using such a model, a control system may then be created to enable autonomous shepherding, with oversight from a farmer.

By developing an autonomous agent that influences outcomes in the physical environment, it will be essential that, in this case, the Sky Shepherd is capable of interpreting the surroundings and basing its future actions on these surroundings, while also achieving a pre-determined objective. Given there is open source evidence of UAVs being used to herd sheep [3], as well as evidence of sound augmented UAVs [3], the first step is to conduct screening tests between sheep and UAVs, to scientifically answer the hypothesis *driving sheep with a UAV, augmented with an audio cue, has the potential to foster the welfare of the animal, while also achieving an efficient outcome*. The first step to developing the screening test was to first understand how sheep will respond to the proximal approach of a UAV, which is based on the reliance sheep place on their visual and aural acuity, as well as general flock dynamics, as discussed in Sect. 9.3.

When determining flock dynamics, such as recording behaviour responses to the presence of a shepherding agent, it is important to consider several alternatives and collect multiple responses to ensure experimental success [10]. In order to understand the behaviour response of sheep to the presence of a Sky Shepherd, the responses presented in Table 9.2 are considered.

The selected behaviours are based on a review of a collection of sheep behaviour, covering improvement to agricultural practices and welfare, with an outline to capture sheep perception and response to a new stimulus [10]. By capturing flocking

Table 9.2 Expected behaviour of sheep to presence of a Sky Shepherd

Behaviour	Measure	Expected occurrence
Visual alertness	Sheep looks at Sky Shepherd	Initial approach
Aural alertness	Sheep ears rotate towards Sky Shepherd	Initial approach
Physiological response	Heart rate increase	Initial approach and increasing as Sky Shepherd moves closer
Social clustering	Social clustering	As Sky Shepherd approaches
Flock flight zone	Flock movement, without fragmentation	When Sky Shepherd exceeds flight zone

Fig. 9.3 Dorper sheep used for Sky Shepherd screening tests

dynamics, it will also be possible to more accurately reflect the response of sheep to the presence of a Sky Shepherd or perceived predator. As detailed by Morrell et al. in [23], flocking agents are more successful at preventing attack by flocking to close neighbours first and joining the central flock as the attack continues. Consequently, flock dynamics in a model used to train an autonomous Sky Shepherd must also reflect this behaviour.

The screening plan used is a high throughput testing (HTT) design, developed using orthogonal array design using rdExpert [27] and is presented in Table 9.3. This screening design allowed for tests to be completed over a total of six days.

During testing, the behaviours of sheep measured were *alert* and *drive*. The behaviour of *alert* is exhibited either by an individual or collectively by the flock and is defined as when a sheep, or sheep, raise their neck and prick their ears forward [2]. Individual sheep that exhibits alert behaviours quicker are often considered a leader [32].

The *drive* behaviour is exhibited when a flock moves together, towards a specific area, and in the context of shepherding, is a result of the influence induced by the presence of an external agent [33]. However, for the response to induce the behaviour of drive and not fear, the influence vector from the shepherd must be great enough to encourage repulsion from the agent, yet small enough to not induce a predatory response in the flock. While the predatory response was required to herd birds away from an airport [25], inducing such high stress responses in sheep may result in injury [3] or death of an individual sheep [18]. Consequently, in field trials conducted to determine the behaviour responses of sheep to the presence of a Sky Shepherd, a heart rate monitor was used to capture Heart Rate Variation and support end of test measures. A test serial that induced a high stress response of 200 beats per minute was ended, in accordance with Animal Ethics approval ACEC 19/122B.

Table 9.3 HTT screening design for initial sheep response to a single Sky Shepherd

Test	Audio cue	Audio sound	UAV display	UAV height (m)	Max UAV speed (km/h)	Flock size
1	Off	Off	Straight and level	10	4	3
2	Off	Off	Straight and level	2	10	7
3	Off	Off	Zig zag	5	10	5
4	Off	Off	Zig zag	2	25	3
5	Off	Off	Dynamic manoeuvre	10	25	5
6	Off	Off	Dynamic manoeuvre	5	4	7
7	On	Siren	Straight and level	5	10	3
8	On	Siren	Zig zag	2	4	7
9	On	Dog bark	Straight and level	10	25	7
10	On	Dog bark	Zig zag	10	4	5
11	On	Dog bark	Dynamic manoeuvre	2	4	3
12	On	Dog bark	Dynamic manoeuvre	5	25	3
13	On	Motorbike	Straight and level	5	4	5
14	On	Motorbike	Zig zag	5	25	7
15	On	Motorbike	Dynamic manoeuvre	2	10	5
16	On	Motorbike	Dynamic manoeuvre	10	10	7
17	On	Music	Straight and level	2	25	5
18	On	Music	Zig zag	10	10	3

The flocking agent in this research is the Dorper. A picture of the twelve (12) sheep used for duration of screening tests, prepared for the conduct of test, is shown in Fig. 9.3. Dorpers are widely spread throughout the agricultural industry in areas such as Australia, the Middle East and Asian countries [1]. Dorpers like all sheep experience stress [2], display flocking response [11] as well as the social dynamics of leadership [32] uncomfortable follower and resistant follower [34]. These three group dynamics were first presented by Syme and Elphick [34] when investigating how to handle sheep, it was noted that there the individual heart rates as well as their response to individual stresses could identify the dynamic of the individual.

To develop the Sky Shepherd the initial phase is a screening test whereby the presence of the UAV as well as its movement towards the flock will be measured for signs of visual alertness, physiological response, social clustering and the flock flight zone. Augmented will be the perceived size of the UAV which is the dynamic performance of the UAV as well as the aural cue. Aural cues are used by shepherds regularly in order to influence the flock towards the desired goal. This influence is applied with the purpose of aligning the flock towards the goal which is where we could say that Reynolds third behaviour of align is implicit within the behaviours exhibited between Shepherd and flock.

The input factor, *Audio Sound* was designed to be nested with *Audio Cue*, to determine whether sound, including the potential for pitch, had an effect over the influence of flock behaviour. When the input factor *Audio Cue* was set to on, the possible sound files used to influence the sheep are presented in Table 9.4:

The UAV factors considered are presented in Table 9.5 and have been selected based on environmental considerations and aligned with regulations and literature where possible.

The flock size input factors selected were three (3), five (5) and seven (7) sheep, chosen to align with sheep herding competition rules [16].

Prior to the conduct of tests, sheep were herded into holding yards and sheared to enable quality collection of heart rate data. Sheep were also colour coded with a binary coding scheme, using approved sheep marker dye, which enabled identification.

The heart rate monitoring equipment selected for use throughout the Sky Shepherd project is the Zephyr Bioharness 3.0 [41], attached using the chest strap and monitored using the BioHarness Module. This equipment was selected due to the range of heart rate monitoring afforded and ease of attachment, which also

Table 9.4 Audio sound input factor for screening tests

Sound file	Comment
Dog bark	Common shepherd sound. A barking UAV has also been used by farmers [3]
Motorbike	Common alternate shepherd agent to sheep dogs [31]
Siren	An alert siren has been used on UAVs by farmers [3]
Music	Varying pitch. Selected music is *The Imperial March* [37], which has also been used by farmers [3]

enabled sheep to freely move. Attached to the chest strap was a QStarz BT-Q818XT GPS module, synchronised with the Bioharness 3.0 monitoring system, allowing for collection of both the heart rate of the sheep and the distance from the UAV at the time of event. The time of event was collected through observation, including also the leader and flock response. Definition of the leader and flock response are provided in Table 9.6.

The Sky Shepherd used in screening tests was the DJI Mavic 2 Enterprise Dual UAV [7], which enabled the recording of image and production of an audio cue using the integrated speaker. The Sky Shepherd was operated by a licensed UAV pilot throughout the duration of testing.

Shown in Fig. 9.4 are the screening results for the *alert* behaviour. As indicated by the magnitude arrows, augmenting the Sky Shepherd with an audio cue initiates the response earlier. The results are specifically detailed in Table 9.7. Of note, the closest approach occurred when the Sky Shepherd emitted a motorbike sound during the final series of testing, while the furthest occurred while emitting a dog bark. The sound of an alert siren generated the most consistent response within the sheep, equivalent to the average *alert* response of the flock.

Shown in Fig. 9.5 are the screening results for the *alert* behaviour. As indicated by the magnitude arrows, augmenting the Sky Shepherd with an audio cue initiates the response earlier. The results are specifically detailed in Table 9.8. Of note, the closest approach occurred when the Sky Shepherd emitted an alert siren at a speed of 4 km/h, while the furthest occurred while emitting a dog bark. The sound of music generated the most consistent response within the sheep, approximately equivalent to the average *drive* response of the flock.

Results from the screening tests confirmed sheep responded to the presence of a Sky Shepherd as per the expected proximal behaviour responses described in Sect. 9.3. Initially, the flocks responded to the UAV and audio cues with a

Table 9.5 UAV input factors for screening tests

UAV manoeuvre	Comment
Straight and level	*Drive* behaviour of a shepherd [33]
Zig zag	*Collect* behaviour of a shepherd [33]. Pilot was instructed to operate UAV with a deviation of 2 m from centre
Dynamic	*Collect* behaviour of a shepherd [33], with a height component added. Pilot was instructed to operate UAV with a deviation of 2 m from centre, and height variation of 1 m from requested height
UAV height	Comment
2 m (Low)	Simulate height of farmer, while also maintaining UAV operating restrictions
5 m (Medium)	Simulate height of low flying bird
10 m (High)	Simulate height of a flying bird
UAV speed	Comment
4 km/h (Slow)	Walking speed of sheep [19]
10 km/h (Medium)	Australian speed limit in a shared zone [24]
25 km/h (Fast)	Approximate running speed of sheep [13]

Table 9.6 Observation descriptions for leader and flock characteristics

Score	Flock	Definition
1	No clear flock	No obvious attraction towards centre of mass
2	Loose flock, weak attraction	Weak attraction towards centre of mass obvious
3	Loose flock, mild attraction	Mild attraction towards centre of mass obvious
4	Flock with good attraction and repulse. Minimal splitting	Obvious cluster and attraction towards a centre of mass. Within approx 2 m of each other
5	Strong flocking. No splitting	Tight cluster, within approx 0.5 m of each other
Score	Leader	Definition
1	No clear leader	No obvious sheep as leader
2	Uncomfortable, weak leader	Sheep slightly outside of flock
3	Uncomfortable leader	Sheep outside of main cluster, but does not stay there for long before returning to main cluster
4	Leader, sometimes indecisive	Sheep inconsistently outside of main cluster, clustered sheep have a weak attraction to them
5	Strong leader	Sheep constantly outside of main cluster
		Clustered sheep have a large attraction to them

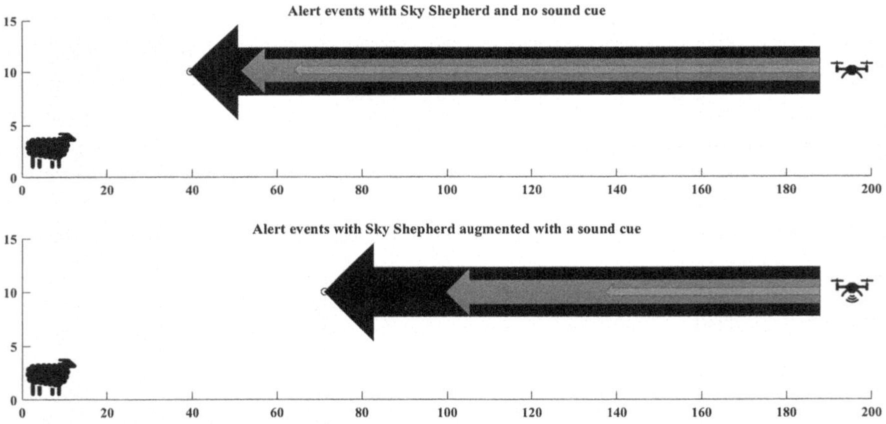

Fig. 9.4 Expected proximal behaviour responses of sheep to a Sky Shepherd

Table 9.7 Absolute *alert* distance response of Dorper sheep to the presence of a Sky Shepherd

	Sky Shepherd	Sky Shepherd with audio cue
Closest approach	39.5 m	71.1 m
Average approach	51.5 m	99.7 m
Furthest approach	63.7 m	137.3 m

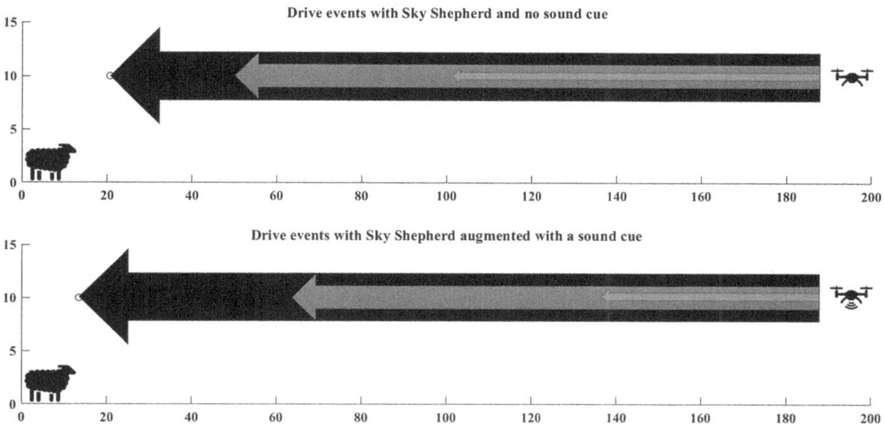

Fig. 9.5 Expected proximal behaviour responses of sheep to a Sky Shepherd

Table 9.8 Absolute *drive* distance response of Dorper sheep to the presence of a Sky Shepherd

	Sky Shepherd	Sky Shepherd with audio cue
Closest approach	20.7 m	13.5 m
Average approach	49.9 m	63.4 m
Furthest approach	101.2 m	136.5 m

higher heart rate response for both *alert* and *drive* responses. This is consistent with findings by Baldock et al. [2], who demonstrated sheep will initially respond to changes in environment with higher heart rate and stress responses. Consistent with these experiments, Sky Shepherds screening tests also indicated aspects of familiarisation with the environment, with *alert* behaviour occasionally becoming synonymous with *drive* behaviour. Overall, the alert siren was the most consistent at stimulating an *alert* response, while the sound of music enabled a consistent *drive* response.

Post-video analysis of Sky Shepherd screening tests reveals that the presence of a UAV is able to influence sheep away from a provided food source, hypothesised to be due to predation risk response. Using a siren audio cue, the flock of sheep will exhibit an alert response earlier, while also driving in a more predictable (flocking) manner. Leaders within the flock will periodically display curiosity towards a shepherding UAV also. The use of a dog barking audio cue results in less flock attraction during alert and drive behaviour responses, with higher concern for sheep welfare during shepherding. While a revving motorbike elicits a simultaneous alert behaviour within a flock, the predation risk response is more closely aligned with a flock and flee response, with sheep driving with no strong attraction to the flock.

9.5 Concluding Comments

By developing an autonomous system capable of responding to animal welfare cues, inducing the correct shepherding influence vectors and teaming with the owner/operator, it is possible to improve agricultural practices in Australia and overseas. A Farmer and Sky Shepherd Team (FaSST) has the potential to improve animal welfare on farms, and relieve labour shortages for future generations.

Sky Shepherd may be singular or multiagent to improve efficiency with flocks of larger sizes or varying dynamics. The initial screening tests discussed in this chapter indicate that there is a need to further investigate the effect of aural cues. While some evidence of heart rate variation was seen during testing, it is unclear whether this was a result of the frequencies present in the audio cues or the locus fidelity provided by enabling the flock to hear the shepherd approach. Consequently, it will be necessary to further investigate the response of sheep to the presence of a Sky Shepherd augmented with an audio cue.

Further, the screening tests only investigated a fraction of the existing flock dynamics, with little understanding of sheep when influenced by the shepherding *collect* behaviour. The flock dynamics observed indicate that there are both leader and follower dynamics to consider, which has implications for improving efficiency in current shepherding models. Further testing is required to support further development.

References

1. American Dorper Sheep Breeders' Society: Dorper Sheep History (2019). https://dorpersheep.org/dorper-history/
2. Baldock, N.M., Sibly, R.M.: Effects of handling and transportation on the heart rate and behaviour of sheep. Appl. Animal Behaviour Sci. **28**, 15–39 (1990)
3. Burry, M.: Barking drones used on farms instead of sheep dogs (2019). https://www.rnz.co.nz/national/programmes/checkpoint/audio/2018685575/barking-drones-used-on-farms-instead-of-sheep-dogs
4. Casterton Kelpie Association: Auction Records (2019). http://www.castertonkelpieassociation.com.au/auction-records/
5. Civil Aviation Safety Authority: Sector Risk Profile for the aerial mustering sector (2015). https://www.casa.gov.au/sites/default/files/Sector_Risk_Profile_%20aerial_mustering_sector.pdf
6. Civil Aviation Safety Authority: Flying drones or model aircraft recreationally (2019). https://www.casa.gov.au/modelaircraft
7. DJI: Mavic 2 enterprise series specs (2019). https://www.dji.com/au/mavic-2-enterprise/specs
8. Fujioka, K., Hayashi, S.: Effective Shepherding Behaviours Using Multi-Agent Systems, pp. 3179–3182. Institute of Electrical and Electronics Engineers, Piscataway (2017)
9. Gade, S., Paranjape, A.A., Chung, S.J.: Herding a flock of birds approaching an airport using an unmanned aerial vehicle. In: AIAA Guidance, Navigation and Control Conference, 2013. American Institute of Aeronautics and Astronautics, AIAA, Reston (2015). https://doi.org/10.2514/6.2015-1540

10. Gonyou, H.W.: Behavioral methods to answer questions about sheep. J. Animal Sci. **69**(10), 4155–4160 (1991)
11. Hamilton, W.D.: Geometry for the selfish herd. J. Theoret. Biol. **31**(2), 295–311 (1971)
12. Harvey, J., Merrick, K., Abbass, H.: Application of chaos measures to a simplified boids flocking model. Swarm Intell. **9**(1), 23–41 (2015)
13. Hitchcock, D.K., Hutson, G.D.: The movement of sheep on inclines. Aust. J. Exp. Agric. **19**(97), 176–182 (1979)
14. Hutson, G.D.: Visual field, restricted vision and sheep movement in laneways. Appl. Animal Ethol. **6**(2), 175–187 (1980)
15. Hutson, G.D.: Flight distance in merino sheep. Animal Sci. **35**(2), 231–235 (1982). https://doi.org/10.1017/S0003356100027409
16. International Sheep Dog Society 2018 Directors: Rules for trials (2018). https://www.isds.org.uk/trials/sheepdog-trials/rules-for-trials/
17. Jackson, T., Zammit, K., Hatfield-Dodds, S.: Snapshot of Australian Agriculture (2018). https://www.agriculture.gov.au/abares/publications/insights/snapshot-of-australian-agriculture#australian-farmers-manage-significant-risk-and-variability
18. Kilgour, R., de Langen, H.: Stress in sheep resulting from management practices. In: Proceedings of the New Zealand Society of Animal Production, vol. 30, pp. 65–76. New Zealand Society of Animal Production (1970)
19. Kim, J., Breur, G.J.: Temporospatial and kinetic characteristics of sheep walking on a pressure sensing walker. Canadian J. Veterinary Res. **72**(1), 50–55 (2008)
20. Lee, W., Kim, D.: Autonomous shepherding behaviors of multiple target steering robots. Sensors **17**(12), 2729 (2017)
21. Lien, J.-M., Rodriguez, S., Malric, J., Amato, N.M.: Shepherding behaviors with multiple shepherds. In: Proceedings of the 2005 IEEE International Conference on Robotics and Automation, pp. 3402–3407 (2005). https://doi.org/10.1109/ROBOT.2005.1570636
22. Lower, T., Temperley, J.: Preventing death and serious injury caused by quad rollovers on Australian farms - Policy Paper (2016). http://sydney.edu.au/medicine/aghealth/uploaded/Quad%20Bike/Position_Statement_Quads_Feb%202016.pdf
23. Morrell, L.J., Ruxton, G.D., James, R.: The temporal selfish herd: predation risk while aggregations form. Proc. R. Soc. B **278**(1705) (2010). https://doi.org/10.1098/rspb.2010.1605
24. National Transport Commission (Road Transport Legislation - Australian Road Rules) Regulations 2006: REG 24 - Speed limit in a shared zone (2016). https://www.legislation.gov.au/Details/F2016C00716
25. Paranjape, A.A., Chung, S.J., Kim, K., Shim, D.H.: Robotic herding of a flock of birds using an unmanned aerial vehicle. IEEE Trans. Rob. **34**(4), 901–915 (2018)
26. Pearson, M., Lennon, J.: Pastoral Australia : Fortunes, Failures & Hard Yakka: A Historical Overview 1788–1967. CSIRO, Melbourne (2010)
27. Phadke Associates, Inc: rdExpert. http://phadkeassociates.com/index_files/rdexperttestplanning.htm
28. Reynolds, C.: Flocks, herds and schools: A distributed behavioral model. In: Proceedings of the 14th Annual Conference on Computer Graphics and Interactive Techniques, SIGGRAPH '87, pp. 25–34. ACM, New Yrok (1987)
29. Rolland, D.: Aerial Agriculture in Australia: A history of the Use of Aircraft in Agriculture and Forestry. Australian Print Group, Sydney (1996)
30. Roussel, S., Hemsworth, P., Boissy, A., Duvaux-Ponter, C.: Effects of repeated stress during pregnancy in ewes on the behavioural and physiological responses to stressful events and birth weight of their offspring. Appl. Animal Behav. Sci. **85**(3–4), 259–276 (2004)
31. SafeWork NSW: Quad bikes. https://www.safework.nsw.gov.au/hazards-a-z/quad-bikes
32. Squires, V., Daws, G.: Leadership and dominance relationships on Merino and Border Leicester sheep. Appl. Animal Ethol. **1**(3), 263–274 (1975)
33. Strömbom, D., Mann, R.P., Wilson, A.M., Hailes, S., Morton, A.J., Sumpter, D.J.T., King, A.J.: Solving the shepherding problem: heuristics for herding autonomous, interacting agents. J. R. Soc. Interf. **11**(100) (2014). https://browzine.com/articles/52614503

34. Syme, L.A., Elphick, G.R.: Heart-rate and the behaviour of sheep in yards. Appl. Animal Ethol. **9**(1), 31–35 (1982)
35. TAFE NSW: Statement of attainment in drone essentials (2019). https://www.tafensw.edu.au/course/-/c/c/164-60035V01/Statement-of-Attainment-in-Drone-Essentials
36. Vaughan, R., Sumpter, N., Frost, A., Cameron, S.: Robot Sheepdog Project Achieves Automatic Flock Control, pp. 489–493. MITP, Bonn (1998)
37. Williams, J., London Symphony Orchestra: Imperial March - Darth Vader's Theme (1980)
38. Williams, T.: Working Sheep Dogs : A Practical Guide to Breeding, Training and Handling. CSIRO Publishing, Melbourne (2007)
39. Yang, W.C., Schmickl, T.: Collective motion as an ultimate effect in crowded selfish herds. Sci. Rep. **9**(6618) (2019). https://doi.org/10.1038/s4158-019-43179-6
40. Yaxley, K.J., Joiner, K.F., Bogais, J., Abbass, H.A.: Life-learning of smart autonomous systems for meaningful human-autonomy teaming. In: Handley, H., Tolk, A. (eds.) A Framework for Human System Engineering Applications and Case Studies. IEEE Wiley (In-print)
41. Zephyr technology: PSM Training ECHO User Guide (2014). https://www.zephyranywhere.com/media/download/psm-training-user-guide.pdf

Chapter 10
Apprenticeship Bootstrapping Reinforcement Learning for Sky Shepherding of a Ground Swarm in Gazebo

Hung Nguyen, Matthew Garratt, and Hussein A. Abbass

Abstract The coordination of unmanned air–ground vehicles has been an active area due to the significant advantages of this coordination wherein unmanned air vehicles (UAVs) have a wide field of view, enabling them to effectively guide a swarm of unmanned ground vehicles (UGVs). Due to significant recent advances in artificial intelligence (AI), autonomous agents are being used to design more robust coordination of air–ground systems, reducing the intervention load of human operators and increasing the autonomy of unmanned air–ground systems. A guidance and control shepherding system design allows for single learning agent to influence and manage a larger swarm of rule-based entities. In this chapter, we present a learning algorithm for a sky shepherd-guiding rule-based AI-driven UGVs. The apprenticeship bootstrapping learning algorithm is introduced and is applied to the aerial shepherding task.

Keywords Apprenticeship learning · Apprenticeship bootstrapping · Swarm guidance · Reinforcement learning · Shepherding · UAV · UGVs

The coordination of unmanned air–ground vehicles has been an active area due to the significant advantages of this coordination wherein unmanned air vehicles (UAVs) have a wide field of view, enabling them to effectively guide a swarm of unmanned ground vehicles (UGVs). Due to significant recent advances in artificial

---

Electronic Supplementary Material: The online version of this chapter (https://doi.org/10.1007/978-3-030-60898-9_10) contains supplementary material, which is available to authorised users.

H. Nguyen (✉) · M. Garratt · H. A. Abbass
School of Engineering and Information Technology, University of New South Wales, Canberra, ACT, Australia
e-mail: hung.nguyen@student.adfa.edu.au; m.garratt@adfa.edu.au; h.abbass@adfa.edu.au

H. A. Abbass, R. A. Hunjet (eds.), *Shepherding UxVs for Human-Swarm Teaming*, Unmanned System Technologies,
https://doi.org/10.1007/978-3-030-60898-9_10

intelligence (AI), autonomous agents are being used to design more robust coordination of air–ground systems, reducing the intervention load of human operators and increasing the autonomy of unmanned air–ground systems. A guidance and control shepherding system design allows for single learning agent to influence and manage a larger swarm of rule-based entities. In this chapter, we present a learning algorithm for a sky shepherd-guiding rule-based AI-driven UGVs. The apprenticeship bootstrapping learning algorithm is introduced and is applied to the aerial shepherding task.

10.1 Introduction

The air coordination of unmanned ground vehicles (UGVs) has been an active area of research due to the significant advantages offered by the situation awareness picture due to a wider field of view (FoV) of a sky shepherd using unmanned air vehicles (UAVs). Recently, with significant advances in artificial intelligence (AI), autonomous agents are being used to improve the robustness of air–ground coordination systems, reducing the intervention of human operators and increasing the independent operating ability of the multi-agent system.

Shepherding approaches [75, 76, 90] use each learning agent to influence a larger swarm of rule-based entities. These scalable hierarchical multi-robot shepherding solutions offer more robust air–ground coordination systems.

In this chapter, we learn a control strategy for the UAV, while the UGVs follow predefined response rules [90]. The apprenticeship bootstrapping (ABS) learning algorithm uses human-collected lessons/data to learn an autonomous sky shepherd.

The chapter is organised as follows. In the first section, we review related research on the air–ground coordination problem, including recent architectures for aerial/sky shepherding of ground swarms. We then define the aerial shepherding task for single and multi-agents. We give an overview of the state-of-the-art learning approaches, including reinforcement learning, apprenticeship learning, and then present ABS learning. Finally, initial results of the implementation of a sky shepherd in Gazebo [51] are presented. The learning task of the sky shepherd agent is decomposed into two primitive skills: collecting and driving. ABS is used to learn a composite skill from human data of these primitive skills.

10.2 Aerial/Sky Shepherding of Ground Swarm

10.2.1 Unmanned Air–Ground Vehicles Coordination

Recently, multi-agent systems (MASs) have been used as a level of abstraction for many real-world problems [81], [52] such as in military missions [9] and search

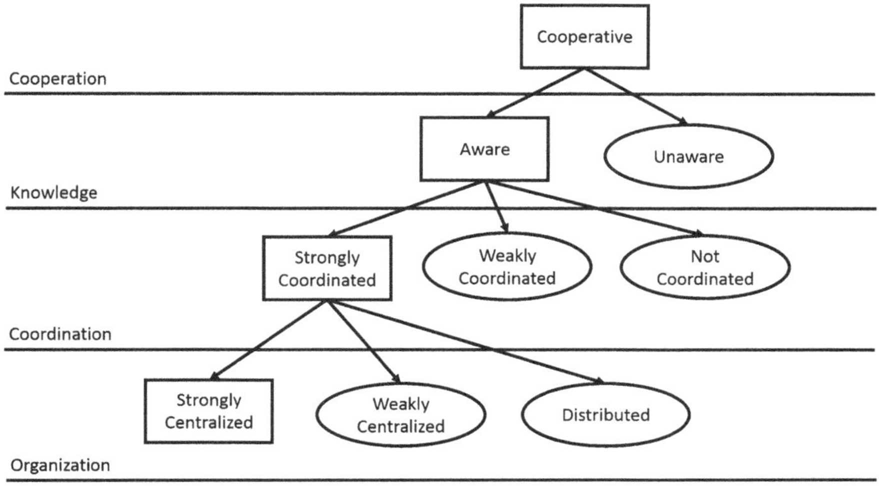

Fig. 10.1 Multi-agent system taxonomy [26]

and rescue operations [8]. Meanwhile, the problem of coordination among agents in MASs is an active topic [83] investigating, for example, how to improve the quality of information coming from different autonomous systems, and how to avoid communication failures [27, 28, 41]. Aiming to evaluate the advantages of coordination in MASs, Farinelli et al. [26] proposed a classification of MAS in which there are four main classes: Cooperation, Knowledge, Coordination, and Organisation as described in Fig. 10.1.

- Cooperation: All parties need to work together to accomplish a global task.
- Knowledge: Each agent in the team should have enough information about its teammates.
- Coordination: First, each agent in the team is aware of the other agents. Then, the actions performed by each agent should help those executed by another agent in order to successfully achieve a task. The degree of strength in coordination (strong or weak) relies on the coordination protocol, which is comprised of predefined rules that the agents are required to follow.
- Organisation: This mechanism requires strong coordination among agents in MASs. There are two key types of organisation: Centralised and Distributed. In the former type [25], leaders decide actions of other agents in the team. If there is only one leader, the organisation mechanism is considered as a strong centralisation; in contrast, when there are two leaders or more, the level of centralisation starts decreasing. In distributed setups, agents are completely autonomous during the decision process; there is no leader.

The decision on which design characteristic a MAS should have relies on the specific task to be carried out by the MAS as Farinelli et al. explained [26]. Focusing

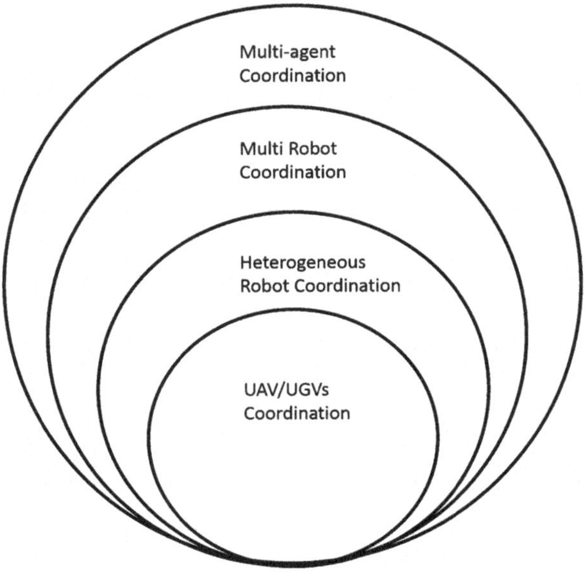

Fig. 10.2 Research fields regarding coordination of autonomous agents [18]

more on the coordination mechanism, Chen et al. [18] provided an overview of research findings focusing on coordinating among agents as can be seen in Fig. 10.2. The coordination of MASs covers the problem of coordinating among multiple homogeneous or heterogeneous robots.

In the case of heterogeneous robots [18, 24], the problem of coordinating Unmanned Aerial Vehicles (UAVs) and Unmanned Ground Vehicles (UGVs) is an active research area due to its significance in real-world situations, including search and rescue missions [3, 18], and crowd control activities [47].

Coordination helps reduce the limitation of each vehicle and leverage their unique advantages [17]. A UAV is assumed to have a large field of view (FoV) [40, 102] that enables it to support UGVs with a continuous feed of real-time situational awareness imagery (RT-SAI), which helps UGVs to better plan and navigate in the environment and coordinate their activities. Additionally, the UAV can fly over obstacles such as buildings and still maintain UGVs in their sensory ranges. The global picture can be understood as the FoV of the UAV's camera. In contexts whereby the UAVs and the UGVs are allowed to communicate with each other, this coordination will help to effectively solve the simultaneous localisation and mapping problem [17] and address failures of sensors such as loss of GPS signals by the UGVs.

According to the report by Chen et al. [18], however, the amount of research regarding UAV/UGV coordination is not significant compared with those related to individual UAVs or UGVs. There is a variety of research focusing on building autonomous aerial agents [16, 62], but development of effective AI agents for air–ground coordination is an ongoing challenge. In the next section, we provide a brief overview about the recent development of autonomous aerial agents in the context of the air–ground vehicles coordination.

10.2.2 A Brief Review of Coordination in Unmanned Air–Ground Vehicles

In recent years, the number of civilian applications using UAVs has drastically increased due to their mobility, automation, and the low cost associated with acquiring UAVs [16, 62]. Their use has received vast attention in the research community towards the development of artificial intelligence agents for the UAVs [16, 71]. In cluttered environments, Ross et al. [85] attempted to design autonomous reactive controllers for UAVs that could fly while avoiding static objects such as trees in a forest. Guo et al. [36] proposed a UAV that has the capability to detect a safe area, which is free of people or obstacles, to land. The attempt used Gaussian mixture models (GMMs) and support vector machines (SVMs), while more recent attempts considered DNNs for designing control policies for UAVs [103]. More recently, with the success of deep learning (DL) in various practical problems, Carrio et al. [16] provided a comprehensive review of applications of DL for UAVs.

The previous literature demonstrate that research on building AI agents for the UAV in domains requiring the coordination between air and ground vehicles is lacking. There is a variety of missions requiring a high level of coordination in the air–ground interaction problem, such as search and rescue, target tacking, site inspection, persistent surveillance, and area mapping [96]. For autonomous UAV–UGV systems, attempts by researchers [63, 95] were made to consider the problem as one of a distributed AI (DAI) and/or multi-agent systems (MASs). Howitt and Richards [38] examine the workload issues of humans during their interaction with the command and control system of a UAV focusing mainly on teleoperations. Semi-autonomous operations in contested environments have also gained attention through the work of Trentini and Beckman [93]. Khaleghi et al. [48] compared different UAV–UGV coordination control systems.

A common aim of these pieces of work is to increase the autonomous ability of the aerial vehicles in order to reduce the complexity caused by their aerodynamic effects in controlling them [96]. It is important and necessary to design an autonomous controller for UAVs to support UGVs in their mission's objective. This controller can either replace or augment human control of the UAVs and extend its horizon to supervisory control of the UAVs, or fully autonomous operations. However, developing autonomous UAVs for this coordination is a non-trivial endeavour. The main reason is due to the complexity of coordination, which demands complex control decisions as described in [3]. Moreover, the asymmetry between the behavioural characteristics and dynamics of UAVs and UGVs makes the learning problem challenging. The difficulty inherent in the problem of coordinating airborne and ground-borne vehicles is multi-faceted. One of the main issues is determining the impact of communication-borne latency on the capability of these vehicles to coordinate their actions.

The focus when designing autonomous aerial vehicles (AAVs) has been mainly placed on contexts lacking coordination between AAVs and other autonomous vehicles, such as UGVs; thus, autonomous coordination of AAVs and UGVs remains an open challenge due to the complexity associated with tasks in these scenarios; especially when the behaviours of UAVs are highly dependent on those of moving targets, such as UGVs on the ground. Although the behaviour of these UGVs can be predefined in advance, the movement of the UGVs is not trivial to be predicted.

When the UGVs act both independently and as a group, which increases the degree of uncertainty in dynamic environments, they can face obstacles or previously unknown situations that will cause unexpected behaviours of the UGVs. Thus, it is challenging for the UAVs when they need to adapt to the uncertain and unpredictable behaviour of the UGVs.

These challenges raise a research question on how to design a new and robust coordination framework. This coordination must be scalable as the number of UAVs and the number of UGVs increase. Fortunately, the shepherding problem [64] could offer an approach to scale up such system to a larger number of vehicles. In the shepherding context, there could be a single UAV or multiple UAVs, which act as shepherds and a swarm of UGVs, which act similar to sheep. The shepherds and sheep act as a biological swarm so that they can be scalable for both the UAVs and UGVs.

10.2.3 Autonomous Aerial/Sky Shepherding in Air–Ground Coordination

The bio-inspired swarm control problem brings novel perspectives from a variety of fields including control theory and computational intelligence [66]. Studies in multi-agent systems and swarm robotics attempt to address the question of how groups of simple agents/robots can be designed such that their collective behaviour emerges from their interaction with one another and their environment [15, 79].

Approaches to control a swarm of agents can be divided into two categories: rule-based algorithms and learning-based algorithms [6, 78]. The former introduces a set of fixed rules or predefined equations to compute the dynamics of the system based on global or local states of agents in the swarm [35, 86, 98]. While they achieve simple design, fast implementation, and scalability in relation with the number of agents, they often suffer from a lack of adaptability and generalisability to different contexts.

The latter approach relies on machine learning algorithms, offers flexibility, and eliminates the need for a large amount of knowledge about the model being used for swarm dynamics, or the need to validate the rules. Various methods applying reinforcement learning or deep reinforcement learning combined with team communication or a shared mental model have been proposed to learn decentralised policies for swarm control [74, 80, 89]. Nevertheless, these approaches do not scale up with an increasing number of swarm members due to the significant increase in the computational resources required to train multiple agents simultaneously.

Fortunately, the shepherding could offer practical solutions to these challenges. The shepherding problem is inspired by the real behaviour of herding sheep using a sheepdog in agriculture. Among a number of researchers, Strömbom et al. [90] developed a heuristic model to explain the interaction between one intelligent individual, the sheepdog, and a swarm of agents/sheep, which can be abstracted to many other problems in human–swarm interaction.

Shepherding-inspired methods for swarm control might employ one learning agent to influence a larger swarm of rule-based entities. Applying the shepherding problem as an application of the UAV–UGV coordination problem, the shepherding agents are analogous to UAVs, and the sheep to UGVs. Solutions to the shepherding problem show promise for the robust coordination of autonomous air–ground vehicles [17]. In this chapter, we propose an approach to developing an effective shepherding agent (aerial/sky shepherd).

10.3 The Aerial/Sky Shepherding Task

In this section, the aerial shepherding task is introduced in a multi-agent system wherein there are UAVs acting as shepherding agents and UGVs playing a similar role to sheep. The UAVs' goal is to herd the UGVs to a predefined target area as described in Fig. 10.3.

In the remainder of this section, a formal description of the aerial shepherding task is introduced, and then we describe the aerial shepherding task in the multi-agent system wherein there might be one shepherd or multiple guiding a large swarm of UGVs acting like sheep.

10.3.1 Description of the Aerial/Sky Shepherding Task

We consider the shepherding task introduced by Baumann and Büning [7] where n UAVs (aerial/sky shepherding agents) are to guide a swarm of m reactive UGVs (sheep) into a predefined target area in as a few time steps for the shepherding agents as possible. The UGVs have a limited range of view and try to move away from the UAVs if the distance between them is less than the UGVs' viewing range. No limitation on the view range of the UAVs is considered. An optimal solution is deemed when the number of steps needed by the UAVs is smallest.

In the research of Strömbom et al. [90], the sheep agents are reactive, responding to the sheepdogs. When the distance between the sheep and the dog is less than the sheep's viewing range, the sheep agents will move far away from the dog

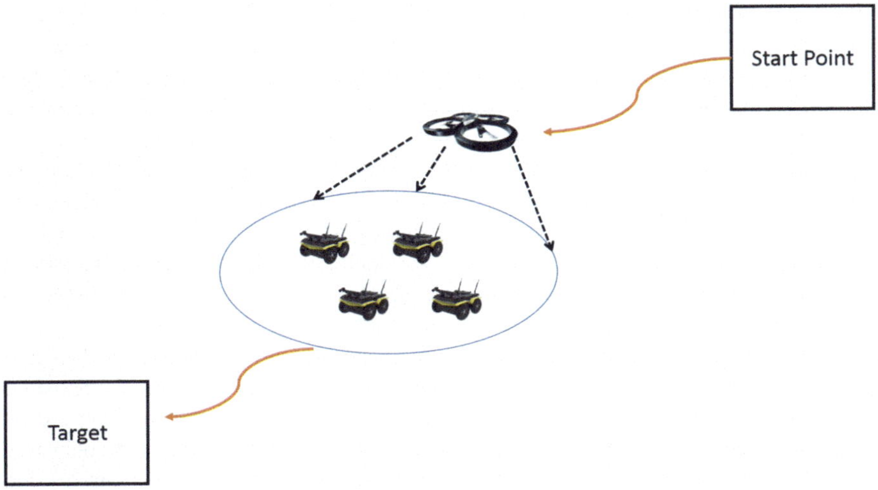

Fig. 10.3 An UAV shepherding a swarm of UGVs

based on predefined moving rules. Simply put, the dogs are smart agents that have sufficient knowledge of the environment and their goals. The UAV–UGV coordination problem we pose here is analogous to this problem. We assume that the UGVs are able to act like sheep described by Strömbom, and the vertical displacement of the UAV does not impact their behaviours in shepherding or influence upon the UGVs.

There are two types of agents initialised in this environment belonging to a set of UGV (influenced agents) $\Pi = \{\pi_1, \ldots, \pi_i, \ldots, \pi_N\}$, or a set of UAVs (influencing agents) $B = \{\beta_1, \ldots, \beta_j, \ldots, \beta_M\}$. Three behaviours of each UAV and four behaviours of each UGV at a time step t are denoted as below.

1. For shepherd β_j:

    - *Driving* behaviour σ_1: If a UAV is aware that all UGVs have been collected in a cluster, the UAV is triggered to adopt a driving behaviour by moving towards a driving point located behind the UGVs' cluster on the line formed between the centre of mass of the UGVs and the target position. Equation 10.1 represents the threshold to decide if the UGVs are considered to form a single cluster or not. If all distances of observed UGVs to their centre of mass are lower than this threshold, the UAVs then calculate their normalised force vectors, $F^t_{\beta_j cd}$, towards the driving position.

$$f(N) = R_{\pi\pi} N^{\frac{2}{3}} \tag{10.1}$$

    - *Collecting behaviour* σ_2: If there is an outlier UGV whose distance to the centre of mass is greater than the threshold specified above, the UAV switches

to collecting behaviour and computes its normalised force vector, $F^t_{\beta_j cd}$, towards a collecting position located behind the furthest UGV to the cluster, on the line between the centre of mass of sheep and the furthest UGV.
- *Jittering behaviour* σ_3: To avoid moving impasse, a random noise is presented $F^t_{\beta_j \epsilon}$ with weight $W_{e\beta_j}$ and summed into the total force.

UAV β_j total force $F^t_{\beta_j}$ (total force behaviour σ_8) is a weighted summation of the forces generated by driving/collecting behaviour and jittering behaviour.

$$F^t_{\beta_j} = F^t_{\beta_j cd} + W_{e\beta_j} F^t_{\beta_j \epsilon} \tag{10.2}$$

2. For UGV π_i:

- *Escaping behaviour* σ_4: This behaviour is represented by a repulsive force $F^t_{\pi_i \beta_j}$ if the distance between UGV π_i at position $P^t_{\pi_i}$ and UAV β_j at position $P^t_{\beta_j}$ is less than $R_{\pi\beta}$; that is,

$$\| P^t_{\pi_i} - P^t_{\beta_j} \| \leq R_{\pi\beta} \tag{10.3}$$

- *Collision avoidance* σ_5: Repulsion of a UGV π_i from another UGV $\pi_{k \neq i}$. The condition for the force to be able to exist between a pair of UGV is that the distance between them is less than $R_{\pi\pi}$; that is,

$$\exists k, \text{ such that } \| P^t_{\pi_i} - P^t_{\pi_k} \| \leq R_{\pi\pi} \tag{10.4}$$

  We then denote $F^t_{\pi_i \pi_{-i}}$ to represent the summed force vectors from all other UGV within the threshold distance applied onto UGV π_i.
- *Grouping behaviour* σ_6: UGV π_i is attracted to the centre of mass of its neighbours $\Lambda^t_{\pi_i}$ by a force $F^t_{\pi_i \Lambda^t_{\pi_i}}$.
- *Jittering behaviour* σ_7: To avoid moving impasse, a random noise is presented $F^t_{\pi_i \epsilon}$ with weight $W_{e\pi_i}$ and summed into the total force.

UGV π_i total force behaviour σ_9 is represented by the total force $F^t_{\pi_i}$, which is a weighted sum of force vectors $F^t_{\pi_i \beta_j}$, $F^t_{\pi_i \pi_{-i}}$, $F^t_{\pi_i \Lambda^t_{\pi_i}}$, and $F^t_{\pi_i \epsilon}$; that is,

$$\begin{aligned} F^t_{\pi_i} = {} & W_{\pi_\upsilon} F^{t-1}_{\pi_i} + W_{\pi\Lambda} F^t_{\pi_i \Lambda^t_{\pi_i}} + W_{\pi\beta} F^t_{\pi_i \beta_j} \\ & + W_{\pi\pi} F^t_{\pi_i \pi_{-i}} + W_{e\pi_i} F^t_{\pi_i \epsilon} \end{aligned} \tag{10.5}$$

The positions of UAVs and UGVs are updated according to Eqs. 10.6 and 10.7 given $S^t_{\beta_j}$ and $S^t_{\pi_i}$ to be the speed of β_j and the speed of π_i at time t. However, in the original Strömbom model, the speed of agents are constant.

$$P^{t+1}_{\beta_j} = P^t_{\beta_j} + S^t_{\beta_j} F^t_{\beta_j} \tag{10.6}$$

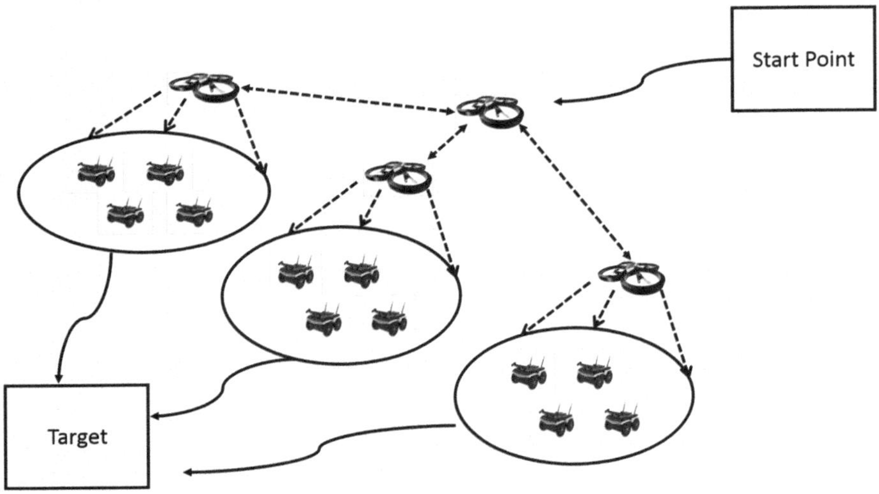

Fig. 10.4 Multiple UAVs shepherding a swarm of UGVs in centralised multi-agent context. There is a UAV leader guiding the remaining UAV aerial shepherding agents

$$P_{\pi_i}^{t+1} = P_{\pi_i}^{t} + S_{\pi_i}^{t} F_{\pi_i}^{t} \tag{10.7}$$

These behaviours of a single UAV at a time step t are used to drive the UGVs within their monitoring region towards the predefined target area. When there are multiple UAVs, the set of their behaviours needs to include factors or dependencies of other UAVs in order to perform the task in as few steps as possible. The context of multiple UAVs acting as aerial shepherding agents is introduced in the next Sect. 10.3.2 with the UAVs described as a multi-agent system.

10.3.2 The Aerial/Sky Shepherding Task as a Multi-Agent System

As mentioned in Sect. 10.2.1, the problem of air–ground vehicle coordination presents itself as a heterogeneous multi-agent coordination problem; therefore, we model the aerial/sky shepherding task as such. Approaching this multi-agent coordination problem, there are two possible models for the aerial/sky shepherding task: centralised and distributed. In the first model, a single sky shepherd is required to monitor and guide the remaining UAVs to perform the shepherding task. The illustration of this model is shown in Fig. 10.4.

This centralised approach has advantages in reducing the amount of interaction among the UAVs. Additionally, it also decreases the complexity in learning as only two types of UAVs models need to be learnt, the leader (who controls the

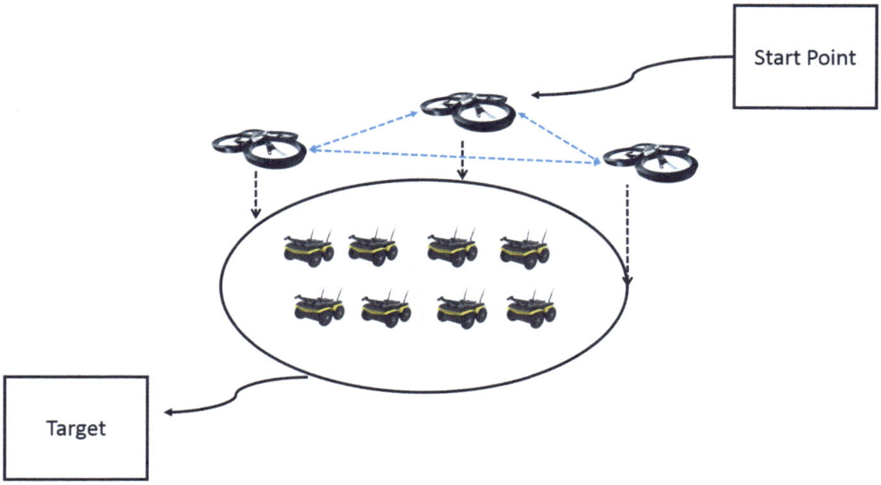

Fig. 10.5 Multiple UAVs shepherding a swarm of UGVs in distributed multi-agent context. The UAV aerial-shepherding agents communicate with each other to produce their actions

sky shepherd fleet) and the followers (the remainder of the fleet). This allows for scalability in the learning process when the number of UAVs increases. However, the centralised model has an underlying assumption that the leader is able to sense or at the least receive communication from the other shepherds. In the distributed model, the shepherds communicate among each other before producing their actions. The illustration of the distributed model is shown in Fig. 10.5.

While the shepherd model can limit the impact of errors due to sensing or loss in peer-to-peer communication, it provides additional challenges when viewing the problem from a learning perspective. First, it requires learning models for UAVs, and these models affect each other in training; that is, models are being created iteratively, which have inter-dependencies on each other. This requires significant knowledge of the model and significant information exchange in the learning process. Second, scalability is a challenge for this approach because when the number of UAVs and/or UGVs increases, the state and action space for the learning process also does.

Developing aerial agents for the shepherding task either in a centralised or in a distributed form is non-trivial when unexpected situations in the environment might cause unexpected behaviours of the UGVs as well as dynamic effects of both the UAVs and UGVs. There exist in the literature various learning approaches to design artificial intelligence agents for autonomous systems. In the next section, we introduce two main learning approaches: reinforcement learning and apprenticeship learning (also known as imitation learning). We then introduce a new learning approach, namely apprenticeship bootstrapping that allows us to address complex tasks such as sky shepherding.

10.4 Learning Approaches

In recent years, dramatic advances in AI have led to greatly improved autonomous systems [104]. It is appealing to attempt developing autonomous aerial agents in the context of the aerial shepherding of ground swarm. In this chapter, we discuss two main learning approaches: reinforcement learning (RL) and apprenticeship learning (AL) and then introduce a new learning approach aiming to bootstrap the learning process in complex tasks where the efficient application of traditional RL or AL approaches is challenging.

10.4.1 Reinforcement Learning

Reinforcement learning (RL) [91] has long been an active research area in the fields of AI and robotics. RL allows AI agents to learn optimal policies through interacting with a given environment. In classic RL algorithms [91], such as Q-learning and SARSA, linear approximation functions are used to address the problem of generalisation in machine learning methods in continuous and dynamic environments [33, 53].

Recently, deep learning approaches have gained traction and achieved significant success in various research domains such as vision and natural language processing applications [58]. These DL methods utilise a significant number of auto-learned parameters and a variety of different network architectures, such as convolutional neural networks (CNNs) [56], and long-short-term memory (LSTM) [31] networks. These deep neural networks (DNNs) are able to extract and operate on unknown features from the data provided and improve the generalisation ability of the machine learner model.

Previously, RL approaches using non-linear approximation functions such as neural networks did not guarantee convergence in training agents [94]. However, when combining with a DL model, convergence of the so-called deep reinforcement learning (DRL) is achieved in training, and thus, resulting in a significantly improved overall performance. Well-known applications of the DRL approach include [68] and [69] in playing Atari and Go games, respectively. In this chapter, we provide background about the Markov decision process, Q-learning, and deep Q-learning, before giving an overview of applying RL for the context of multi-agent systems.

10.4.1.1 Markov Decision Process

A Markov decision process (MDP) [91] represents a sequential decision-making problem for which an agent must select sequential actions to maximise its optimisation criteria based on reward values. The finite MDP is formally presented as a tuple $M = \{\mathbb{S}, \mathbb{A}, T, r, \gamma\}$, where:

- $\mathbb{S} = \{s_1, s_2, \ldots, s_n\}$ is a finite set of n states describing the dynamic world.
- $\mathbb{A} = \{a_1, a_2, \ldots, a_m\}$ a set of m actions executed by the agent.
- $T : S \times A \times S \mapsto [0, 1]$ a transition function or transition model, where $T(s, a, s')$ is the transition probability when the agent takes action $(a \in A)$ in a state $(s \in S)$, which leads to the next state $s' \in S$, i.e., $T(s, a, s') = P(s'|s, a)$.
- $r : S \times A \mapsto \mathbb{R}$ is a reward function, where $r(s, a)$ is the immediate reward received by the agent when it takes action $(a \in A)$ in state $(s \in S)$.
- $\gamma \in [0, 1]$ a discount parameter.

A control policy is defined as a mapping $(\tau : S \times A \rightarrow [0, 1])$, where $\tau(s, a)$ is the probability of selecting action $(a \in A)$ in state$(s \in S)$. When the agent follows a given policy, τ, the value function V^τ and the action-value function Q^τ are defined respectively, by:

$$V^\tau(s) = \mathbb{E}_\tau \left[\sum_{i=0}^{\infty} \gamma^t r(s_t, a_t) | s_0 = s \right]. \tag{10.8}$$

$$Q^\tau(s, a) = \mathbb{E}_\tau \left[\sum_{i=0}^{\infty} \gamma^t r(s_t, a_t) | s_0 = s, a_0 = a \right], \tag{10.9}$$

where (s_t, a_t) is a state–action pair generated under policy τ in time step t, the value function $V^\tau : \mathbb{S} \rightarrow \mathbb{R}$, and action-value one $Q^\tau : \mathbb{S} \times \mathbb{A} \rightarrow \mathbb{R}$.

The aim of the agent is to discover an optimal control policy (τ^*) that maximises the totally discounted reward accumulated over all states. The optimal value function is denoted by $V^*(s)$, and $V^*(s) = sup_\tau V^\tau(s)$. Similarly, the optimal action-value function is denoted by $Q^*(s, a)$, and $Q^*(s, a) = sup_\tau Q^\tau(s, a)$ with τ identified as an optimal policy in the MDP framework if and only if:

$$\tau(s) \in arg \max_{a \in \mathbb{A}} Q^\tau(s, a), \forall s \in \mathbb{S}. \tag{10.10}$$

Based on Bellman equations, $V^*(s)$ and $Q^*(s, a)$ are defined. For all $a \in \mathbb{A}$ and $s \in \mathbb{S}$, respectively, as

$$V^*(s) = \max_{a \in \mathbb{A}} Q^*(s, a). \tag{10.11}$$

$$Q^*(s, a) = r(s, a) + \gamma \sum_{s' \in \mathbb{S}} P(s'|s, a) V^*(s'). \tag{10.12}$$

10.4.1.2 Q-Learning

Various algorithms can be used to solve an MDP through computing the optimal value functions [92]. The most common of these is *Q-learning*, which is a model-free RL approach. A model-free approach directly learns action values through trial-error experiences without a need to formulate a model of the operating environment; thus, it does not need to know the transition probabilities among states [22]. The classic Q-learning approach [91] is shown in Algorithm 1. The learning is denoted as α, and γ is the discount factor for the value of future rewards.

Algorithm 1 The Q-learning algorithm

1: Initialise Q[`num_states,num_actions`]
2: Observe initial state (s)
3: **repeat**
4: Select and perform action a
5: Observe reward r and new state s'
6: $Q[s,a] = Q[s,a] + \alpha(r + \gamma \max_{a'} Q[s',a'] - Q[s,a])$.
7: $s = s'$
8: **until** terminated

In many real-world scenarios, Q-Learning is impractical because of its limited state and action spaces, and inability to generalise to unseen states. The approach is innately discrete and has difficulty in coping with continuous state spaces. Therefore, in practical Q-learning systems, function approximators [33, 53], generally linear ones, are used to approximate the Q-function as

$$Q(s,a;\theta) \approx Q^*(s,a) \tag{10.13}$$

However, these functions are limited to manually crafted features and low-dimensional state space. On the contrary, deep neural networks (DNNs) [58] have been shown to be capable of solving low- and high-dimensional data in real-word scenarios. Hence, combining DNNs and RL, deep reinforcement learning (DRL) is a logical progression and such approaches show advances in addressing continuous environments, while still guaranteeing convergence [69, 99].

10.4.1.3 Deep Q-Network

Recent successes in deep reinforcement learning (DRL) models, which even surpass top human competitors in some particular tasks, such as playing Atari games or developing autonomous robots [69, 99], have attracted researchers and technologists in an effort to address various practical problems, such as in video games [46], robotics [59, 67, 76], natural language processing (NLP) [60, 61], intelligent transportation systems [13, 14], and smart grid [30].

In Deep Q-network (DQN) [68], the authors utilise an experience replay approach when training the DNNs. The AI agents trained through DQN outperformed others that learned from classic RL algorithms and even reached a professional human level. In this technique, experiences are stored in a sufficiently large memory over episodes, and then a certain number, called a mini-batch, is randomly chosen from the memory and updated to the DNN in every step. The advantages of this technique are clearly analysed in [68]. The original DQN with experience replay [68] is shown in Algorithm 2.

Algorithm 2 DQN with experience replay

Initialise replay memory D to capacity N
Initialise action-value function Q with random weights θ
Initialise target action-value function Q^* with weights $\theta^- = \theta$
for episode=1, M **do**
    Initialise sequence $s_1 = x_1$ and preprocessed sequence $\phi_1 = \phi(s_1)$
    for t=1, T **do**
        With probability ϵ select a random action a_t
        otherwise select $a_t = arg\max_a Q(\phi(s_t), a; \theta)$
        Execute action a_t in emulator and observed reward r_t and image x_{t+1}
        Set $s_{t+1} = s_t, a_t, x_{t+1}$ and pre-process $\phi_{t+1} = \phi(s_{t+1})$
        Store transition $(\phi_t, a_t, r_t, \phi_{t+1})$ in D
        Sample random mini-batch of transitions $(\phi_j, a_j, r_j, \phi_{j+1})$ from D
        Set $y_j = \begin{cases} r_j \text{ if episode terminates at step j+1} \\ r_j + \gamma \max_{a'} Q^*(\phi_{j+1}, a'; \theta^-) \text{ otherwise} \end{cases}$
        Perform a gradient descent step on $(y_j - Q(\phi_j, a_j; \theta))^2$ with respect to the network parameters θ
        Every C steps reset $Q^* = Q$
    end for
end for

10.4.1.4 Multi-Agent Reinforcement Learning

In the literature, RL has been used in the context of the multi-agent or swarm systems [37] to learn an appropriate policy at both entity level and swarm level focusing on addressing distributed or decentralised multi-agent coordination. An example of the RL framework in the multi-agent context is mentioned by Iima and Kuroe [44]. In this research, the Q-learning algorithm for a swarm scenario with discrete state spaces was investigated. Each agent is an independent learner, but a swarm reward function is used to reflect the overall performance of the entire swarm. A similar effort on UAV formation control [89] has been performed when multiple look-up tables of state–action values are used with the SARSA algorithm. However, these above approaches do not address the scalability of the approach when the state space is continuous or less trivial (i.e., larger); in such cases, the approach requires a large computational resource without guaranteeing convergence.

Recently, with advances of deep reinforcement learning (DRL) [68], researchers [80] proposed a multi-agent DRL approach, namely Lenient Multi-Agent Deep Reinforcement Learning (LDQN). The research aims to address coordination between multi-agents in a distributed fashion. However, it too faces computational resource challenges due to the complexity of simultaneously training multiple agents. Thus, the method is not scalable with an increase in agent numbers. In terms of centralised multi-agent coordination, the RL approaches might be used to address components of behaviour sequentially. For example, models of the monitored agents are firstly built, and then a high-level model of the leader is trained.

In this chapter, we focus on the latter approach when adopting the shepherding problem since we aim for robustness in the coordination among ground and air vehicles. In our approach, the aerial shepherd acts as a leader to guide the UGVs. However, transferring RL approaches to autonomous systems for the aerial shepherding of ground swarm problem poses challenges in terms of dealing with the large state space and designing effective reward functions. Therefore, another learning approach partially able to deal with such challenges, namely apprenticeship learning or imitation learning[42], is used.

10.4.2 Apprenticeship Learning

Apprenticeship learning (AL), also called imitation learning (IL), learning from demonstration (LfD) or programming by demonstration (PbD), is a learning approach in which learners obtain knowledge based on watching teacher's demonstrations. Described in [4, 11, 12, 42], agents' behaviours are learned based on teacher's demonstrations, each of which is formed as a set of state–action pairs. AL learning methods are categorised into two major classes: supervised learning (SL) and inverse reinforcement learning (IRL), which we describe below. In addition, a combination of SL with RL or genetic algorithms to bootstrap the learning process has also been researched [43, 45, 55].

10.4.2.1 Supervised Learning

In SL, the data-set is a set of demonstrations from which the most suitable action is chosen through applying function approximations, for a given state. SL comes in two main forms: label prediction or classification, and point prediction or regression. Classification could be used for any type of actions, that is, low-level, high-level, and complex ones. Regression produces continuous output values and is generally applied to low-level actions, such as velocity control of robots' wheels.

Recently, there have been dramatic advances in DL for solving a large variety of problems in both computer vision and natural language processing [58]. Therefore, the idea of using DNNs to directly learn from human demonstrations is appealing [59]. However, learning by directly matching states and actions as in the SL approach might not be effective for obtaining the desired behaviour [42] because of the inability to obtain expert demonstrations or errors induced by poor demonstrations caused by the limitations of the collection methods used. To address these issues, an additional step is frequently required in AL, which is for a learner to perform the learned action and re-optimise the learned model. This self-improvement can be achieved through the IRL approach.

10.4.2.2 Inverse Reinforcement Learning

In RL [91], an agent executes actions and receives feedback signals in the form of a *reward*, which indicates the degree of success of an action. The agent attempts to learn a policy (a state–action map) that maximises its own accumulated reward function over time. However, the success of an RL agent depends on an appropriate manual design of the reward function that can guide the search process towards optimal policies.

Based on MDP, IRL [5] learns the reward function from human-generated policies based on the underlying assumption that, as these policies are produced by experts, they are (near) optimal. This machine-induced reward function is then used by an RL algorithm to find a near-optimal machine policy that is closest to that of the expert. There are various IRL approaches within the literature [5], such as maximum margin programming [1], entropy [97, 101, 106], and Bayesian [84]. In the chapter, we describe the maximum margin programming IRL approach introduced by Abbeel et al. [1], where it demonstrated success in designing an autonomous agent within a helicopter [2] in the real environment.

In the RL framework, the agent needs to discover an optimal control policy that maximises a reward-based criterion. However, in real-world scenarios, defining the reward function is challenging. Abbeel et al. [1] proposed using AL via IRL to find an approximate policy sufficiently close to the observed one. This approach avoids the need to explicitly recover the reward function as a state's reward is represented by a sum of its linearly weighted features. The algorithm, called AL via IRL, recovers the reward function by learning it from human demonstrations.

The MDP framework without the reward function is denoted as the MDP\R model in which an expert's demonstrations are assumed to be an optimal policy. The set of expert demonstrations, denoted D_E, consists of m trajectories $d^{(i)}$: $D_E = \{d^{(i)}\}_{1 \le i \le m}$, each of which $d^{(i)}$ is composed of k state–action pairs: $d^{(i)} = \{(s_t^{(i)}, a_t^{(i)})\}_{1 \le i \le k}$. Therefore, the expert's demonstrations D_E might be formulated as: $D_E = \{(s_1, a_1), (s_2, a_2), \ldots, (s_M, a_M)\}$, where $M = m \times k$.

AL via IRL [1] assumes that the reward function, $R(s)$, is a linear function represented as a weighted, w, sum of a state feature vector $\phi(s) = \{\phi(s_1), \ldots, \phi(s_i), \ldots, \phi(s_K)\}$ as

$$R(s) = w^T \cdot \phi(s) = \sum_{k=1}^{K} w_k \phi_k(s), \forall s \in \mathbb{S} \tag{10.14}$$

As IRL assumes that the human expert provides the optimal policy, there is an associated optimal weight vector w^* such that the optimal reward $R^*(s)$ is

$$R^*(s) = w^* \cdot \phi(s), \forall s \in \mathbb{S} \tag{10.15}$$

Then, IRL uses human demonstrations to find the weight vector w that approximates w^*.

For a fixed policy, its value is defined as

$$V^\tau(s_0) = w \cdot E\left[\sum_{t=0}^{\infty} \gamma^t \phi(s_t) \mid \tau\right] \tag{10.16}$$

Also, the feature expectations vector under a policy τ is defined as

$$\mu(\tau) = E\left[\sum_{t=0}^{\infty} \gamma^t \phi(s_t) \mid \tau\right] \tag{10.17}$$

If the dynamic model is unknown, the feature expectations vector $\mu(\tau)$ can be

$$\mu(\tau) = \frac{1}{m} \sum_{i=0}^{m} \sum_{t=0}^{\infty} \gamma^t \phi(s_t^{(i)}) \tag{10.18}$$

where m is the number of trajectories.

The AL via IRL algorithm is described in Algorithm 3.

10.4.2.3 Hybrid Methods

Other approaches focus on how to combine AL with other learning methods, such as reinforcement learning and genetic algorithms. In the supervised approach to imitation learning, the learning process is relatively fast, but naturally, the agent may only learn from what it observes. In contrast, reinforcement learning enables the agent to find new policies through free and wide exploration. Therefore, a combination of supervised learning and reinforcement learning can provide improvements over both algorithms. The approach can be understood as the exploration of reinforcement learning being guided by demonstrations. This can reduce the searching time for reinforcement learning algorithms to find an adequate policy and still enable the agent to depart from the demonstrated behaviours. Early research in combining imitation learning and reinforcement learning is described in [10, 50, 55, 82, 100].

Algorithm 3 The AL via IRL algorithm introduced in [1]

Require: An MDP\R, a feature mapping ϕ, and the expert feature expectations vector $\mu(\tau_E)$ calculated by expert trajectories.
Ensure: A number of policies $\{\tau^{(i)} : i = 0..n\}$.
Randomly choose policy $\tau^{(0)}$, estimate $\mu^{(0)} = \mu(\tau^{(0)})$ via Monte Carlo, and set $i = 1$.
Set $w^{(1)} = \mu_E - \mu^{(0)}$ and $\bar{\mu}^{(0)} = \mu^{(0)}$.
Set $t^{(1)} = \|\mu_E - \mu^{(0)}\|_2$.
if $t^{(i)} \leq \epsilon$ **then**
    terminate
end if
while $t^{(i)} > \epsilon$ **do**
    Using RL algorithm, compute optimal policy $\tau^{(i)}$ for MDP with $R = (w^{(i)})^T\phi$.
    Compute $\mu^{(i)} = \mu(\tau^{(i)})$ and set $i = i + 1$.
    Set $x = \mu^{(i-1)} - \bar{\mu}^{(i-2)}$.
    Set $y = \mu_E - \bar{\mu}^{(i-2)}$.
    Set $\mu^{(i-1)} = \bar{\mu}^{(i-2)} + \frac{x^T y}{x^T x} x$.
    Set $w^{(i)} = \mu_E - \bar{\mu}^{(i-1)}$.
    Set $t^{(i)} = \|\mu_E - \bar{\mu}^{(i-1)}\|_2$.
end while

An example of using demonstration to bootstrap reinforcement learning is to generate an initial list of primitive skills [54]; reinforcement learning is used to select these primitives. Another example is that demonstrations are used to reduce the search space of reinforcement learning [34]. Besides reinforcement learning, AL can also be integrated with genetic algorithms such as learning manipulation skills [43], or navigation strategies [45].

10.4.2.4 Multi-Agent Apprenticeship Learning/Imitation Learning

In the context of multi-agent systems, most of the apprenticeship learning approaches are developed based on IRL [88]. An exception to this is the research by Zhan et al. [105] in which a generative multi-agent policy is learned from demonstration data based on deep generative models. The latest work in the multi-agent context includes two main approaches [88]: fully cooperative agents and non-cooperative settings. For the first, the main factors, which are considered to decide the reward function, are the size of swarm and the communication ability among agents; for instance, partial or global. The reward function of the latter is decided by Bayesian inference or a linear combination of predefined state features.

10.4.3 Apprenticeship Bootstrapping

In practice, when applying apprenticeship learning approaches for building autonomous agents in complex tasks like aerial shepherding, a human expert may not be available due to novelty of the task or the cost of access to someone with appropriate skills. In the review made by Billard et al. [11], the clarification of tasks as well as learning approaches are introduced. In this sub-section, we describe these key points before introducing a learning approach. We introduce *apprenticeship bootstrapping*, which focuses on addressing how to collect human data for complex tasks, where it is expensive (or impractical due to task novelty) to access a human expert.

10.4.3.1 Clarifying Tasks

Tasks can be can divided into two classes [11, 42]. Sequential Tasks are composed of a sequence of individual skills or motions. Composite tasks are composed of a combination of individual skills. It can be seen that a composite task is composed of a combination of different sequential tasks.

For sequential tasks, the human data collection can easily be obtained by requiring the human to separately teach individual skills. In contrast, for composite tasks, the problem of collecting human data is non-trivial when this requires the human to combine many individual skills simultaneously while performing these tasks; that is, there is interdependence between the required skills. Thus, collecting human data is difficult. Different collection methods will lead to different learning approaches. The apprenticeship bootstrapping approach aims to address these composite tasks.

10.4.3.2 Learning Tasks

Billard et al. [11] provided a broad review of how to learn sequential and composite tasks as introduced below.

1. *Learning Sequential Tasks:*
   Sequential tasks are composed of a sequence of individual skills or motions/actions; for example, the juicing task might be considered as a sequential task in which there are two main skills: juicing an orange and pouring liquid into a cup. In the learning process, the human data of these individual skills can be collected separately and sequentially. After obtaining the human data, the sequential tasks can be learned by various learning methods, such as reinforcement learning, supervised learning, inverse reinforcement learning, or a combination of supervised learning and reinforcement learning as described in the previous approaches.
2. *Learning Composite Tasks:*
   For the composite tasks, the learning process is highly complex, but it is also an open area for the research community. The composite tasks are composed of a combination of individual skills. According to Billard et al. [11], there are two major approaches to address the problem of learning composite tasks as follows:

- *Approach 1:* There are two stages: the first stage is to learn all individual skills independently [21, 65] using demonstrations of these skills, and then the second aims to learn the correct order and combination of these skills through observing human demonstrations of the entire task [23, 87] or through the use of reinforcement learning [70]. This approach assumes that there is a known list of all necessary individual skills (termed primitives). This assumption might be true for specific composite tasks, but regarding general-purpose individual skills, the list does not exist. Additionally, the finite list may not sufficiently represent all human skills required, thus leading to the learning of a sub-optimal approach to the task.
- *Approach 2:* Instead of manually segmenting the finite primitive skills, this approach automatically segments the composite task to extract a set of possible primitive skills by observing a human demonstrating the whole task [57, 77]. The approach has the advantage of learning a combination of these primitive skills because of automatic segmentation. However, one problem that appears is that the number of necessary primitive skills is unknown, and the leaned combination might be sub-optimal because the learning approach is ineffective or the demonstration is poor [32].

10.4.3.3 Apprenticeship Bootstrapping Approach

In the discussion above, it is clear that human experts and their knowledge is a critical component of a successful system. Interestingly, although there has been much research conducted on RL approaches used to find the appropriate combination of skill primitives, a gap exists in addressing how best to collect data from human experts in non-trivial cases.

Apprenticeship bootstrapping (ABS) when it was first introduced [72], the approach was used to learn from human policies tasks in a UAV–UGV coordination context. The approach builds on apprenticeship learning via inverse reinforcement learning (AL via IRL) proposed by Abbeel et al. [1]. In ABS via inverse reinforcement learning, sub-tasks are used to approximate the reward signal needed for a

---

Algorithm 4 Apprenticeship bootstrapping (ABS) via deep learning algorithm

Require: Sub-task demonstrations $\{D_1, D_2, D_3, \ldots, D_H\}$; AF_{state} - the function fusing a sub-state into a composite state; AF_{action} - the function fusing a sub-action into a composite action
Ensure: A DNN trained model outputting composite actions.
1: Initialising a composite set.
2: Initialising a DNN model.
3: **for** each sub-task demonstration (s, a) **do**
4: $s_{composite} = AF_{state}(s)$
5: $a_{composite} = AF_{action}(a)$
6: Adding the composite demonstration $(s_{composite}, a_{composite})$ to the composite set.
7: **end for**
8: Training the DNN model using the composite set.

---

subset of the state vector. These incomplete approximations are then fused through an expectation function in order to approximate the overall reward function for an RL to discover a policy for the composite task. The primary assumption made is that there exists a human who can perform the sub-tasks. Each sub-task encodes sub-skills for the composite task. However, the fusion of these sub-skills is left to the RL agent to learn how to appropriately switch and combine.

Another application of ABS is shown in [73], where a deep network is used to train the composite set of states and actions. Sub-state space values are input for the network, and an action space covering all possible primitives is the output. This approach is used to build an autonomous agent for a UAV in an air–ground coordination scenario and was tested on a physical robotic test-bed. The high-level algorithmic description of ABS-DL is shown in Algorithm 4.

Additionally, recent ABS research [29] relies on a curriculum learning technique to decompose the shepherding task into primitive sub-tasks, and then lessons of these sub-tasks are learned from human data.

The results of the ABS approaches described above show promise for using ABS for the aerial shepherding of a ground swarm. However, the two first ABS approaches [72, 73] just work on the basic air–ground coordination in which according to the leader–follower rules, the only three allowed moves are forward, turn right, and left in a short distance while maintaining a predefined formation. Our air–ground shepherding task is more complex when the number of UGVs increases, and they act as a swarm in a larger operating environment. Thus, the behaviour of the UAVs will need to be investigated by high-level actions or primitive sub-tasks, such as collecting and driving sub-tasks. Requesting humans to perform demonstrations requiring multiple skills simultaneously could be very challenging [29].

Moreover, in this research, the shepherding task is only conducted in a two-dimensional environment but considering the aerodynamic effects of the UAVs operating in three-dimensional environment. These effects might also cause challenges for the human operator while controlling the UAV as investigated in our previous ABS research [72, 73]. Therefore, it can be seen that it is non-trivial for a human, even for an expert, to produce the data for successful control of one UAV or multiple UAVs in the context of a single shepherd or multi-shepherds, respectively. It is appealing to learn the aerial shepherding task from human data of primitive sub-tasks based on the ABS approach. In the next section, we initially investigate our ABS approach for the aerial shepherding task in a three-dimensional realistic simulation environment.

10.5 Initial Results

In this section, we aim to investigate our ABS approach for the aerial shepherding task. We apply the methodology proposed by Nguyen et al. [73] to bootstrap the learning process for the task. We create a robotic test-bed with an interface for control that allows a human to guide a group of UGVs with a UAV. ABS allows the human's approaches to be learnt and improved upon as discussed below.

10.5.1 Proposed Methodology

As mentioned in Sect. 10.4.3, the challenges of learning shepherding behaviours come from the complexity of the task's objectives including the multiple sub-tasks needed to be addressed. This presents a large state space, which may prove problematic for a deep network. Our ABS approach decomposes the search space of the aerial shepherding task into two problem sub-spaces: driving and collecting sub-tasks; we argue that this assists in managing the complexity of learning as previously discussed [76].

Figure 10.6 shows the learning process of the ABS approach. There are two demonstration scenarios to collect human data for two primitive behaviours: collecting outlier UGVs to create a cluster of UGVs and driving this cluster of UGVs towards a given target. The human demonstration scenario for the driving behaviour is used when all UGVs are clustered in a single cluster. The collecting scenario is utilised when there is one or more UGVs outside a threshold distance from the flock global centre of mass, which is calculated by Eq. 10.1. After collecting two human data-sets of the two sub-tasks, there are two training methods: hierarchical apprenticeship bootstrapping (HABS) and aggregation apprenticeship bootstrapping (AABS). First, we train the collecting and driving data-sets individually, called DNN-Collecting and DNN-Driving, respectively. In the second method, the two data-sets are aggregated to produce a composite data-set, which will be used for training a deep neural network (DNN), namely DNN-Composite. In the actions and states aggregation component, wherein the composite set is created, we initially conduct a straightforward fusion of these demonstrations. The high-level algorithmic description of this learning approach is described in Algorithm 4.

We utilise a state space representation of two vectors; namely, (1) the UAV to the UGV's centre of mass, and (2) the UAV to the sub-goal, which is collecting or driving point. This representation holds for both the collecting and driving sub-tasks. The action space of the UAV comprises a two real-valued vector, $V(x, y)$, representing the linear velocities while performing the drive or collect sub-tasks, respectively.

In testing phases of the HABS method, we combine the two trained networks (DNN-Collecting and DNN-Driving) in a framework shown in Fig. 10.7, where a behaviour selection component at each time step chooses to trigger one of the networks based on the threshold specified in Eq. 10.1. We test the performance of the system by continuously activating the two deep networks in a full shepherding problem scenario.

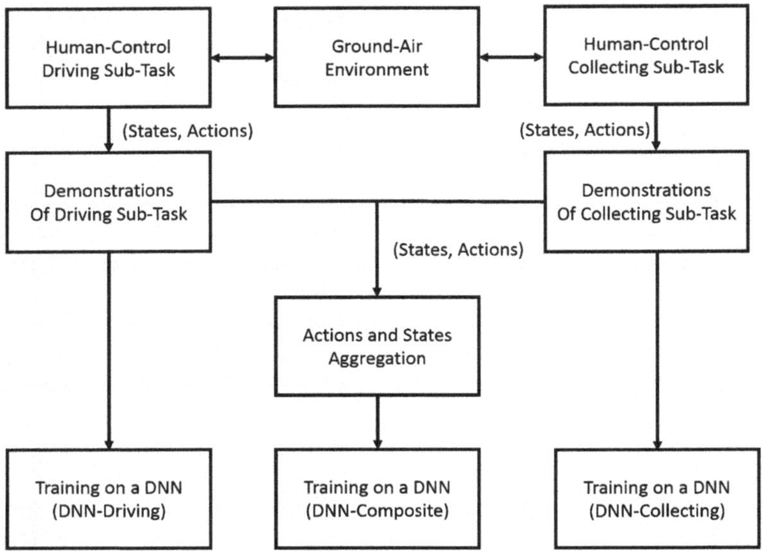

Fig. 10.6 The training phases of the ABS

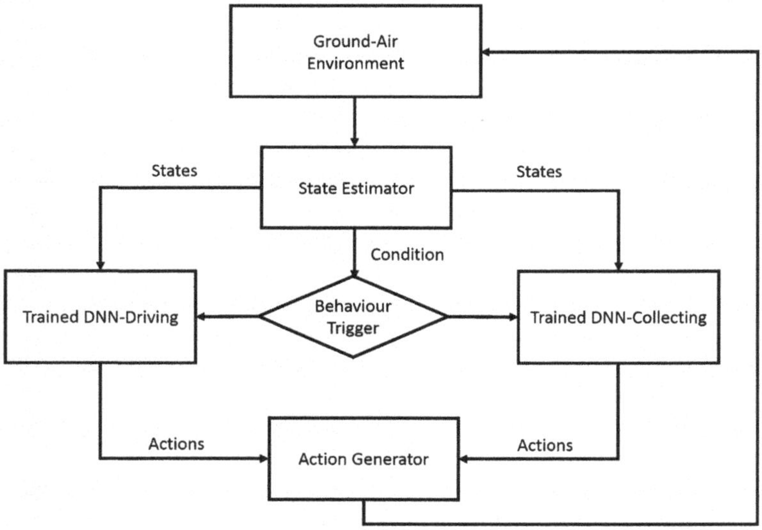

Fig. 10.7 The testing phases of the hierarchical apprenticeship bootstrapping (HABS)

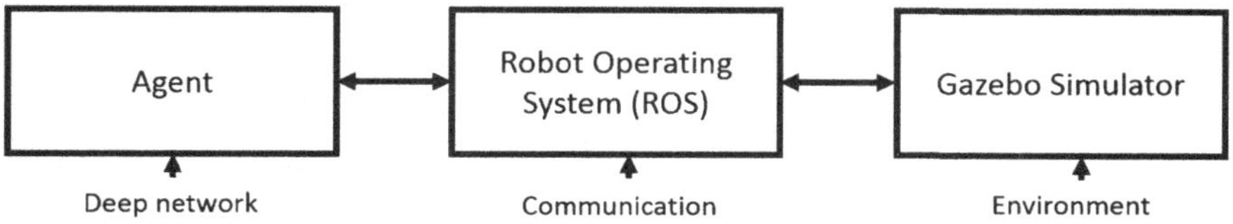

Fig. 10.8 The communication protocol of the simulation environment

10.5.1.1 Evaluation Metrics

To evaluate the effectiveness of an autonomous controller of the UAV, we introduce three metrics: the number of steps/actions to complete the task, the distance travelled, and success rate. Moreover, we visualise the footprints of the UAV and the UGVs to show the efficiency of our learning approach.

10.5.2 Experimental Design

In this section, we conduct the aerial shepherding task in simulation with one UAV herding four UGVs. The mission of the UAV is to herd the swarm of UGVs to a given target. A clear illustration of the UAV–UGVs shepherding mission is shown in Fig. 10.3.

The Gazebo simulator is used to design the environment. We use the Tum-Simulator [39] UAV package to simulate a Parrot AR UAV 2, and the Husky [20] package for the UGVs. For communications, ROS is used as an interface between the DNN agent and the Gazebo environment. Actions and states are sent and received through ROS messages. We use ROS Indigo installed on Ubuntu 14.04. The communication protocol is shown in Fig. 10.8.

10.5.2.1 Demonstration Interface

The human operator is supplied a *global-view* screen to control the UAV towards the sub-goal, which is calculated based on the methodology of Strömbom et al. [90]; that is, based on the locations of UGVs, we calculate the driving or collecting point and provide this to the operator, who must then determine how to get to these locations. There are two demonstration scenarios for the operator: collecting and driving sub-tasks. The operator is required to minimise the number of steps to reach the sub-goal or minimise the distance between the UAV and the sub-goal (Fig. 10.9).

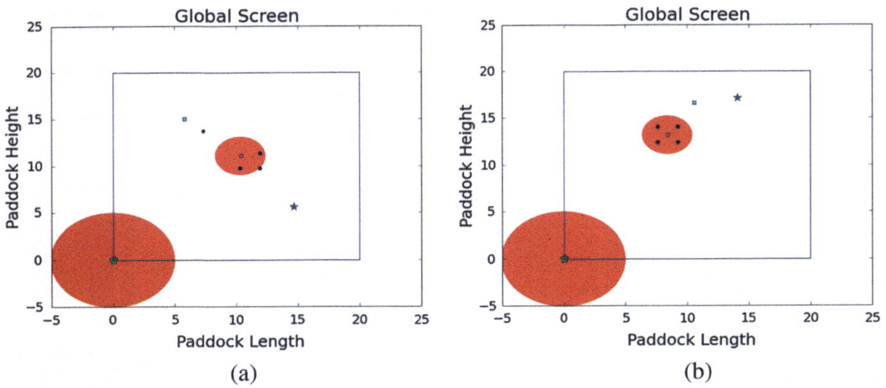

Fig. 10.9 Global screen for human operator in task demonstrations. Blue, green, and black dots represent the positions of the UAV, sub-goal, and UGVs, respectively. (**a**) Collecting. (**b**) Driving

Table 10.1 The state spaces

State ID	State name	State description
1–2	$GCMtoD(x, y)$	Vector from GCM to shepherd
3–4	$SubtoD(x, y)$	Vector from the sub-goal to shepherd

10.5.2.2 Actions and States Space

The UGVs' action space is composed of two real-valued actions representing the linear velocity, $V(x, y)$, in which this velocity is set to 0.5 (m/s), and the angular velocity (yaw rate), ω. Similarly, the UAV's action space consists of two continuous values representing the linear velocity in the longitudinal and lateral directions, and these are denoted using the values (p, r), respectively. Each episode in collecting the human data or testing, the UAV ascends to a height $2m$ with vertical velocity set to $1(m/s)$ and maintains its height till the end of the episode.

The state spaces presented in Table 10.1 are used as inputs, while the outputs, as discussed above, are four discrete actions of the UAV.

The Tensorflow and Keras libraries [19] were used to create a deep network architecture comprising an input layer, two fully connected hidden layers, and a fully connected output layer. The number of nodes of the two hidden layers are 32 and 64, respectively. The optimiser method is Adam [49]. The DNN is trained in 50,000 epochs, and the mini-batch size is set to equal to the size of the data-set. The deep network was trained on an NVIDIA GeForce GTX 1080 GPU.

10.5.2.3 Experimental Setups

In the experiment, we collect the human data from two scenarios: human-combined and human-primitive. For the first, the human operator is required to perform the

Table 10.2 Experimental setups

ID	Name	Meaning
1	DNN-Combined	Training on human-combined
2	DNN-Composite	Training on human-composite by our AABS
3	DNN-Primitive	Training on two human-primitive data-sets by our HABS

Table 10.3 Parameters in setups

Name	Value	Description
N	4 (UGVs)	Number of UGVs or sheep
$W_{\pi\pi}$	2 (m)	Weight repel from other sheep
$R_{\pi\pi}$	1 (m)	Sheep radius
$R_{\pi\beta}$	5 (m)	Shepherd radius

entire aerial shepherding task; that is, the UAV needs to herd the swarm of the UGVs towards the target successfully. For the second, we decompose into sub-tasks: the operator is required to perform either collecting or driving based on the dispersion of the UGV swarm. The mission of the human is to control the UAV towards the sub-goal. This provides three demonstration data-sets: human-combined, two human-primitive data-sets of the two sub-tasks, and human-composite data-set such that the two human-primitive data-sets are aggregated.

Therefore, to investigate our ABS approach for the task, we conduct three experimental setups: DNN-Combined, DNN-Composite, and DNN-Primitive. First, for DNN-Combined, a DNN is used to directly learn from the human-combined data. In the second method, our AABS approach is used to train a DNN on the human-composite data-set. Third, our HABS approach is used to train the two human-primitive data-sets. The three configurations are described in Table 10.2.

In terms of parameters' setup, we adopt the parameters in the shepherding model of Strömbom et al. [90] except for the size, the sheep step, and the shepherd radius. The step time of the UAV and the UGVs is set to the simulate time of the Gazebo simulator framework. In our Gazebo simulation, the size of the square field is 20 × 20 (m × m). Table 10.3 shows the parameters used in all the setups.

10.5.3 Results and Discussion

We collect 9567 instances for *human-combined* in 30 episodes, and 10,288 instances for *human-primitive* in 100 episodes for collecting and driving sub-tasks.

10.5.3.1 Training

We fully train four DNNs for the human-combined, human-composite, and two human-primitive data-sets till convergence. Figure 10.10 shows the value of the

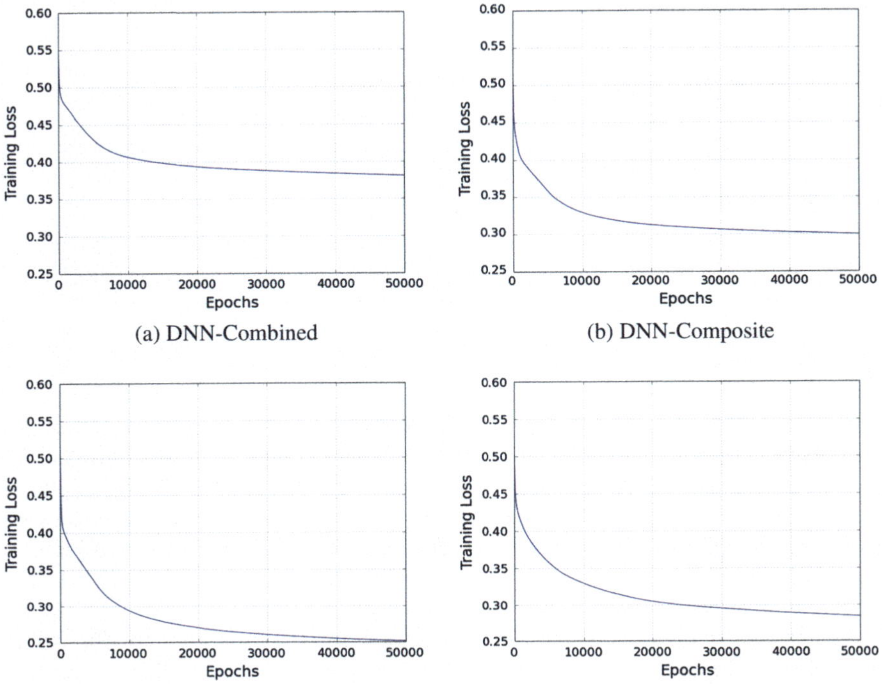

(a) DNN-Combined
(b) DNN-Composite
(c) DNN-Primitive in Driving Sub-Task
(d) DNN-Primitive in Collecting Sub-Task

Fig. 10.10 Training loss of four deep neural networks on human-combined, human-composite, and two human-primitive data-sets on the two collecting and driving sub-tasks. (**a**) DNN-Combined. (**b**) DNN-Composite. (**c**) DNN-Primitive in driving sub-task. (**d**) DNN-Primitive in collecting sub-task

Mean Squared Error (MSE) in the four experimental scenarios. In all the scenarios, the errors cease to decline at approximately 10,000 epochs. This tendency shows the success of the training process on both data-sets.

10.5.3.2 Testing

After the DNN-Combined, DNN-Composite, and DNN-Primitive models are fully trained, we conduct 20 different testing cases on these trained models as well as on the Strömbom model [90]. For testing the Strömbom model, we decrease the velocity of the UAV by multiplying with 0.5 (m/s) in order to reduce the fluctuation of the UAV that might cause significant changes of the direction of the UAV and leads to the unexpected actions. With each testing case, the positions of the four UGVs are differently initialised in the environment. The position $(UGV1_x, UGV1_y)$ of the UGV1 is set by Eq. 10.19.

Table 10.4 Success rate, averages, and standard deviations of number of steps and distance travelled of success testing cases of 20 testing cases in setups

	Success rate	Number of steps	Distance travelled (m)
Experiment ID	(%)	MSE $\mu \pm \sigma$	MSE $\mu \pm \sigma$
DNN-Combined	85	1745 ± 140	**45** ± 6.2
DNN-Composite	70	1723 ± 257	50 ± 17
DNN-Primitive	95	**1596** ± 225	60.7 ± 13.1
Strömbom	**100**	1730 ± 278	101.4 ± 17

Best values when comparing among four trained DNN agents.

$$\begin{cases} UGV1_x = 10 + \delta_t \\ UGV1_y = 20 - UGV1_x \\ \delta_t = t \times 0.2 \end{cases} \tag{10.19}$$

with t, which indicates the testing case tth, $\in [1, 20]$. Meanwhile, the position of the UGV2 is considered as the furthest UGV being out of the mass, and $UGV2_x$ and $UGV2_y$ are set by $UGV1_x - 3$ and $UGV1_y + 4$, respectively. For the remaining UGV3 and UGV4, their positions are set close to the UGV1 in order to create the mass of the UGVs. In the chapter, the position $(UGV3_x, UGV3_y)$ of the UGV3 is set by $UGV1_x + 1.6$ and $UGV1_y$, respectively, and the UGV4 position is set by $UGV1_x + 1.6$ and $UGV1_y + 1.6$. We calculate average and standard deviations of the number of steps, distance travelled, and success rate in the 20 testing cases shown in Table 10.4 in order to investigate the effectiveness of our ABS approach.

We can see from the results of HABS that DNN-Primitive is able to outperform DNN-Combined in both the success rate and the duration taken (number of steps). Additionally, compared to the Strömbom model, the HABS surpasses in both the number of steps and the travelled distance and is close in the success rate. These results show that our HABS approach is able to overcome the problem of learning a large space in terms of the success rate and produce the more effective behaviour than that of the Strömbom model, according to the number of steps and the travelled distance.

Besides these promises, these results show that the travelled distance of the DNN-Primitive UAV is longer than that of the DNN-Combined UAV. The DNN-Primitive UAV needs to move to sub-goals precisely by switching between the collecting and driving DNN-Primitive models that can also be seen in the Strömbom model. Meanwhile, the DNN-Combined model produces the behaviour of the UAV being close to the human demonstrations. Thus, it does not move to sub-goals precisely and then reduces the UAV travelled distance significantly. This difference can be seen in Fig. 10.11. This sub-optimal movement of the proposed HABS approach leads to an open question of finding an alternative design for producing the behaviour of the UAV to replace switching between the trained DNN-Primitive models.

For the second approach (AABS), although DNN-Composite has the lowest success rate, and the large number of steps approximating that of DNN-Combined

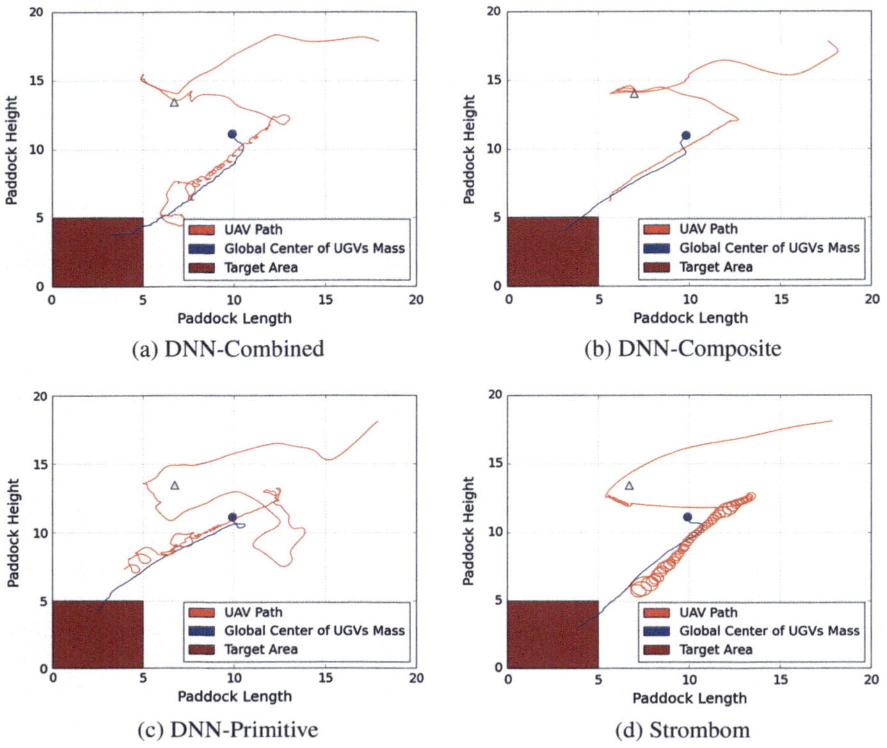

Fig. 10.11 Trajectories of the UAV and the four UGVs in the 15th testing case. (**a**) DNN-Combined. (**b**) DNN-Composite. (**c**) DNN-Primitive. (**d**) Strömbom

and the Strömbom model, its travelled distance is reduced considerably when comparing its results with the first approach. Figure 10.11 shows that the behaviour of DNN-Composite is closest with DNN-Combined. The clue provides promises of the AABS approach. The low success rate of AABS might be understood that the training composite data-set is not sufficient, or the straight-forward fusion method of producing the data-set is not effective. Both of them are open questions for our future work.

10.6 Conclusions and Open Issues

In this work, we introduce various learning approaches including reinforcement learning (RL) and apprenticeship learning (AL). While the RL algorithms have still been challenging in developing autonomous systems, AL seems to be more practical and promising.

In practice, when applying these AL approaches for building autonomous agents in complex tasks or composite tasks, which are composed of a combination of individual skills, a human expert may not be available due to novelty of the task or the cost of access to someone with appropriate skills. We introduced *apprenticeship bootstrapping* (ABS), which focuses on addressing how to collect human data for complex tasks, where it is expensive (or impractical due to task novelty) to access a human expert.

Taking advantage of the shepherding problem[90], the approach of air–ground vehicles coordination seems to be more practical when limiting the complexity and dynamic of the interaction among these vehicles. In the context of the sky/aerial shepherding of ground swarm, the UGVs acting like sheep are rule-based agents. The behaviours of the UGVs are homogeneous, i.e., every UGV has the same behaviour and will identically react in exact situations, so the level of the uncertainty due to response variations of UGVs has been significantly reduced. When decreasing these effects considerably, learning-based approaches for the sky shepherd might show promise to produce effective autonomous aerial agents.

We conducted two ABS approaches: hierarchical apprenticeship bootstrapping (HABS) and aggregation apprenticeship bootstrapping (AABS). First, we trained the collecting and driving data-sets individually, called DNN-Primitive. In the second method, the two data-sets were aggregated to produce a composite data-set, which was used for training a deep neural network (DNN), namely DNN-Composite. In the actions and states aggregation component, wherein the composite set was created, we initially conducted a straight-forward fusion of these demonstrations. We also collected the human data-set while the human performs the whole task and used these data-sets to train a deep neural network (DNN); this model is called DNN-Combined.

During testing, we compared our ABS approaches with the Strömbom model[90] and DNN-Combined through three evaluation metrics: success rate, the number of steps, and the distance travelled. The results showed that HABS was able to overcome the problem of DNN-Combined in learning a large space in terms of the success rate and produces the more effective behaviour than that of the Strömbom model, according to the number of steps and the travelled distance. Additionally, for the AABS approach, the results and visualisations showed promises when the travelled distance was reduced considerably when compared with the first approach, and the behaviour was closest to DNN-Combined cloned from the human acting on the whole task.

In conclusion, this chapter demonstrated that the ABS approach is feasible and effective to produce an autonomous UAV for sky shepherding. Some open questions, which need further investigations, include:

1. How to aggregate the state spaces and the action spaces of primitive sub-tasks into a composite state–action space?
2. How to combine the primitive skills produced from trained individual models?
3. How to apply ABS for the context of multi-shepherds acting as the centralised or distributed multi-agent systems?

Our future work will focus on answering these questions in the context of the aerial shepherding of a ground swarm as well as investigating the behaviour of UAV shepherds and UGVs by adding the UAV altitude as a control variable.

References

1. Abbeel, P., Ng, A.Y.: Apprenticeship learning via inverse reinforcement learning. In: Proceedings of the Twenty-First International Conference on Machine Learning, p. 1. ACM, New York (2004)
2. Abbeel, P., Coates, A., Ng, A.Y.: Autonomous helicopter aerobatics through apprenticeship learning. Int. J. Rob. Res. **29**(13), 1608–1639 (2010)
3. Aghaeeyan, A., Abdollahi, F., Talebi, H.A.: UAV–UGVs cooperation: with a moving center based trajectory. Rob. Auton. Syst. **63**, 1–9 (2015)
4. Argall, B.D., Chernova, S., Veloso, M., Browning, B.: A survey of robot learning from demonstration. Rob. Auton. Syst. **57**(5), 469–483 (2009)
5. Arora, S., Doshi, P.: A survey of inverse reinforcement learning: Challenges, methods and progress (2018). Preprint arXiv:1806.06877
6. Balch, T., Arkin, R.C.: Behavior-based formation control for multirobot teams. IEEE Trans. Robot. Autom. **14**(6), 926–939 (1998)
7. Baumann, M., Büning, H.K.: Learning shepherding behavior. Ph.D. Thesis, University of Paderborn (2016)
8. Baxter, J.L., Burke, E., Garibaldi, J.M., Norman, M.: Multi-robot search and rescue: A potential field based approach. In: Autonomous Robots and Agents, pp. 9–16. Springer, Berlin (2007)
9. Beard, R.W., Lawton, J., Hadaegh, F.Y.: A coordination architecture for spacecraft formation control. IEEE Trans. Control Syst. Technol. **9**(6), 777–790 (2001)
10. Bentivegna, D.C., Atkeson, C.G., Cheng, G.: Learning tasks from observation and practice. Rob. Auton. Syst. **47**(2–3), 163–169 (2004)
11. Billard, A.G., Calinon, S., Dillmann, R.: Learning from humans. In: Springer Handbook of Robotics, pp. 1995–2014. Springer, Berlin (2016)
12. Billing, E.A., Hellström, T.: A formalism for learning from demonstration. Paladyn **1**(1), 1–13 (2010)
13. Bojarski, M., Del Testa, D., Dworakowski, D., Firner, B., Flepp, B., Goyal, P., Jackel, L.D., Monfort, M., Muller, U., Zhang, J., et al.: End to end learning for self-driving cars (2016). Preprint arXiv:1604.07316
14. Bojarski, M., Yeres, P., Choromanska, A., Choromanski, K., Firner, B., Jackel, L., Muller, U.: Explaining how a deep neural network trained with end-to-end learning steers a car (2017). Preprint arXiv:1704.07911
15. Carelli, R., De la Cruz, C., Roberti, F.: Centralized formation control of non-holonomic mobile robots. Lat. Am. Appl. Res. **36**(2), 63–69 (2006)
16. Carrio, A., Sampedro, C., Rodriguez-Ramos, A., Campoy, P.: A review of deep learning methods and applications for unmanned aerial vehicles. J. Sensors **2017**, 3296874 (2017)
17. Chaimowicz, L., Kumar, V.: Aerial shepherds: Coordination among UAVS and swarms of robots. In: Alami, R., Chatila, R., Asama, H. (eds.) Distributed Autonomous Robotic Systems, vol. 6, pp. 243–252. Springer Japan, Tokyo (2007)
18. Chen, J., Zhang, X., Xin, B., Fang, H.: Coordination between unmanned aerial and ground vehicles: A taxonomy and optimization perspective. IEEE Trans. Cybern. **46**(4), 959–972 (2016)
19. Chollet, F.: Keras: Theano-based deep learning library. Code: https://github.com/fchollet. Documentation: http://keras. IO (2015)

20. ClearpathRobotics: ROS husky robot. ROS package at http://wiki.ros.org/Robots/ Husky (2017)
21. Daniel, C., Neumann, G., Peters, J.: Learning concurrent motor skills in versatile solution spaces. In: 2012 IEEE/RSJ International Conference on Intelligent Robots and Systems, pp. 3591–3597. IEEE, Piscataway (2012)
22. Daw, N.D., Niv, Y., Dayan, P.: Uncertainty-based competition between prefrontal and dorsolateral striatal systems for behavioral control. Nat. Neurosci. **8**(12), 1704–1711 (2005)
23. Dillmann, R.: Teaching and learning of robot tasks via observation of human performance. Rob. Auton. Syst. **47**(2–3), 109–116 (2004)
24. Duan, H., Li, P.: Bio-Inspired Computation in Unmanned Aerial Vehicles. Springer, Berlin (2014)
25. Dunk, I., Abbass, H.: Emergence of order in leader-follower boids-inspired systems. In: 2016 IEEE Symposium Series on Computational Intelligence (SSCI), pp. 1–8. IEEE, Piscataway (2016)
26. Farinelli, A., Iocchi, L., Nardi, D.: Multirobot systems: a classification focused on coordination. IEEE Trans. Syst. Man Cybern. Part B (Cybern.) **34**(5), 2015–2028 (2004)
27. Fernandez-Rojas, R., Perry, A., Singh, H., Campbell, B., Elsayed, S., Hunjet, R., Abbass, H.A.: Contextual awareness in human-advanced-vehicle systems: A survey. IEEE Access **7**, 33304–33328 (2019)
28. Fraser, B., Hunjet, R.: Data ferrying in tactical networks using swarm intelligence and stigmergic coordination. In: 2016 26th International Telecommunication Networks and Applications Conference (ITNAC), pp. 1–6. IEEE, Piscataway (2016)
29. Gee, A., Abbass, H.: Transparent machine education of neural networks for swarm shepherding using curriculum design. In: Proceedings of the International Joint Conference on Neural Networks (2019)
30. Glavic, M., Fonteneau, R., Ernst, D.: Reinforcement learning for electric power system decision and control: Past considerations and perspectives. IFAC-PapersOnLine **50**(1), 6918–6927 (2017)
31. Graves, A., Mohamed, A.r., Hinton, G.: Speech recognition with deep recurrent neural networks. In: 2013 IEEE International Conference on Acoustics, Speech and Signal Processing (ICASSP), pp. 6645–6649. IEEE, Piscataway (2013)
32. Grollman, D.H., Jenkins, O.C.: Incremental learning of subtasks from unsegmented demonstration. In: 2010 IEEE/RSJ International Conference on Intelligent Robots and Systems, pp. 261–266. IEEE, Piscataway (2010)
33. Grounds, M., Kudenko, D.: Parallel reinforcement learning with linear function approximation. In: Adaptive Agents and Multi-Agent Systems III. Adaptation and Multi-Agent Learning, pp. 60–74. Springer, Berlin (2008)
34. Guenter, F., Hersch, M., Calinon, S., Billard, A.: Reinforcement learning for imitating constrained reaching movements. Adv. Rob. **21**(13), 1521–1544 (2007)
35. Guillet, A., Lenain, R., Thuilot, B., Rousseau, V.: Formation control of agricultural mobile robots: A bidirectional weighted constraints approach. J. Field Rob. **34**, 1260–1274 (2017)
36. Guo, X., Denman, S., Fookes, C., Mejias, L., Sridharan, S.: Automatic UAV forced landing site detection using machine learning. In: 2014 International Conference on Digital Image Computing: Techniques and Applications (DICTA), pp. 1–7. IEEE, Piscataway (2014)
37. Gupta, J.K., Egorov, M., Kochenderfer, M.: Cooperative multi-agent control using deep reinforcement learning. In: International Conference on Autonomous Agents and Multiagent Systems, pp. 66–83. Springer, Berlin (2017)
38. Howitt, S., Richards, D.: The human machine interface for airborne control of UAVS. In: 2nd AIAA "Unmanned Unlimited" Conference and Workshop & Exhibit, p. 6593 (2003)
39. Huang, H., Sturm, J.: Tum simulator. ROS package at http://wiki.ros.org/tum_simulator (2014)
40. Hudjakov, R., Tamre, M.: Aerial imagery terrain classification for long-range autonomous navigation. In: 2009 International Symposium on Optomechatronic Technologies, pp. 88–91. IEEE, Piscataway (2009)

41. Hunjet, R., Stevens, T., Elliot, M., Fraser, B., George, P.: Survivable communications and autonomous delivery service a generic swarming framework enabling communications in contested environments. In: MILCOM 2017–2017 IEEE Military Communications Conference (MILCOM), pp. 788–793. IEEE, Piscataway (2017)
42. Hussein, A., Gaber, M.M., Elyan, E., Jayne, C.: Imitation learning: A survey of learning methods. ACM Comput. Surv. (CSUR) **50**(2), 21 (2017)
43. Hwang, Y.K., Choi, K.J., Hong, D.S.: Self-learning control of cooperative motion for a humanoid robot. In: Proceedings 2006 IEEE International Conference on Robotics and Automation, 2006. ICRA 2006, pp. 475–480. IEEE, Piscataway (2006)
44. Iima, H., Kuroe, Y.: Swarm reinforcement learning method for a multi-robot formation problem. In: 2013 IEEE International Conference on Systems, Man, and Cybernetics (SMC), pp. 2298–2303. IEEE, Piscataway (2013)
45. Jansen, B., Belpaeme, T.: A computational model of intention reading in imitation. Rob. Auton. Syst. **54**(5), 394–402 (2006)
46. Justesen, N., Risi, S.: Learning macromanagement in starcraft from replays using deep learning. In: 2017 IEEE Conference on Computational Intelligence and Games (CIG), pp. 162–169. IEEE, Piscataway (2017)
47. Khaleghi, A.M., Xu, D., Minaeian, S., Li, M., Yuan, Y., Liu, J., Son, Y.J., Vo, C., Lien, J.M.: A dddams-based UAV and UGV team formation approach for surveillance and crowd control. In: Proceedings of the 2014 Winter Simulation Conference, pp. 2907–2918. IEEE Press, Piscataway (2014)
48. Khaleghi, A.M., Xu, D., Minaeian, S., Li, M., Yuan, Y., Liu, J., Son, Y.J., Vo, C., Mousavian, A., Lien, J.M.: A comparative study of control architectures in UAV/UGV-based surveillance system. In: IIE Annual Conference. Proceedings. Institute of Industrial and Systems Engineers (IISE), p. 3455 (2014)
49. Kingma, D.P., Ba, J.: Adam: A method for stochastic optimization (2014). Preprint arXiv:1412.6980
50. Kober, J., Peters, J.R.: Policy search for motor primitives in robotics. In: Advances in Neural Information Processing Systems, pp. 849–856 (2009)
51. Koenig, N., Howard, A.: Gazebo-3d multiple robot simulator with dynamics (2006)
52. Kolling, A., Walker, P., Chakraborty, N., Sycara, K., Lewis, M.: Human interaction with robot swarms: A survey. IEEE Trans. Human-Mach. Syst. **46**(1), 9–26 (2015)
53. Konidaris, G., Osentoski, S., Thomas, P.S.: Value function approximation in reinforcement learning using the fourier basis. In: Association for the Advancement of Artificial Intelligence, vol. 6, p. 7 (2011)
54. Kormushev, P., Calinon, S., Caldwell, D.G.: Robot motor skill coordination with em-based reinforcement learning. In: 2010 IEEE/RSJ International Conference on Intelligent Robots and Systems, pp. 3232–3237. IEEE, Piscataway (2010)
55. Kormushev, P., Calinon, S., Saegusa, R., Metta, G.: Learning the skill of archery by a humanoid robot ICub. In: 2010 10th IEEE-RAS International Conference on Humanoid Robots, pp. 417–423. IEEE, Piscataway (2010)
56. Krizhevsky, A., Sutskever, I., Hinton, G.E.: Imagenet classification with deep convolutional neural networks. In: Advances in Neural Information Processing Systems, pp. 1097–1105 (2012)
57. Kulić, D., Ott, C., Lee, D., Ishikawa, J., Nakamura, Y.: Incremental learning of full body motion primitives and their sequencing through human motion observation. Int. J. Rob. Res. **31**(3), 330–345 (2012)
58. LeCun, Y., Bengio, Y., Hinton, G.: Deep learning. Nature **521**(7553), 436 (2015)
59. Levine, S., Finn, C., Darrell, T., Abbeel, P.: End-to-end training of deep visuomotor policies. J. Mach. Learn. Res. **17**(1), 1334–1373 (2016)
60. Li, J., Monroe, W., Ritter, A., Galley, M., Gao, J., Jurafsky, D.: Deep reinforcement learning for dialogue generation (2016). Preprint arXiv:1606.01541
61. Li, X., Chen, Y.N., Li, L., Gao, J., Celikyilmaz, A.: End-to-end task-completion neural dialogue systems (2017). Preprint arXiv:1703.01008

62. Lin, S., Garratt, M.A., Lambert, A.J.: Monocular vision-based real-time target recognition and tracking for autonomously landing an uav in a cluttered shipboard environment. Autonom. Rob. **41**(4), 881–901 (2017)
63. Liu, M., Amato, C., Anesta, E.P., Griffith, J.D., How, J.P.: Learning for decentralized control of multiagent systems in large, partially-observable stochastic environments. In: Thirtieth AAAI Conference on Artificial Intelligence (2016)
64. Long, N.K., Sammut, K., Sgarioto, D., Garratt, M., Abbass, H.A.: A comprehensive review of shepherding as a bio-inspired swarm-robotics guidance approach. IEEE Trans. Emer. Topics Comput. Intell. **4**, 523–537 (2020)
65. Mangin, O., Oudeyer, P.Y.: Unsupervised learning of simultaneous motor primitives through imitation. In: Frontiers in Computational Neuroscience Conference Abstract: IEEE ICDL-EPIROB 2011 (2011)
66. Martinez, S., Cortes, J., Bullo, F.: Motion coordination with distributed information. IEEE Control Syst. Mag. **27**(4), 75–88 (2007)
67. Mirowski, P., Pascanu, R., Viola, F., Soyer, H., Ballard, A.J., Banino, A., Denil, M., Goroshin, R., Sifre, L., Kavukcuoglu, K., et al.: Learning to navigate in complex environments (2016). Preprint arXiv:1611.03673
68. Mnih, V., Kavukcuoglu, K., Silver, D., Rusu, A.A., Veness, J., Bellemare, M.G., Graves, A., Riedmiller, M., Fidjeland, A.K., Ostrovski, G., et al.: Human-level control through deep reinforcement learning. Nature **518**(7540), 529 (2015)
69. Mnih, V., Badia, A.P., Mirza, M., Graves, A., Lillicrap, T., Harley, T., Silver, D., Kavukcuoglu, K.: Asynchronous methods for deep reinforcement learning. In: International Conference on Machine Learning, pp. 1928–1937 (2016)
70. Mülling, K., Kober, J., Kroemer, O., Peters, J.: Learning to select and generalize striking movements in robot table tennis. Int. J. Rob. Res. **32**(3), 263–279 (2013)
71. Nguyen, H.T., Garratt, M., Bui, L.T., Abbass, H.: Supervised deep actor network for imitation learning in a ground-air UAV-UGVs coordination task. In: 2017 IEEE Symposium Series on Computational Intelligence (SSCI), pp. 1–8. IEEE, Piscataway (2017)
72. Nguyen, H., Garratt, M., Abbass, H.: Apprenticeship bootstrapping. In: 2018 International Joint Conference on Neural Networks (IJCNN), pp. 1–8. IEEE, Piscataway (2018)
73. Nguyen, H., Tran, V., Nguyen, T., Garratt, M., Kasmarik, K., Barlow, M., Anavatti, S., Abbass, H.: Apprenticeship bootstrapping via deep learning with a safety net for UAV-UGV interaction (2018). Preprint arXiv:1810.04344
74. Nguyen, T., Nguyen, H., Debie, E., Kasmarik, K., Garratt, M., Abbass, H.: Swarm Q-Leaming with knowledge sharing within environments for formation control. In: 2018 International Joint Conference on Neural Networks (IJCNN), pp. 1–8. IEEE, Piscataway (2018)
75. Nguyen, H.T., Garratt, M., Bui, L.T., Abbass, H.: Apprenticeship learning for continuous state spaces and actions in a swarm-guidance shepherding task. In: 2019 IEEE Symposium Series on Computational Intelligence (SSCI), pp. 102–109. IEEE, Piscataway (2019)
76. Nguyen, H.T., Nguyen, T.D., Garratt, M., Kasmarik, K., Anavatti, S., Barlow, M., Abbass, H.A.: A deep hierarchical reinforcement learner for aerial shepherding of ground swarms. In: International Conference on Neural Information Processing, pp. 658–669. Springer, Berlin (2019)
77. Niekum, S., Osentoski, S., Konidaris, G., Barto, A.G.: Learning and generalization of complex tasks from unstructured demonstrations. In: 2012 IEEE/RSJ International Conference on Intelligent Robots and Systems, pp. 5239–5246. IEEE, Piscataway (2012)
78. Oh, K.K., Park, M.C., Ahn, H.S.: A survey of multi-agent formation control. Automatica **53**, 424–440 (2015)
79. Oh, H., Shirazi, A.R., Sun, C., Jin, Y.: Bio-inspired self-organising multi-robot pattern formation: a review. Rob. Auton. Syst. **91**, 83–100 (2017)
80. Palmer, G., Tuyls, K., Bloembergen, D., Savani, R.: Lenient multi-agent deep reinforcement learning. In: Proceedings of the 17th International Conference on Autonomous Agents and MultiAgent Systems, pp. 443–451. International Foundation for Autonomous Agents and Multiagent Systems (2018)

81. Parker, L.: Multiple Mobile Robot Systems, pp. 921–941. Springer, Berlin (2008). https://doi.org/10.1007/978-3-540-30301-5_41
82. Pastor, P., Kalakrishnan, M., Chitta, S., Theodorou, E., Schaal, S.: Skill learning and task outcome prediction for manipulation. In: 2011 IEEE International Conference on Robotics and Automation, pp. 3828–3834. IEEE, Piscataway (2011)
83. Pendleton, S.D., Andersen, H., Du, X., Shen, X., Meghjani, M., Eng, Y.H., Rus, D., Ang, M.H.: Perception, planning, control, and coordination for autonomous vehicles. Machines **5**(1), 6 (2017)
84. Ramachandran, D., Amir, E.: Bayesian inverse reinforcement learning. In: International Joint Conferences on Artificial Intelligence, vol. 7, pp. 2586–2591 (2007)
85. Ross, S., Melik-Barkhudarov, N., Shankar, K.S., Wendel, A., Dey, D., Bagnell, J.A., Hebert, M.: Learning monocular reactive uav control in cluttered natural environments. In: 2013 IEEE International Conference on Robotics and Automation, pp. 1765–1772. IEEE, Piscataway (2013)
86. Sen, A., Sahoo, S.R., Kothari, M.: Cooperative formation control strategy in heterogeneous network with bounded acceleration. In: 2017 Indian Control Conference (ICC), pp. 344–349. IEEE, Piscataway (2017)
87. Skoglund, A., Iliev, B., Kadmiry, B., Palm, R.: Programming by demonstration of pick-and-place tasks for industrial manipulators using task primitives. In: 2007 International Symposium on Computational Intelligence in Robotics and Automation, pp. 368–373. IEEE, Piscataway (2007)
88. Song, J., Ren, H., Sadigh, D., Ermon, S.: Multi-agent generative adversarial imitation learning. In: Advances in Neural Information Processing Systems, pp. 7461–7472 (2018)
89. Speck, C., Bucci, D.J.: Distributed UAV swarm formation control via object-focused, multi-objective SARSA. In: 2018 Annual American Control Conference (ACC), pp. 6596–6601. IEEE, Piscataway (2018)
90. Strömbom, D., Mann, R.P., Wilson, A.M., Hailes, S., Morton, A.J., Sumpter, D.J.T., King, A.J.: Solving the shepherding problem: heuristics for herding autonomous, interacting agents. J. R. Soc. Interf. **11**(100) (2014). https://browzine.com/articles/52614503
91. Sutton, R.S., Barto, A.G.: Reinforcement Learning: An Introduction, vol. 1. MIT Press, Cambridge (1998)
92. Szepesvári, C.: Algorithms for reinforcement learning. Synth. Lect. Artif. Intell. Mach. Learn. **4**(1), 1–103 (2010)
93. Trentini, M., Beckman, B.: Semi-autonomous UAV/UGV for dismounted urban operations. In: Unmanned Systems Technology XII, vol. 7692, p. 76921C. International Society for Optics and Photonics (2010)
94. Tsitsiklis, J.N., Van Roy, B.: Analysis of temporal-diffference learning with function approximation. In: Advances in Neural Information Processing Systems, pp. 1075–1081 (1997)
95. Vidal, R., Rashid, S., Sharp, C., Shakernia, O., Kim, J., Sastry, S.: Pursuit-evasion games with unmanned ground and aerial vehicles. In: Proceedings 2001 ICRA. IEEE International Conference on Robotics and Automation (Cat. No. 01CH37164), vol. 3, pp. 2948–2955. IEEE, Piscataway (2001)
96. Waslander, S.L.: Unmanned aerial and ground vehicle teams: Recent work and open problems. In: Autonomous Control Systems and Vehicles, pp. 21–36. Springer, Berlin (2013)
97. Wulfmeier, M., Ondruska, P., Posner, I.: Maximum entropy deep inverse reinforcement learning (2015). Preprint arXiv:1507.04888
98. Xu, D., Zhang, X., Zhu, Z., Chen, C., Yang, P.: Behavior-based formation control of swarm robots. Math. Problems Eng. **2014**, 205759 (2014)
99. Yang, Z., Merrick, K., Jin, L., Abbass, H.A.: Hierarchical deep reinforcement learning for continuous action control. IEEE Trans. Neur. Netw. Learn. Syst. (99), 1–11 (2018)
100. Yoshikai, T., Otake, N., Mizuuchi, I., Inaba, M., Inoue, H.: Development of an imitation behavior in humanoid kenta with reinforcement learning algorithm based on the attention during imitation. In: 2004 IEEE/RSJ International Conference on Intelligent Robots and Systems (IROS) (IEEE Cat. No. 04CH37566), vol. 2, pp. 1192–1197. IEEE, Piscataway (2004)

101. You, C., Lu, J., Filev, D., Tsiotras, P.: Advanced planning for autonomous vehicles using reinforcement learning and deep inverse reinforcement learning. Rob. Auton. Syst. **114**, 1–18 (2019)
102. Yu, H., Beard, R.W., Argyle, M., Chamberlain, C.: Probabilistic path planning for cooperative target tracking using aerial and ground vehicles. In: Proceedings of the 2011 American Control Conference, pp. 4673–4678. IEEE, Piscataway (2011)
103. Zhang, T., Kahn, G., Levine, S., Abbeel, P.: Learning deep control policies for autonomous aerial vehicles with mpc-guided policy search. In: 2016 IEEE International Conference on Robotics and Automation (ICRA), pp. 528–535. IEEE, Piscataway (2016)
104. Zhang, T., Li, Q., Zhang, C.s., Liang, H.w., Li, P., Wang, T.m., Li, S., Zhu, Y.l., Wu, C.: Current trends in the development of intelligent unmanned autonomous systems. Front. Inf. Technol. Electron. Eng. **18**(1), 68–85 (2017)
105. Zhan, E., Zheng, S., Yue, Y., Sha, L., Lucey, P.: Generative multi-agent behavioral cloning. In: Proceedings of the 35th International Conference on Machine Learning (2018)
106. Ziebart, B.D., Maas, A.L., Bagnell, J.A., Dey, A.K.: Maximum entropy inverse reinforcement learning. In: Proceedings of the Twenty-Third AAAI Conference on Artificial Intelligence, vol. 8, pp. 1433–1438. Chicago (2008)

Chapter 11
Logical Shepherd Assisting Air Traffic Controllers for Swarm UAV Traffic Control Systems

Heba El-Fiqi, Kathryn Kasmarik, and Hussein A. Abbass

In recent years, Unmanned Aerial Vehicles (UAVs) have attracted attentions from almost every industry. Their low cost, high accessibility, and low-risk compared to human-operated vehicles, created a unique opportunity for a variety of use cases in many application domains. The addition of these tele-operated, and sometimes autonomous, vehicles to the air traffic control environment imposed significant challenges and has been calling for appropriate UAV traffic control systems. The complexity of this situation increases manyfold when the UAVs need to work together as a swarm. Air traffic controllers are used to manage 20 or so aircraft separated according to strict guidelines. A highly dynamic, adaptive, fast, and large swarm of UAVs present unprecedented complexity. Shepherding offers an opportunity to provide a concept of a single sheepdog simultaneously guiding a large sheep flock. We present a logical shepherd that could act both in an autonomous mode or in a tele-operation mode by simply sitting in the hands of a swarm traffic controller. Due to the safety critical nature of the environment, we modified the concept of shepherding by designing an asynchronous shepherding algorithm coupled with a digital twin environment to assess consequences. Once the logical shepherd location and orientation is chosen by the human operator, the influence force vectors start to propagate asynchronously from one aircraft to another, maintaining separation assurance and safety constraints. The updated trajectory intent information of the UAVs gets displayed on the screen for the human operator to see if the change is acceptable or not. If acceptable, the recommendation is made and the UAVs commence to follow the new path.

H. El-Fiqi (✉) · K. Kasmarik · H. A. Abbass
School of Engineering and Information Technology, University of New South Wales, Canberra, ACT, Australia
e-mail: h.el-fiqi@adfa.edu.au; k.kasmarik@adfa.edu.au; h.abbass@unsw.edu.au

H. A. Abbass, R. A. Hunjet (eds.), *Shepherding UxVs for Human-Swarm Teaming*, Unmanned System Technologies,
https://doi.org/10.1007/978-3-030-60898-9_11

11.1 Introduction

Unmanned Aerial Vehicles (UAVs) have been used for many interesting tasks in recent years including package delivery, agriculture, mapping, surveillance, military operations, and search and rescue missions. With these growing trends in commercial and personal usages, the U.S. Federal Aviation Administration (FAA) developed new safety regulations including UAV registration and requiring UAV pilots to obtain a remote pilot certificate licence [7]. Commercial package delivery alone as per Amazon, United Parcel Service, Google, and Deutsche Post DHL [10, 14, 16] created a new industry that relies on UAV management that is going to continue to grow with other companies wanting to follow and to compete. The impact of releasing a large number of UAVs requiring concurrent control and management will exceed the capacity of a single UAV-owner company.

Moreover, these companies will not have a monopoly on the airspace, which needs to be shared among all companies as well as individual users. Relying on each UAV manufacturer for automatic separation assurance, collision avoidance and automated path planning will unlikely be enough to manage the flow smoothly and offer safety assurance and operational worthiness of the airspace. When Air-traffic control (ATC) is required to manage UAVs, its current approach will not work due to the significant differences between UAV operations and classic air traffic management, let alone if these UAVs are operating as a swarm.

Shepherding provides a communication resource-efficient concurrent control approach for a large number of agents. The shepherd is the only agent who needs to know the goal. In our discussion, the UAV Traffic Controller (UTC) will control a logical shepherd on the UTC's screen that is autonomously controlling and guiding the swarm.

In this chapter, we propose an asynchronous shepherding algorithm, which transforms the location of the logical shepherd into a sequence of asynchronous force vectors that propagate from one UAV to another, causing the UAV swarm to adjust its route. The asynchronous shepherding provides a UTC with a single point of control for a large UAV fleet or a swarm. The concept reduces the communication load between a UTC and all UAVs.

11.2 Background

Unmanned Aircraft System Traffic Management (UTM) systems emerged in the last few years. In 2015, the National Aeronautics and Space Administration (NASA) started a project on UTM to enable the safe integration of UAVs to the airspace. NASA aims at developing prototypes that enable sharing the low-altitude air space in a way similar to air traffic management. However, the challenge with UAVs is their large numbers. The solutions that NASA are developing focuses on advancing UAV autonomy by improving automation skills such as self-configuration, self-optimisation, and self-protection [11], and only allowing identified UAVs that match specific automation characteristics to operate. Furthermore, the human role will be limited to major and strategic decisions. As the operational assumption for NASA's UTM is "Air traffic controllers are not required to actively *control* every UAS in uncontrolled airspace or uncontrolled operations inside controlled airspace" [8]. Most research focused on the challenges associated with enabling UAVs with the required autonomous capabilities.

A number of trajectory modelling and path planning frameworks for UAVs have been proposed to address the specific dynamics of UAVs [4, 6, 13, 19]. Examples of topics that have been explored in the literature include: conflict detection and consequence identification [12] and collision avoidance for UAVs [9]. Bekkouche et al. [3] evaluated the UAV to UTM communication latency and associated reliability impact. They proposed using Multi-access Edge Computing (MEC) to overcome potential high latency, a concept well suited to the communication protocols applicable to the shepherding problem.

Various approaches have been proposed to describe the *level* of autonomy within UAVs; these are referred to as their Autonomous Control Level (ACL). The U.S. Air Force Research Laboratory (AFRL) identifies ten levels. Simple remote-controlled vehicles are classified as level zero, and fully autonomous vehicles are classified as level ten. Clough used three measures to evaluate the autonomy of a UAV: (1) Perception/Situational Awareness, (2) Analysis/Decision Making, (3) Communication/Cooperation [5]. The air traffic controller role in UTM is on the side of supervisory control. The UTM system allows the air traffic controller the rights of initiating, allowing, and/or terminating a UAV operation. However, as in air traffic management, the air traffic controller may need to make some strategic decisions to resolve risky situations that evolved within the monitored sector. The assumption that assured autonomy will always work imposes very high risk on the acceptance of this operation concept.

If fully automated UAVs are considered as human-operated aircraft from an autonomy perspective (i.e. they are able to make independent decisions), the need for the air traffic controller for continued risk assessment within the observed sector continues. This need persists whether or not the role of an air traffic controller has been partially or fully automated.

The need to act on large numbers of UAVs in the case of an emerging risk situation is not addressed in the current UTM literature. This is the gap the research in this chapter is addressing. However, while shepherding is a promising method for crowd control, it cannot be applied directly to UTM due to various reasons. Advantages and challenges of adapting shepherding for UTM are discussed below.

11.2.1 Advantages of Shepherding in Air Traffic Control

One of the main challenges of maximising utilisation of available airspace to accommodate an adequately large number of aircraft is the human resource, specifically:

the availability of enough well-trained air traffic controllers. With shepherding, a single air traffic controller can simultaneously control and update intended paths for a large number of UAVs. As the air traffic controller moves a logical shepherd in the logical airspace (the information displayed on the controller's screen), the shepherd exerts a "force" on the UAV "sheep". The system calculates how this force will update the intended path of this UAV. Then, the system automatically calculates how the changes for this UAV will propagate, through the propagation of the shepherd force, to other UAVs. This process continues until all targeted UAVs have been updated based on that single logical shepherd move. There is no need to use the old sequence of select a UAV, initiate an order, send, wait for a confirmation, and repeat. Thus, shepherding offers a mechanism to propagate influence on a dynamic graph that both impacts the graph structure and the flow on the graph.

As the ATC controls the UAVs via forces, these forces will create the desired effect on *all* intended UAV's paths. In this manner, the control message from the ATC need only be sent *once* rather than continually to multiple UAVs.

11.2.2 Challenges for Shepherding in Air Traffic Control

The constant application of force vectors causes continuous update of aircraft location. This behaviour may create high instability risk for the aircraft if it is required to update its location at each time step. It is therefore crucial that the shepherding algorithm for UTM takes this into account.

In a high-risk environment, there is a need to eliminate any factor that contributes to unnecessary complexity in the system. In air traffic management, some of these shepherding based forces become unnecessary as they are handled by the autopilot, such as standards on minimum separation distances and technologies for tactical conflict detection and resolution. Similarly, there is no need to include randomness in ATC. The genesis of this random perturbation in the original shepherding algorithm is to avoid deadlock situations [15]. In a safety critical domain, if randomness is used, it will always be within strict bounds that guarantee the continuity of safe environment and operations. However, if these bounds are established, randomness is no longer needed, and a deterministic algorithm could decide on safe decision making without a random element.

Using the classic shepherding algorithm, which assumes concurrent propagation of forces, the change in a single aircraft position could cause propagation of conflicts, or cascading failures, that are near impossible to overcome; leading to very high uncertainty in this safety critical domain.

The above challenges have called for designing a new algorithm for shepherding suitable for safety critical domains. This is the subject of our next section.

11.3 Asynchronous Shepherding

In this section, we propose *Asynchronous Shepherding* algorithm to address the challenges mentioned above.

In a classic shepherding algorithm, a sheep agent is always affected by five forces as presented in Fig. 11.1a: (1) repulsion from other sheep agents to avoid collisions, (2) repulsion from shepherd, (3) attraction to the fleet centre, (4) previous heading, and (5) mild randomness (jittering). A sheepdog agent that herds the fleet attempts to drive them to a pre-defined goal, known only by the sheepdog. If the fleet is not grouped, the sheepdog agent collects them first. The dynamics of these forces are used by the sheepdog to guide the sheep to the desired goal.

With *Asynchronous Shepherding*, a UAV's *intended path* is updated as a result of the shepherding command. Thus, there is no continuous update of UAVs' locations at each time step. Model complexity is also reduced with sequential shepherding. The previous heading force and attraction to the fleet centre are managed by the autopilot. Next waypoint information is updated via updating the intended path. The autopilot has access to the current location, heading, speed, and next waypoint location required heading and speed. To avoid the unexpected results caused by propagating the forces, it is important to have a high fidelity response model for the sheep/aircraft to anticipate the effect of force propagation on each aircraft. We use a high-fidelity simulation environment in fast mode to do this form of prediction in this paper. This allows the user to evaluate the resultant system-level state of their shepherding action before accepting and executing the recommendation.

For our purpose, the shepherding agent is controlled by the UTC. The total force that affects a UAV agent (π_i) is still composed of the five forces mentioned earlier. However, as shown in Fig. 11.1:

- The attraction force is not used due to the UAV following its flight plan, which is designed to maintain group cohesion.
- Repulsion between two UAVs is removed because each of our UAVs is equipped with an internal embedded Conflict Detection and Resolution (CD&R) system. This system is also part of the consequence analysis module that utilises a digital twin of the UAVs in the sky. In other words, we assume that there is at minimum a digital twin of the CD&R module on the UAV on the ground. However, a digital twin of each UAV would offer a more powerful ground capability for safety critical domains.
- The inclusion of the direction at the previous timestep in classic shepherding to ensure a smoother trajectory is not needed in ATC. The digital twin of the UAV autopilot on the ground does the job by taking the intent information from the shepherding algorithm and use the particular aircraft model to identify the immediate feasible sequence of velocity vectors to achieve the manoeuvre required to meet the intent.
- Jittering is also removed due to being problematic in a safety critical domain as we discussed in the previous section.

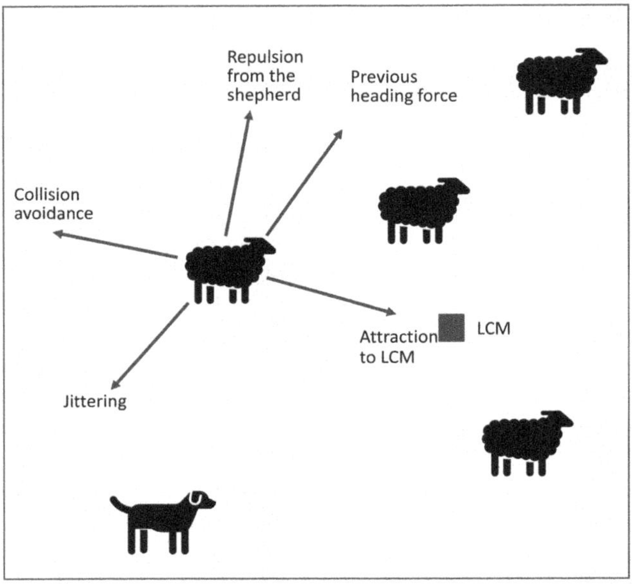

(a)

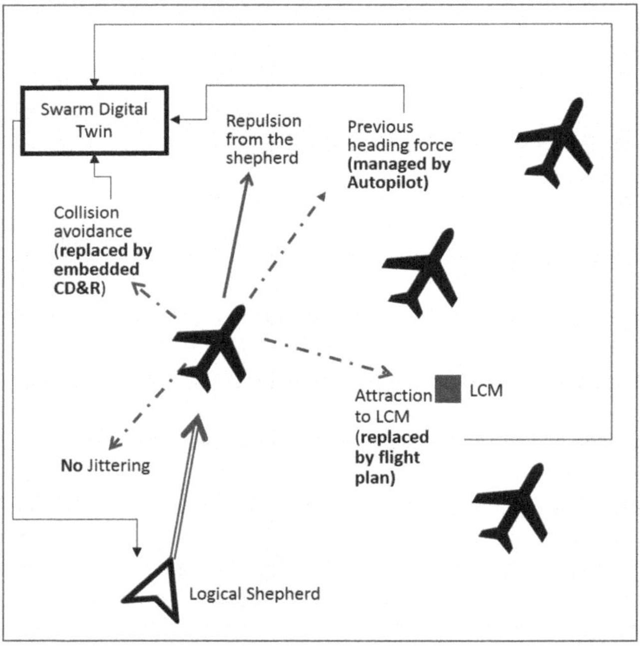

(b)

Fig. 11.1 The shepherding concept in agriculture and the corresponding concept in air traffic control. (**a**) Classic shepherding. (**b**) Air traffic controller shepherding

In classic shepherding, the agent decides on its immediate velocity vector. This reactive one-step approach is appropriate for some domains, but it is inadequate for a safety critical domain such as air traffic control. As such, we do not update the immediate position, π_i, of a UAV. Instead, we update intent information (the remaining portion of the flight plan). This update continuously executes one of three rules based on the state of a UAV agent.

1. Current Influencing Agent (CIA): Identify which agent to influence.
   (a) Sense: Detect if the agent π_i is selected as CIA.
   (b) Decide: If the agent is selected, find m where

$$d^t_{\pi_i \pi_m} = min(D^t_{\pi_i \pi_j}) \quad \forall \pi_j \in \Omega^t_{\pi_i \pi} \tag{11.1}$$

   where $(D^t_{\pi_i \pi_j})$ is the distance between an agent π_i andπ_j, and $\Omega^t_{\pi_i \pi}$ is the set of UAV agents that are within Agent π_i sensing range.
   (c) Act: Set π_m as the influenced agent.
2. Influenced Agent (IA): Update its own path based on the propagated force.
   (a) Sense: Detect if the agent π_i is selected as the IA.
   (b) Decide: If the agent is selected, then
      (i) Calculate the distance between this agent π_i and the CIA π_j. The CIA can be a shepherding or UAV agent.

$$d^t_{\pi_i \pi_j} = \| P^t_{\pi_i} - P^t_{\pi_j} \| \tag{11.2}$$

      where $P^t_{\pi_i}$ is the position of agent π_i at time t.
      (ii) Calculate the UAV $P^t_{\pi_i}$ to CIA $P^t_{\pi_j}$ Force Vector $F^t_{\pi_i \pi_j}$ as follows: If the CIA is a shepherding agent, apply classic shepherd to sheep repulsion force

$$F^t_{\pi_i \pi_j} = \frac{P^t_{\pi_i} - P^t_{\pi_j}}{\| P^t_{\pi_i} - P^t_{\pi_j} \|} \tag{11.3}$$

      else, propagate the carried force

$$F^t_{\pi_i \pi_j} = f(C_{\pi_j}, P^t_{\pi_i}, d^t_{\pi_i \pi_j}) \tag{11.4}$$

      where C_{π_j} is the force carried by the CIA to propagate.
      (iii) Calculate the first waypoint $P^t_{I^1_{\pi_i}}$ to UAV position $P^t_{\pi_i}$ Force Vector

$$F^t_{I^1_{\pi_i} \pi_i} = \frac{P^t_{I^1_{\pi_i}} - P^t_{\pi_i}}{\| P^t_{I^1_{\pi_i}} - P^t_{\pi_i} \|} \tag{11.5}$$

where $P^t_{I^1_{\pi_i}}$ is the position of the first waypoint in the intended path of agent π_i.

(iv) Calculate a rotation matrix R that can be used to rotate $F^t_{I^1_{\pi_i}\pi_i}$ to $F^t_{\pi_i\pi_j}$ around the agent's current position $P^t_{\pi_i}$ using

$$R^t_{\pi_i\pi_j} = T(P^t_{\pi_j}, P^t_{\pi_i}) \tag{11.6}$$

(c) Act:

(i) Update the intended path I_{π_i} using

$$P^{t+1}_{I^{\pi_i}} = \frac{d^t_{\pi_i\pi_j}}{a} \odot R^t_{\pi_i\pi_j} \times (P^t_{I_{\pi_i}} - P^t_{\pi_i}) + P^t_{\pi_i} \tag{11.7}$$

where parameter a is used to control the magnitude of the repulsion force.

(ii) Set the carried force that this agent will propagate when being selected as CIA to $C_{\pi_i} = F^t_{\pi_i\pi_j}$.

3. Waiting Agent (WA): Wait until being selected as the influenced agent.
   (a) Sense: Detect if the selected agent π is selected as the influencing agent or the influenced agent.
   (b) Decide: If not selected yet, then no further action is required and continue the current intended path.
   (c) Act: execute the decision from the "Decide" outcome.

The approach above allows us to address the challenges of shepherding discussed in the previous section. Specifically, risk instability is reduced by updating the intended path rather than continuously updating the aircraft location. Model complexity is addressed by reducing the number of forces considered in *Asynchronous Shepherding* and leveraging the digital twin technology we have adopted within our architecture. Figure 11.1b describes which forces can be eliminated to reduce the model complexity. Cascading effects are minimised by utilising (1) the sequential propagation order, (2) the consequence analysis using digital twinning, and (3) the visualisation of the propagation effect, which allows the ATC to make informed decision.

11.4 The Digital Twin

11.4.1 ATOMS

Air Traffic Operations and Management Simulator (ATOMS) [1] is a high fidelity model for air traffic control and national airspace. The system was originally proposed to model and analyse free-flight concepts. Multiple aspects of ATM are considered in the design including flight aerodynamics, airspace configurations, aviation emission, Cockpit Display of Traffic Information (CDTI), data communications, and weather, in order to evaluate advanced ATM concepts such as free flight, severe weather avoidance, and conflict detection and resolution. ATOMS has been used in different studies to evaluate air traffic concepts. Zhao et al. [20] used it to evaluate aircraft landing sequencing algorithms. Optimisation of dynamic airspace sectorisation was evaluated using ATOMS [17]. Alam et al. [2] used ATOMS to evaluate conflict resolution algorithms using evolutionary computation for risk assessment. Aviation emission was also analysed by Pham et al. [18] using ATOMS.

We built on top of an ATOMS the concept of digital twins, whereby we designed and implemented a digital twin with a new ATC interface that uses the *Asynchronous Shepherding Algorithm* to allow concurrent control of a number of aircraft.

ATOMS has two interfaces: The first is the simulation interface, shown in Fig. 11.2, which allows the control of the scenario, weather, and other simulation data. The second interface is the ATC interface.

11.4.2 UTC Interface

ATOMs' original ATC interface only allows a single control command to be sent to a single aircraft, as shown in Fig. 11.3. Our ATC interface for the digital twin design is shown in Fig. 11.4. There is a "Display" command that the user can use to show the shepherding agent location on the screen (Fig. 11.4b). Then, to move the shepherd, the user selects the "Move and Select" button. This enables selection of the shepherding agent's new location and the UAV to be targeted by the shepherd agent (Fig. 11.4d). An arrow visually displays force direction, allowing the user

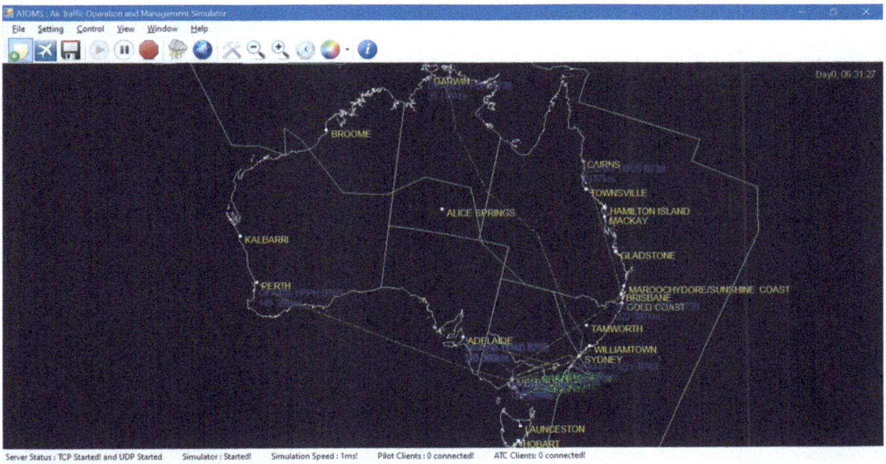

Fig. 11.2 ATOMS simulation interface

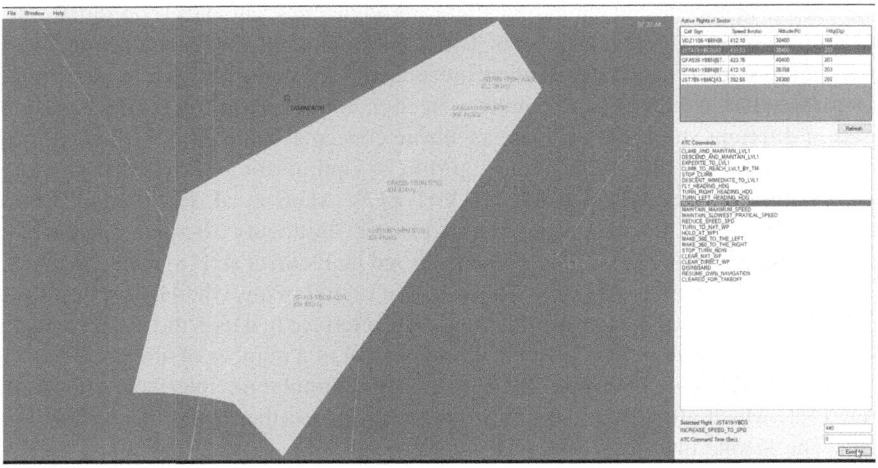

Fig. 11.3 Original ATC in ATOMS: single aircraft control

to explore its potential propagation effect on all UAVs via the "Predict" command (Fig. 11.4e). The proposed updated path is depicted by a yellow line. The user can accept the path by selecting the "Accept" button, or ignore it and wait. We used two timers, S1 and S2, described in Algorithm 5, to control the user actions to eliminate continuous exploration of the shepherding effect and to allow logging the timing of the events for analysis in our future human-autonomy teaming studies.

11.4.3 Applying the Asynchronous Shepherding Rules

In order to apply the *Asynchronous Shepherding* rules into an algorithm, there are a number of decisions that need be made and can be used to customise the algorithm.

1. The propagation of the shepherding forces between the influencing aircraft and subsequently selected aircraft: There are many ways of propagating the forces based on the desired effect. If we would like to maintain the original alignment, the initial force received from the shepherd to the first influenced aircraft can be passed as it is to the subsequent aircraft. However, if we would like the force to be reduced as it propagates, we can discount the strength of the force as it travels and propagate further in the system. The force propagation strategy can be customised to different settings and opens a whole space of possibilities for future work in this area.
2. The Meaning/Interpretation of the repulsion from the shepherding agent force within ATC: The decision to choose a rigid motion to reflect the desired repulsion effect needs to be made. This rigid motion can be rotating the intended path of

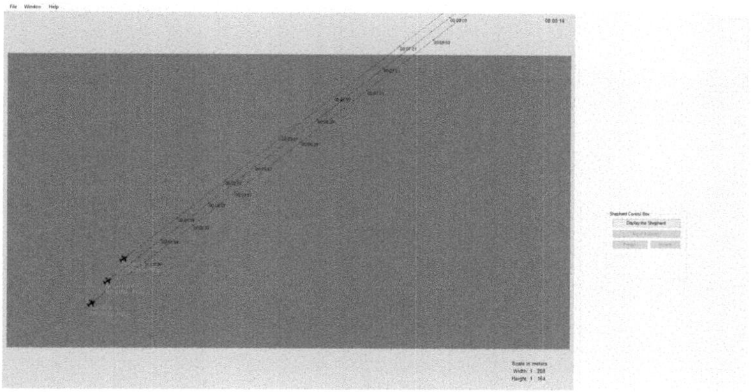

(a)

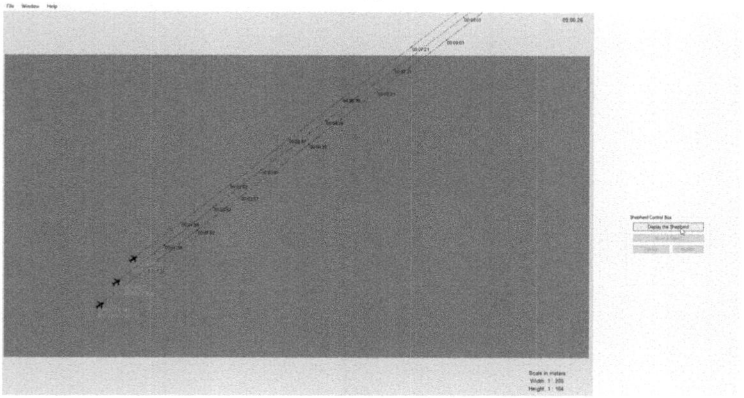

(b)

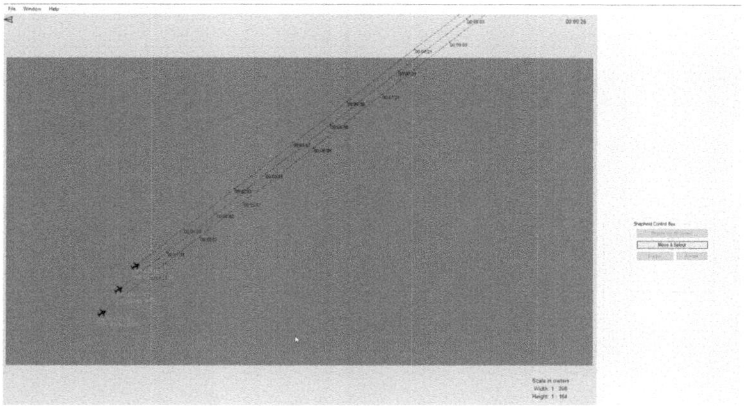

(c)

Fig. 11.4 New air traffic control interface for ATOMS using asynchronous shepherding algorithm. (**a**) Initial state. (**b**) The user clicks on display the shepherd. (**c**) The logical Shepherd is displayed at the top left corner. (**d**) The user uses move and select to move the shepherd and selects the first aircraft to influence. (**e**) The user selects predict to forecast the propagation of the forces. (**f**) The user accepts the proposed forces propagation. The aircraft updates their intended path

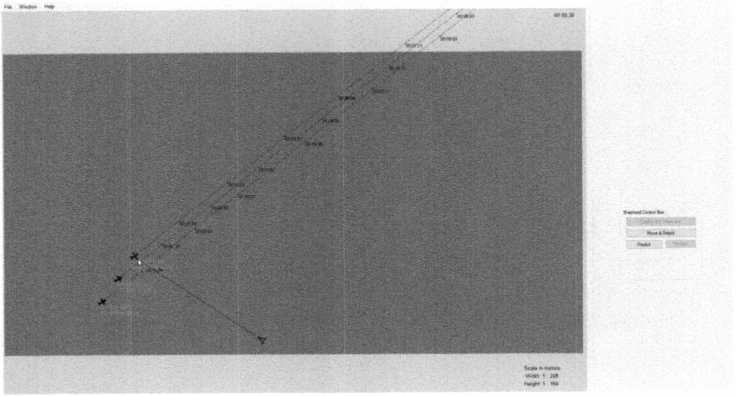

(d)

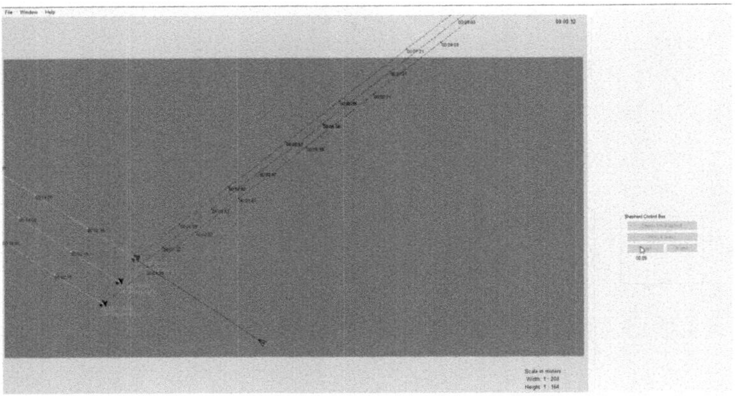

(e)

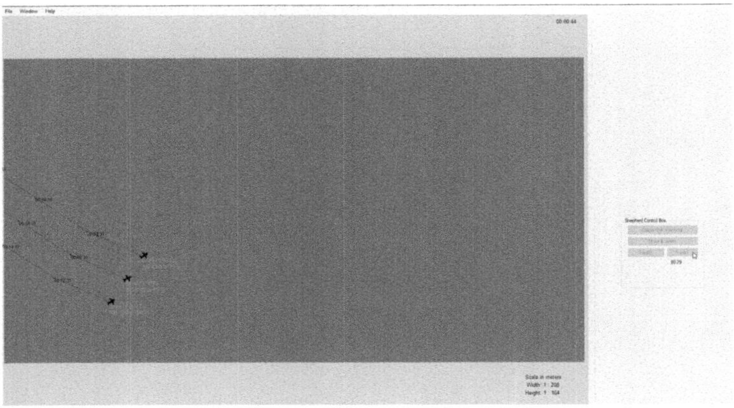

(f)

Fig. 11.4 (continued)

the aircraft versus translating (shifting) the intended path. However, using the shifting idea will result in extended path and duration compared to rotating the path. Furthermore, with rotating the path, the repulsion force can be interpreted as either pulling toward the shepherding force direction or escaping that force (a reflection of the force path).

3. Propagating the force to all aircraft within the sector or considering aircraft within a certain diameter or distance to the shepherding agent only. While the target is to maximise the number of aircraft controlled by a single ATC, a decision on the shepherding effect can be made according to the ATC skills and the overall mission objectives.
4. Controlling the speed of UAVs can be modelled through shepherding via considering the distance between the shepherd and the first UAV within its sight. For example, if the shepherd is too close to an aircraft, it could be handled as a more prominent situation where a fast turn is needed, but then the propagation strategy needs to decide whether this fast turn should cause subsequent aircraft to increase or reduce speed. Some aspects of this configuration space need to be explored further beyond the scope of the current chapter. In general, speed control is an essential strategy for different application domains.

11.4.4 Asynchronous Shepherding Algorithm

In this section, we propose the Asynchronous Shepherding Algorithm. The Shepherding Procedure and the Asynchronous Shepherding Algorithm are described in Algorithms 5 and 6. A visualisation of the algorithm is shown in Fig. 11.5.

First, we choose to propagate the received force as it is, unchanged, to maintain the spatial proximity between the UAVs. Therefore, $f(C_{\pi_j}, P^t_{\pi_i}, d^t_{\pi_i \pi_j})$ is set to C_{π_j}.

Second, the rigid transformation was set to rotation to maintain the planned trip distance. Figure 11.5 shows how rotation can be used to represent escaping from the shepherd attitude.

Third, we chose to add all UAVs within the sector to the Waiting Agent List. Thus, the force will be propagated to all UAVs, which will minimise the effort of managing different clusters of UAVs if not required by the task. It also assumes a broadcast communication protocol. If peer-to-peer communication is more desirable with a need to reduce the amount of transmissions in the environment, the adopted strategy in this chapter will need to change.

Fourth, for the speed control based on the distance, we chose to eliminate the speed factor for this algorithm to minimise the complexity of our initial phase of testing.

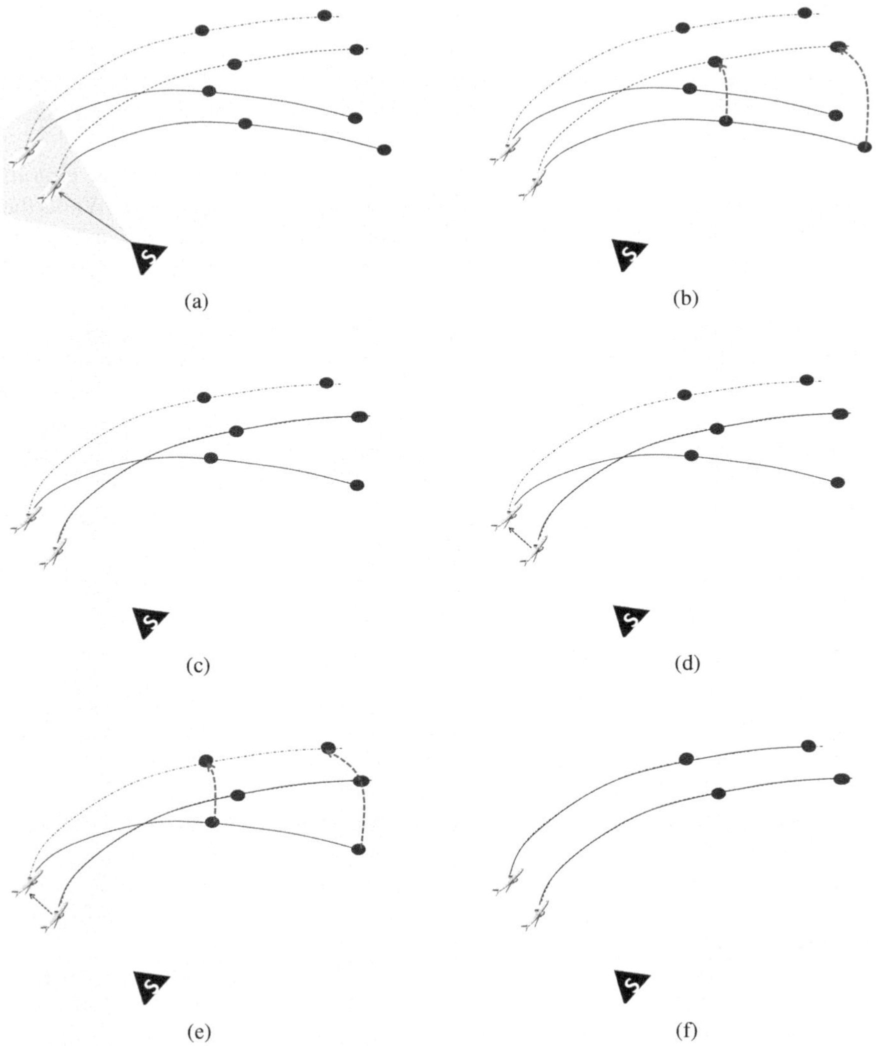

Fig. 11.5 Visualisation of asynchronous shepherding algorithm. (**a**) Line 2 (Algorithm 6): Let CIA be the closest aircraft in *WaitingAgentList* in the cone of sight of the shepherd. (**b**) Line 3 (Algorithm 6): Use the force vector to rotate the original flight plan of the CIA around the CIA current position. (**c**) Line 3 (Algorithm 6) cont.: Use the force vector to rotate the original flight plan of the CIA around the CIA current position. (**d**) Line 8 (Algorithm 6): Let CIA be the closest aircraft with the smallest distance to any aircraft in *InfluencingAgentList*. (**e**) Line 9 (Algorithm 6): Use the force vector to rotate the flight plan of the CIA Agent. (**f**) Line 10 (Algorithm 6): Add CIA to *InfluencingAgentList* Line11: Remove CIA from *WaitingAgentList*

Algorithm 5 Shepherding procedure

Click to display the shepherd
repeat
    Move and rotate the shepherd while displaying the influence vector
    Click to show the effect of influence on updated flight plans and updated trajectories using the ***Asynchronous Shepherding Algorithm***
    Wait $s1$ seconds
until click to accept Shepherd's position
Call the ***Asynchronous Shepherding Algorithm*** to execute the change
Wait $s2$ seconds
End

Algorithm 6 Asynchronous shepherding algorithm

Add all aircraft to a list *WaitingAgentList*
Let *CurrentInfluencingAgent* be the closest aircraft in *WaitingAgentList* in the cone of sight of the shepherd
Use the force vector F to rotate the flight plan of *CurrentInfluencingAgent*
Add *CurrentInfluencingAgent* to *InfluencingAgentList*
Remove *CurrentInfluencingAgent* from *WaitingAgentList*
while WaitingAgentList is not empty **do**
    Calculate the distance between every aircraft in *WaitingAgentList* and every aircraft in the *InfluencingAgentList*
    Let *CurrentInfluencingAgent* be the closest aircraft with the smallest distance to any aircraft in *InfluencingAgentList*
    Use the force vector to rotate the flight plan of *CurrentInfluencingAgent*
    Add *CurrentInfluencingAgent* to *InfluencingAgentList*
    Remove *CurrentInfluencingAgent* from *WaitingAgentList*
end while

A flow diagram that represents the process required to implement the Asynchronous Shepherding algorithm is shown in Figs. 11.6 and 11.7.

11.4.5 Issues for Future Research

Integration of UAVs speed control has been discussed in this chapter using the distance between the shepherding agent and the UAV. However, changing the speed of a UAV based on the propagation of forces offers another avenue of interesting research. Additionally, future work will investigate and analyse alternative methods of force propagation and efficacy. The impact of the spatial distribution of UAVs is likely to be an important factor to decide which of these design choices need to be adopted in which context. We are also planning human experiments to prototype the concept and evaluate its impact on human cognitive load. This will allow us to compare the proposed shepherding concept with classic ATC operations. Last, but not least, we will examine the scalability of the concept by understanding the sources of complexity in shepherding. For example, while it is trivial to expect that the size of the flock will impact mission success, our initial investigations have identified that this is incorrect. What matters most is the interaction between

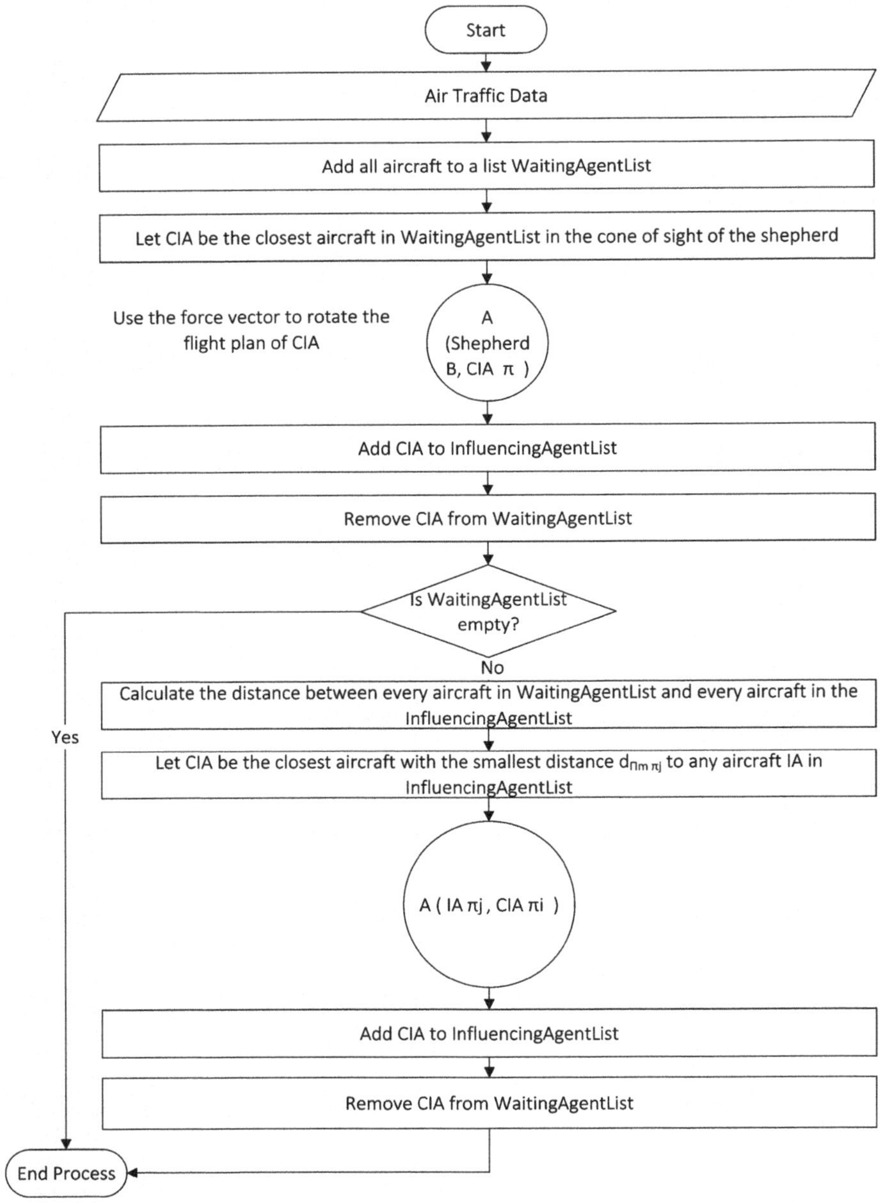

Fig. 11.6 Main algorithm

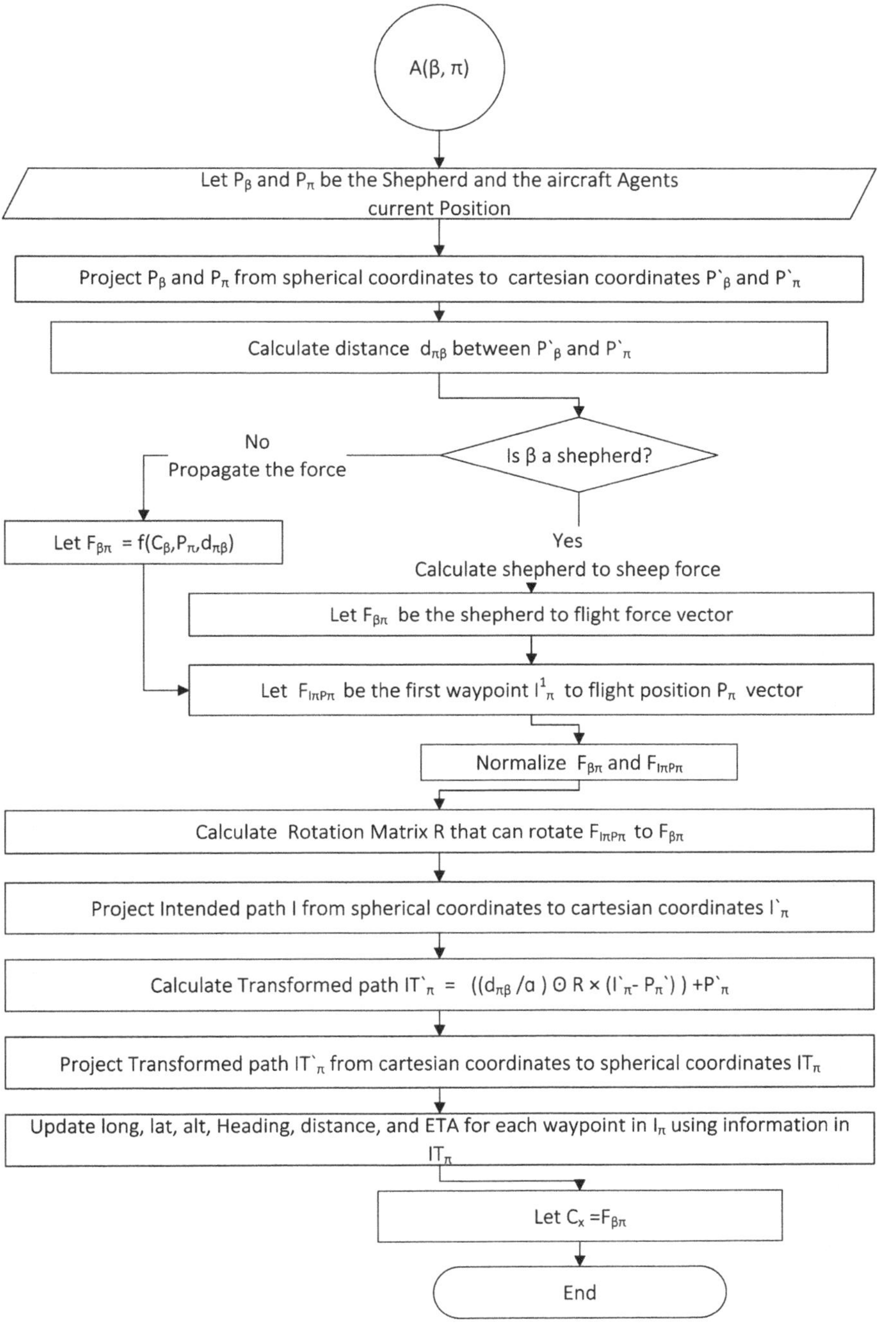

Fig. 11.7 Asynchronous shepherding flow diagram: procedure A(β,π): apply agent β shepherding force to rotate agent π path

flock size and the features in the environment. An infinitely large obstacle-free environment with an infinitely large number of very well-behaved agents can still be managed by a single sheepdog. Throwing in this environment an obstacle or increasing the jittering of an agent will quickly transform the problem from being solvable to unsolvable. These factors are crucial for classic synchronous shepherding and our proposed asynchronous shepherding in this chapter.

11.5 Conclusion

Unmanned Aerial Vehicles (UAVs) based industries are growing. The simultaneous control of multiple UAVs is a pertinent issue. With the limited growth in the number of air traffic controllers compared to UAVs, and the significant cognitive load the problem of UAVs imposes on human air traffic controller, there is an air traffic management challenge that needs to be addressed.

In this chapter, we introduced Asynchronous Shepherding for ATC. This algorithm allows a single air traffic controller to manage a large number of UAVs concurrently. It also reduces the number of control commands communicated between the ATC and UAVs. We presented on how a digital twin of the UAV environment, together with the proposed asynchronous shepherding algorithm, can offer safe implementation of the concept.

A variety of extensions and questions emerged from this research that lays out the foundations for future research in this area. Examples of future work include the impact of speed control and different force propagation strategies on the asynchronous shepherding algorithm. Understanding the sources of task complexity that impact how hard it is for the shepherding algorithm to resolve the swarm traffic management problem is crucial, both because these sources of task complexity identify the important dimensions for understanding context and offering the swarm traffic controller with the appropriate information to form their situation awareness picture, and for the autonomous artificial intelligence agent that automates some of the problem solving functions traditionally carried out by the human air traffic controller to reduce load on the human.

Acknowledgments This project is partially funded by an Australian Research Council Discovery Grant DP160102037 and partially funded by the Office of Naval Research Global.

References

1. Alam, S., Abbass, H.A., Barlow, M.: Atoms: Air traffic operations and management simulator. IEEE Trans. Intell. Transp. Syst. **9**(2), 209–225 (2008). https://doi.org/10.1109/TITS.2008.922877
2. Alam, S., Tang, J., Abbass, H.A., Lokan, C.: The effect of symmetry in representation on scenario-based risk assessment for air-traffic conflict resolution strategies. In: 2009 IEEE Congress on Evolutionary Computation, pp. 2180–2187. IEEE, Piscataway (2009)

3. Bekkouche, O., Taleb, T., Bagaa, M.: UAVs traffic control based on multi-access edge computing. In: 2018 IEEE Global Communications Conference (GLOBECOM), pp. 1–6. IEEE, Piscataway (2018)
4. Cardei, M., Cardei, I., Steinberg, A.: UAS trajectory scheduling system. In: 2018 Annual IEEE International Systems Conference (SysCon), pp. 1–8. IEEE, Piscataway (2018)
5. Clough, B.T.: Metrics, schmetrics! how the heck do you determine a UAV's autonomy anyway. Technical Report, Air Force Research Lab Wright-Patterson AFB OH (2002)
6. Devasia, S., Lee, A.: A scalable low-cost-UAV traffic network (uNet). CoRR **abs/1601.01952** (2016). http://arxiv.org/abs/1601.01952
7. FAA, U.S.F.A.A.: Fact sheet – small unmanned aircraft regulations (part 107). https://www.faa.gov/news/fact_sheets/news_story.cfm?newsId=22615 (2019). Accessed 2019 Oct 17
8. Johnson, R.D.: Unmanned aircraft system traffic management (UTM) project (2018). https://ntrs.nasa.gov/archive/nasa/casi.ntrs.nasa.gov/20180002542.pdf
9. Liu, M., Wan, Y.: Analysis of random mobility model with sense and avoid protocols for uav traffic management. In: 2018 AIAA Information Systems-AIAA Infotech@ Aerospace, p. 0076 (2018)
10. Murray, C.C., Chu", A.G.: The flying sidekick traveling salesman problem: optimization of drone-assisted parcel delivery. Transp. Res. Part C Emerg. Technol. **54**, 86–109 (2015). https://doi.org/10.1016/j.trc.2015.03.005. http://www.sciencedirect.com/science/article/pii/S0968090X15000844
11. NASA: Unmanned Aircraft System (UAS) Traffic Management (UTM). https://utm.arc.nasa.gov/index.shtml (2019). Accessed 2019 Aug 19
12. Rattanagraikanakorn, B., Sharpanskykh, A., Schuurman, M.J., Gransden, D., Blom, H., Wagter, C.D.: Characterizing UAS collision consequences in future UTM. In: 2018 Aviation Technology, Integration, and Operations Conference, p. 3031 (2018)
13. Ren, L., Castillo-Effen, M., Yu, H., Yoon, Y., Nakamura, T., Johnson, E.N., Ippolito, C.A.: Small unmanned aircraft system (SUAS) trajectory modeling in support of UAS traffic management (UTM). In: 17th AIAA Aviation Technology, Integration, and Operations Conference, p. 4268 (2017)
14. Stolaroff, J.K., Samaras, C., O' Neill, E.R., Lubers, A., Mitchell, A.S., Ceperley, D.: Energy use and life cycle greenhouse gas emissions of drones for commercial package delivery. Nat. Commun. **9**(1), 409 (2018)
15. Strömbom, D., Mann, R.P., Wilson, A.M., Hailes, S., Morton, A.J., Sumpter, D.J.T., King, A.J.: Solving the shepherding problem: heuristics for herding autonomous, interacting agents. J. R. Soc. Interf. **11**(100) (2014). https://browzine.com/articles/52614503
16. Sudbury, A.W., Hutchinson, E.B.: A cost analysis of amazon prime air (drone delivery). J. Econ. Educ. **16**(1), 1–12 (2016)
17. Tang, J., Alam, S., Lokan, C., Abbass, H.A.: A multi-objective approach for dynamic airspace sectorization using agent based and geometric models. Transp. Res. Part C Emerg. Technol. **21**(1), 89–121 (2012)
18. Van Pham, V., Tang, J., Alam, S., Lokan, C., Abbass, H.A.: Aviation emission inventory development and analysis. Envir. Model. Softw. **25**(12), 1738–1753 (2010)
19. Wang, B., Xie, J., Wan, Y., Guijarro Reyes, G.A., Garcia Carrillo, L.R.: 3-d trajectory modeling for unmanned aerial vehicles. In: AIAA Scitech 2019 Forum, p. 1061 (2019)
20. Zhao, W., Tang, J., Alam, S., Bender, A., Abbass, H.A.: Evolutionary-computation based risk assessment of aircraft landing sequencing algorithms. In: Distributed, Parallel and Biologically Inspired Systems, pp. 254–265. Springer, Berlin (2010)

Part IV
Human-Shepherding Integration

Chapter 12
Transparent Shepherding: A Rule-Based Learning Shepherd for Human Swarm Teaming

Essam Debie, Raul Fernandes Rojas, Justin Fidock, Michael Barlow, Kathryn Kasmarik, Sreenatha Anavatti, Matthew Garratt, and Hussein A. Abbass

This chapter aims to demonstrate how rule-based Artificial Intelligence algorithms can address a few human swarm teaming challenges. We will start from the challenges identified by the cognitive engineering community for building human autonomy teaming and how they scale to human swarm teaming. The discussion will follow with a description of rule-based machine learning with a focus on learning classifier systems as a representative for these algorithms and their benefits for human swarm teaming. Shepherding affords a human to manage a swarm by teaming with an autonomous single shepherd. A learning classifier system is designed to learn behaviour needed to be exhibited by the shepherd. Results demonstrate the effectiveness of the rule-based XCS model to capture shepherding behaviour, where the XCS model achieves comparable performance to the standard Strömbom's shepherding method as measured by the number of steps needed by a sheep-dog to guide group of sheep to target destination. These results are promising and demonstrate that learning classifier systems could design autonomous shepherds for new type of shepherding tasks and scenarios that we may not have rules for today.

E. Debie (✉) · M. Barlow · K. Kasmarik · S. Anavatti · M. Garratt · H. A. Abbass
School of Engineering and Information Technology, University of New South Wales, Canberra, ACT, Australia
e-mail: e.debie@adfa.edu.au; e.debie@unsw.edu.au; m.barlow@adfa.edu.au; k.merrick@adfa.edu.au; s.anavatti@adfa.edu.au; m.garratt@adfa.edu.au; h.abbass@adfa.edu.au

R. F. Rojas
Faculty of Science and Technology, University of Canberra, Canberra, ACT, Australia
e-mail: raul.fernandezrojas@canberra.edu.au

J. Fidock
Defence Science and Technology Group, Edinburgh, SA, Australia
e-mail: Justin.Fidock@dst.defence.gov.au

H. A. Abbass, R. A. Hunjet (eds.), *Shepherding UxVs for Human-Swarm Teaming*, Unmanned System Technologies,
https://doi.org/10.1007/978-3-030-60898-9_12

12.1 Introduction

Over recent years, the practical importance of autonomous mobile robots has increased in industrial and consumer market applications such as search and rescue missions, logistics, agriculture, unmanned vehicles, and cognitive aid. Such an increase is due to the fact that autonomous mobile robots play a significant role in optimising efficiency, reducing costs and hazards, maintaining safety, security and complying with regulations.

Increased AI and autonomy in these applications require several sophisticated algorithms to handle high level of autonomous behaviour while being transparent as much as possible. Bohren et al. [1] described the capabilities needed by a single robot in a "robot butler" by analysing the task of fetching a specific drink in detail. Despite the simplicity of the scenario, it requires several complex algorithms such as collision avoidance, recognition and identification of the specific drink to fetch. Scaling this scenario to a swarm-level where multiple robots interact together to achieve a common goal poses extra challenges and usually requires a number of sophisticated and scalable artificial intelligence algorithms. On the other hand, autonomous swarms should work seamlessly in collaboration with human operators (either in supervisory control or as team members) to support performance, independently reorganising priorities in response to evolving environments and their members collaborative activities. In achieving these goals, the swarm still needs to be adequately expressive for human supervisors or team members to understand, guide, and intervene in its operations if needed. It is very challenging to achieve this algorithmic complexity and human interpretability through traditional machine learning algorithms.

The Cognitive Engineering (CE) community has developed valuable research and related perspectives to model successful teaming for humans and machines. Research results from these studies suggest that there are some challenges to build effective human autonomy teams [24, 26]. In this chapter, we shall shed light on rule-based artificial intelligence (AI) techniques as a promising approach to mitigate the different challenges identified by the cognitive engineering community for building human autonomy teaming. Rules are accepted means of representing knowledge for many domains. They can be used, for example, in robot control, modelling, classification, and prediction [15].

A review and summary is first presented on the recent challenges identified in the cognitive engineering literature for building effective human autonomy teaming. Although the challenges introduced discuss a single autonomous entity, we scale these to swarms of autonomous systems. A taxonomy is then proposed to group these challenges from AI design perspectives.

The second part of this chapter focuses on rule-based machine learning; an introduction to rule-based learning common in the literature is presented. Then, we discuss how such systems can mitigate the challenges highlighted in our taxonomy and their benefits for systems involving human swarm interactions. Finally, a proposed rule-based reinforcement learning model for the shepherding herding problem is presented and discussed.

12.2 Challenges for Efficient Human Swarm Teaming

In human-robot teaming (HRT), humans and robots concurrently interact with each other to achieve some task. Typically, a human-robot teaming system consists of three components: (1) human partner; (2) robot partner (AI based autonomous system with varying levels of autonomy); and (3) relationship manager. Figure 12.1 illustrates the components of a human autonomy teaming system. Robots (with varying level of autonomy) should work collaboratively with human partners in a seamless way to achieve common goal, aiding performance by alerting human teammate about behaviour that deviates from normal, suggesting alternative solutions that they may not have considered, reorganising priorities autonomously in reaction to evolving objectives, or other cooperative activities. A swarm is a group of artificially intelligent agents whose collective behaviour results from local interactions between the agents and between the agents and the environment in which they work. Human swarm teaming (HST) extends HRT to systems consisting of multiple human/AI partners.

The cognitive engineering (CE) community has produced helpful studies and associated insights to design efficient human autonomy teams. Research findings from these studies suggest that there exists some challenges for designing efficient human autonomy teaming [24, 26]. Improperly constructed human autonomy teaming systems have many adverse effects, such as unstable performance, mis-calibrated human trust in the autonomous partner, and lack of user acceptance. In this section, we adapt these challenges to human swarm teaming and discuss them from AI design perspectives (see Fig. 12.2). These challenges are:

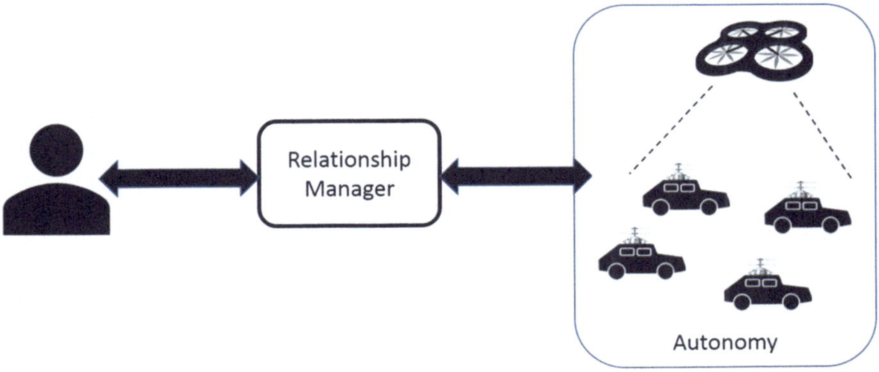

Fig. 12.1 Human autonomy teaming (HAT) architecture

- Converged situational awareness: Despite different agents require different knowledge (based on their capabilities and objectives), a shared situational awareness should always be maintained among swarm members. To achieve converged situational awareness, relevant knowledge and assumptions should be shared among human and swarm team members. This needs to be continually updated and maintained so that members of the human swarm team can maintain a mutual understanding of what is happening in the environment and appropriately react to support one another.
- Observability: Observability refers to the level of transparency into what an autonomous partner is doing with respect to task progress. An autonomous partner is observable when it offers an adequate level of insight into its reasoning and rationale that contributes to the decisions or recommendations it provides to a human partner. Observability supports mutual understanding of the problem to be solved and progress towards goals [12].
- Predictability: Predictability is another dimension of system transparency that reflects on how discernible/understandable the future intentions and activities of the autonomous partners are. It is a complementary component for observability where the latter reflects on the past while the former reflects on the future behaviour of the system.
- Information Presentation: Information should be presented to human partners in a simple and understandable way. Human partners should be able to view and interact with mission-relevant information in order to understand the implications of behaviour of their autonomous partners.
- Calibrated Trust: Calibrated Trust is supported when the AI model offers performance indicators that enable users to know when and how much they can trust an autonomous partner in various situations.
- Exploring the Solution Space: Autonomous swarm partners should be able to leverage multiple views and knowledge bases, in order to jointly reach a reasonable response to the problem space. The swarm then uses these multiple inputs to produce alternative courses of action and provide ways of comparing those alternatives with regard to implications and results.
- Adaptability: Autonomous swarms should be prepared to recognise unexpected situations and adapt their behaviour to address evolving and dynamic environments.
- Directing Attention: Autonomous swarms should proactively interact to direct attention of their own members and their human partners to critical issues whenever information becomes available.
- Directability: Directability is supported when humans can easily guide the behaviour of autonomous partners and re-prioritise their activities. Ultimately, humans are responsible for system results, therefore they must be able to interrupt procedures, alter course, switch between autonomy levels (i.e. switch from high level autonomy where agents act independently to a lower level control such as tele-operation or, if required, override and manually regulate the process).

From an AI design perspective, these challenges can be organised into three categories as shown in Sect. 12.2 and explained below:

1. Transparency: This category includes observability, predictability, information presentation, and calibrated trust. Shared expectations and mutual understanding are critical facets of teamwork. By building more transparent (interpretable and explainable) systems, humans will develop a better understanding of the system behaviour leading to calibrated trust in the intelligent systems [12, 20, 27]. Interpretability and explainability are two complementary approaches to increase transparency and trust of AI systems and will form part of many trusted autonomous systems [28].

   Interpretability refers to the degree to which a human can understand how an AI system works. The higher the machine learning model's interpretability, the simpler it is for people to comprehend how they make choices or predictions. Despite the benefits of transparent AI, it can be difficult to achieve mutual knowledge and shared expectations in cooperative situations of human autonomy teaming, as human beings and autonomous systems are unlikely to share a common language to express intentions, plans, or justifications. Moreover, it can be very difficult to gain insight into the logic or reasoning that regulates the behaviours of autonomous systems, requiring another layer of AI to model the decision making process for the main AI model. This increases the system complexity in terms of computational costs and errors.

   Explainability refers to reasoning about decisions explicitly to human partners. It includes the extraction of causal information from learned models. One approach for providing these explanations is through post-hoc processing (another machine learning layer/algorithm) and includes natural language explanations and visualisations of learned models. Explainable machine learning models are useful for multiple uses such as explaining behaviour of autonomous agents in robotics [20, 27], debugging machine learning models [25], and medical decisions explanation [18].
2. Flexibility: This category includes adaptability, directability, and exploring the solution space. Autonomy is often a required characteristic to respond both flexibly and rapidly to changes in task demand and environment dynamics. An AI partner should be able to respond to momentary changes in the human partner's efficiency and task requirements. A flexible autonomy that is able to communicate with its partners and adapt quickly is of high need.
3. Coordination: This category includes common ground and directing attention. An AI partner can assist the human operator by providing crucial information. However, communicating this information is challenging. The human needs to know where to locate and how to interpret information, particularly in time critical situations where they might be stressed and forget to access them. Furthermore, information should be presented in a format that is readily understood and minimally interferes with the task being conducted. On the other hand, humans have the capability to pre-process, format, and present information according to a given context to other human partners. For instance, a co-pilot

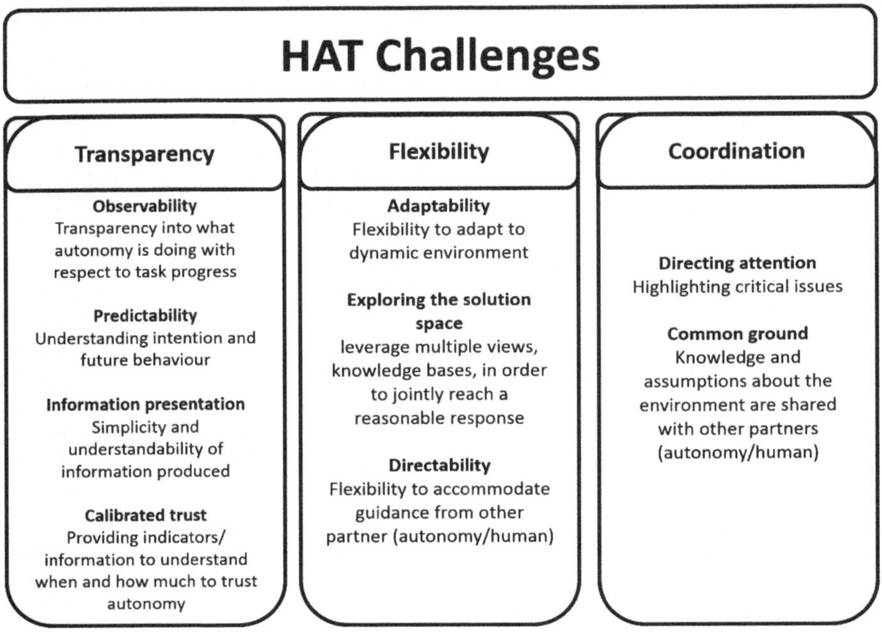

Fig. 12.2 HAT challenges

can correct flight control unit figures, draw attention to an incomplete checklist, or acknowledge volatile performance. Such conduct is also desirable for an autonomous agent.

12.3 Fundamentals of Rule-Based Artificial Intelligence

In this section we outline some of the fundamental aspects of rule-based machine learning algorithms used in artificial intelligence applications. There are many rule-based machine learning systems available. We focus here on the Learning Classifier Systems (LCSs) as an example, since they are popular and well-studied in the literature. Rule-based machine learning can be defined as a type of machine learning which uses an easy to interpret condition-action or stimuli-response (rules) to represent the action production model of an agent. They are normally represented as if. . . then. . . rules.

12.3.1 Structure

Figure 12.3 shows the general structure of a rule-based machine learning model. Data sensed from the environment is fed along with a knowledge base to what is called an inference mechanism. The knowledge stored in this knowledge base is often obtained and validated from continuous interaction with the environment. A reasoning process is conducted by the inference mechanism to decide the right inference through application of the knowledge to the sensed data. Finally, feedback from the environment is used to update and maintain the knowledge base.

12.3.2 Representation of Knowledge: Rules

A mechanism is needed to capture the knowledge extracted from the environment the AI works within. One popular approach to achieve this is to use the so-called production rules. A production rule is a basic mapping from inputs to outputs. In general, each rule has a prerequisite that is either met or not met by the global database. If the precondition is met, the rule is applied. They typically take the form

IF condition **THEN** inference

What is shown here as condition is also referred to as the antecedent because it is the logical expression that precedes the inference step. The inference is called the consequent or decision. The strength of production rules is their interpretability; their IF-THEN structure is semantically similar to natural language and the way humans think. Some simple examples of production rule are

1. **IF** collecting mode is ON **AND** range = [30, 45] **AND** bearing = [325, 55] **THEN** turn = 45°

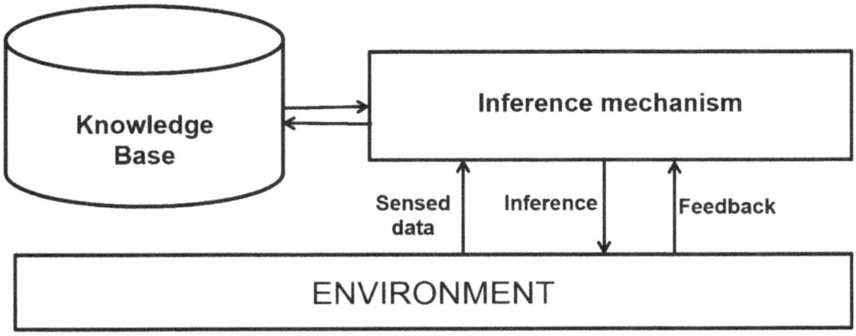

Fig. 12.3 General rule-based machine learning framework

2. **IF** shepherd to sheep distance > threshold1 **AND** farthest sheep to global centre of mass <= threshold2 **AND** target angle = [−10°, 10°] **THEN** Change in steering angle = 15°

The condition in these examples is a logical statement that can either be true or false. If it is true, it is presumed that the inference is true or justified. Otherwise, the rule does not provide any evidence for or against the assessment of the present condition. In addition to simple logical expression with a single logical statement, the condition can be a compound statement in which several components are combined through the logical and/or operations.

In the second rule above, a set of parameters is used (threshold1 and threshold2). These parameters require numerical values which will depend on the environment. The values could be set manually before the system starts the learning process or, alternatively, they could be estimated through a number of training cycles during which the system interacts with the environment and uses the received reward to update them.

Each single rule can be thought of as providing one item of knowledge about how to interact with the environment being considered. A knowledge base might consist of hundreds of rules of these types, obtained from interactions with the environment. When a situation (environment state) is presented to the AI model, the inference mechanism goes through the rule base and selects those rules which are relevant to the given situation. Among those filtered rules, the inference mechanism checks support for or against various decisions. Some rules will offer strong support while others will be weak. Several decisions for a particular situation might be supported among the rules; a conflict resolution mechanism is then needed to decide a final decision.

12.3.3 The Inference Mechanism

The inference mechanism is responsible for maintaining the rules in the knowledge base and spans a number of functions such as generating new rules, and updating rules strengths based on interaction with the environment. It also has to account for redundant reasoning (e.g. disposal of redundant rules). In addition, the inference mechanism monitors all rules that infer a particular inference and those that contradict that inference. By weighing all the evidence from the rules, the inference mechanism makes a decision about the correct inference. In addition, challenges such as uncertainties, data quality, and missing rules are looked after during the inference process.

12.4 Learning Classifier Systems

LCS is a rule-based machine learning paradigm introduced by John Holland in 1976 [21] and was the first implementation of an adaptive rule learning algorithm. A classifier refers to the combination of a rule, represented as a production rule IF *condition* Then *action* form, and a number of performance statistics. LCS employs reinforcement learning and a genetic algorithm to evolve a set of rules. Since its invention, LCS has been applied to a wide range of applications including autonomous robotics [16], traffic control [9], and network routing [11]. Due to their implicitly parallel nature, LCSs are able to explore a large search space to find an optimal solution effectively [31]. They are incremental learners, and being evolutionary based, they are inherently adaptive, making them widely used for robot control problems [30, 40].

12.4.1 Learning Classifier System Components

A basic LCS consists of three major components:

Knowledge Base The knowledge base is a population of classifiers that collectively capture the current knowledge of the LCS. Conditions can be defined in different ways (e.g. exact value equality, ranges of values, fuzzy intervals, etc.) depending on the type of input the environment provides (e.g. binary, nominal, real...). Each classifier condition defines a problem subspace. The population of classifiers as a whole completely covers the entire problem space. Independent classifiers are those that cover completely different subspaces and collectively provide a solution for a problem, while overlapping classifiers compete with each other for activity and reward.

Rule Evaluation Mechanism The evaluation mechanism estimates (and propagates) rule quality/usefulness and is usually implemented as an adapted reinforcement learning technique.

Rule Evolution Mechanism Based on the estimated utilities, the evolutionary mechanism generates offspring classifiers and deletes less useful classifiers.

Figure 12.4 illustrates the interaction of these LCS components with each other and with the environment.

Learning Architecture The learning architecture differs between LCS variants. Here, we describe the learning architecture of the accuracy-based extended classifier system (XCS) [43], one of the most popular and well-studied LCS variants. XCS differs from the original LCS in two main aspects: (1) the evaluation of classifiers is based on their prediction accuracy rather than the prediction itself; and (2) the rule discovery is performed in a *niche-based* manner (e.g. only a subset of the classifiers in the population is used rather than the whole population). Owing

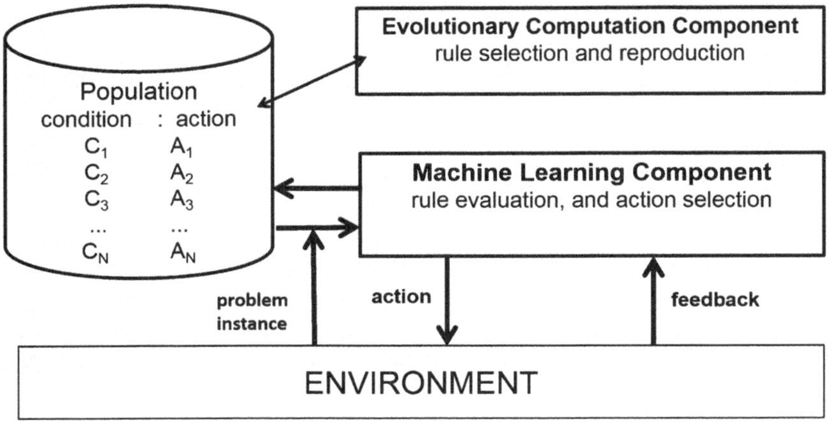

Fig. 12.4 Main components of LCS

to these differences, it has been anticipated that XCS can handle more complex problems, produce compact solutions (smaller number of classifiers), and can also utilise different representations of data, which is useful particularly in fields such as robotics [22].

In this section, we provide an overview of XCS operation in a multi-step problem environment; for a detailed description of the XCS classifier system the reader is directed to [8].

There are two modes of operation in XCS: (1) Exploration mode (acquisition of new knowledge or rules); and (2) Exploitation mode (the efficient use of learned knowledge to influence/control the environment). The system alternates between exploration and exploitation modes.

At any time t, XCS evolves a population $[P]^t$ of N_t classifiers, each denoted cl_n where $n \in 1, 2, \ldots, N$, and N must not exceed the maximum population size N_{max} set by the user. Each classifier has a rule and a set of associated parameters estimating the quality of the rule. Each rule consists of a condition part and an action part. The condition specifies the set of input states where the classifier can be applied. It is a conjunction of predicates, where each predicate can be coded using one of several representations (e.g. binary, real-valued inputs, etc.). The classifiers in XCS are macro-classifiers such that each classifier represents a number num of traditional (micro-classifiers) that have identical conditions and actions. A typical classifier cl_n has the form:

$$cl_n = (c_n, P_n, Perr_n, F_n^{macro}, as_n, num_n, exp_n) \rightarrow a_n^k \tag{12.1}$$

where

c_n	A conjunction of predicates that specifies the set of input states where the classifier can be applied, and has the form: $c_n = \{c_n[1], c_n[2], \ldots, c_n[d], \ldots, c_n[D]\}$
D	The number of input attributes in the problem
P_n	The reward prediction
$Perr_n$	the error in reward prediction estimate
F_n^{macro}	A measure of classifier quality
as_n	Estimated average size of all action sets in which the classifier participates
num_n	The number of copies that this classifier represents
exp_n	The number of times that a classifier has belonged to a match set
a_n	The decision action advocated by the classifier chosen from a list of possible discrete action set A

A classifier applies to a problem instance from the environment if its condition part matches the values provided for this problem instance. Matching the classifier's condition to the problem instance relies on the input type (binary, real, etc.) and the representation used in the condition part. For instance, in binary problem representation, the condition is usually represented in a ternary alphabet where each predicate $c_n[d] \in \{0, 1, \#\}$.). The symbol #, called "don't care", allows the formation of generalisations in the rule's condition. In this case, a condition $c_n = \{c_n[1], c_n[2], \ldots, c_n[d], \ldots, c_n[D]\}$ matches a problem instance $x = (x_1, x_2, .., x_d, \ldots, x_D)$ as follows:

$$match(c_n, x) \Rightarrow (c_n[d] = x_d)\, or\, (c_n[d] = \#)\, \forall d \in \{1, 2, \ldots, D\} \tag{12.2}$$

Various approaches have been proposed to represent real-valued problems such as intervals [44], and fuzzy logic. At any time t, the input to XCS is often a problem instance x^t (characterising the current state of the environment). Input data may be binary, integer, real-valued, or some other custom representation, assuming the XCS algorithm has been coded to handle it. A match set $[M]^t$ is then formed of all classifiers in the population $[P]^t$ whose conditions match the current problem instance according to Eq. (12.2). The classifiers in the match set reflect the system knowledge about the current state of the environment. These classifiers may advocate different inferences (actions) and compete with each other to propose its advocated action.

A covering mechanism is activated when the match set $[M]^t = \phi$, to generate a number of new classifiers $\{cl_{cov1}, cl_{cov2}, \ldots, cl_{covM}\}$. The condition part of every classifier matches the current problem instance x_t and contains "don' t cares" with probability $p\#$ such that

$$C_{cov}[d] = \begin{cases} \# \text{ with probability } p\# \\ x_d \text{ with probability } 1 - p\# \end{cases} \tag{12.3}$$

Each newly created classifier advocates an action from the possible list of actions defined. The newly created classifiers are added to the match set $[M]^t$ and the population of rules $[P]^t$ deleting other classifiers from $[P]^t$ if the population exceeds its bounded size N_{max}. Different strategies may be used to decide which classifier to delete, a common method is to use a roulette wheel selection strategy based on how frequently a given classifier matches problem instances [45].

Once a match set is formed, a decision is made to either do one explore (acquiring new knowledge) or exploit (using current knowledge to influence the environment) step:

Exploration Mode In exploration mode, the system chooses an action randomly from those actions represented by the classifiers in $[M]^t$ and sends it to the Environment.

Exploitation Mode In exploitation mode, XCS uses the classifiers in the match set $[M]^t$ to decide on an action that influences the environment. The system calculates for each possible action $a \in A$ the predicted payoff. In general, each classifier in $[M]^t$ contributes with a payoff weighted by its fitness for the action it advocates. The accumulated payoff for each action is calculated:

$$payoff_a^t = \frac{\sum_{\{cl_n \in [M]^t \wedge cl_n.A=a\}} P_n.F_n}{\sum_{\{cl_n \in [M]^t \wedge cl_n.A=a\}} F_n} \tag{12.4}$$

where $payoff_a^t$ is the predicted payoff (accumulated vote) for action a. P_i and F_i are the reward prediction and fitness values of classifier cl_n, and $\{cl_n \in [M]^t \wedge cl_n.A = a\}$ is the set of all classifiers in the match set that advocate action a. The best action a_*^t is chosen with the highest vote and is chosen from the possible set of actions A :

$$a_*^t = argmax_{a \in A} payoff_a^t \tag{12.5}$$

Once an explore/exploit step is performed, all classifiers in $[M]^t$ with the chosen action form what is called the action set $[A]^t$:

$$[A]^t = \left\{cl \in [M]_t \left| cl.A = a^* \right.\right\} \tag{12.6}$$

The chosen action is sent back to the environment to be executed and an immediate reward r^t is received as a result of executing that action. In multi-step problems, the previous action set $[A]^{t-1}$ is updated using the previous reward r^{t-1} and the maximum predicted payoff obtained at the current learning iteration $maxPayoff_a^t$ calculated using Eq. (12.4).

The update follows an adapted Q-learning mechanism to update the reward prediction of each classifier as follows:

$$P_i^{t-1} = P_i^{t-1} + \beta \left(r^{t-1} + \gamma \, maxPayoff_a^t - P_i^{t-1}\right) \tag{12.7}$$

where β is the learning rate and γ is the discount factor used. Following the reward prediction update, the error in prediction ϵ and classifier fitness F are updated. Now that the reinforcement learning component updated classifiers statistics, it is time for rule evolution to take place. The role of rule evolution is to discover new and promising classifiers. Several approaches have been proposed in the literature to implement this discovery component such as differential evolution [14] and rule combining [19]. However, a steady-state genetic algorithm (GA) is the most common algorithm used as the evolution strategy mechanism in LCSs. GA is applied to $[A]^{t-1}$by selecting two parent classifiers $cl_{p1}$and cl_{p2} with a probability that depends on the classifier's fitness and applies crossover and mutation with probabilities χ and μ respectively (i.e., mutation can change a condition attribute or the action part to one of the other possible values) to yield two offspring classifiers. The newly generated classifiers are inserted into the population unless they are subsumed by their parents; a parent classifier cl_{pn} subsumes its offspring if it is sufficiently accurate (its prediction error ϵ_{pn} is less than a pre-specified threshold ϵ_0, that is $\epsilon_{pn} < \epsilon_0$), and sufficiently experienced (its experience exp_{pn} or the number of times it participated in an action set is higher than a minimum threshold θ_{sub}, that is $exp_{pn} > \theta_{sub}$), and is more general than its offspring (the inputs it matches are a super-set of the inputs matched by the offspring). If one offspring is subsumed by its parent, this offspring is not added to the population and instead the numerosity (num_{pn}) of this parent is incremented by 1. The numerosity is a memory-efficiency strategy to avoid growing the memory with many copies of the same classifier by keeping track of the number of classifier copies that a given classifier represents instead of storing the individual classifiers themselves.

Generalisation Theories Since the development of XCS as an accuracy-based LCS (e.g. guided by rule accuracy instead of strength), a number of theoretical studies were introduced to analyse the behaviour of accuracy-based LCSs. For example, Wilson [43] proposed the generalisation hypothesis which states that a combination of mechanisms (referred to as pressures) applied implicitly by XCS leads the system towards more accurate and maximally general classifiers. Although Wilson did not prove his hypothesis formally, it formed the basis for later formal studies on XCS. For example, Butz [5] introduced the first formal analysis of XCS to explain why it works. He followed what is called a facet-wise approach to analyse each component of XCS on its own in order to understand how the system works as a whole. In his work, Butz defined five pressures that drive XCS towards maximally general, accurate solutions. These are: (1) fitness pressure; (2) set pressure; (3) mutation pressure; (4) deletion pressure; and (5) subsumption pressure. Interested readers are referred to [7] for detailed description of the XCS pressures.

Following Butz's work, several theoretical research studies were introduced into XCS' parameter configurations. Butz et al. introduced learning challenges [4, 6] from which parameter bounds were derived. Most of the theoretical studies on challenges and parameter bounds are based on analysing pressures in XCS. Orriols-Puig et al. [29] analysed XCS performance on learning tasks with imbalances (e.g. niches are not represented with similar number of instances). The authors derive an upper bound on the imbalance ratio between the different niches to ensure system performance. They also derive parameter bounds to guarantee system convergence in such scenarios. Debie and Shafi [13] followed a similar approach to the challenges and bounds developed by Butz et al. to define parameter learning bounds in high-dimensional and real-valued problem spaces. They regarded XCS derivative for supervised learning, the Supervised Learning Classifier System (UCS). However, their findings are applicable to both system.

12.5 Learning Classifier Systems for Human Swarm Teaming

LCS offers a number of capabilities that can be utilised to address or mitigate the challenges discussed in Sect. 12.2.

12.5.1 Transparency: Rules as Justifiers

Since the introduction of the extended classifier system XCS [43], LCS has been commonly recognised as a robust learning technique with the highly desired advantage of transparency owing to its typical use of human-interpretable production rules that operate explicitly in the problem domain. Unlike black-box machine learning algorithms such as artificial neural networks or random forests, LCS generates solutions that are intuitively human interpretable. LCS classifiers act as justifiers since they provide a degree of justification or evidence in favour of a particular action. This evidence is built through a credit assignment process through which classifiers that are responsible for system outcomes are rewarded/penalised according to feedback received from the environment during learning (exploration) mode. These credits are accumulated as a weight/support by the classifiers. During exploitation mode LCS acts as an ensemble learner, no single model is applied to a given instance to yield a decision/action. Instead, a set of relevant classifiers contribute a "vote" based on their accumulated credits. This ensemble-like behaviour tends to make more accurate and reliable predictions. Various LCS versions provide alternative methods to accumulate this credit. In general, stronger classifiers are more likely to influence the system outcomes than weak ones. The combination of rule representation and reinforcement learning mechanism enables the LCS to explain its reasoning, characterise its strengths and weaknesses, and provide an easy way for humans to comprehend how LCS will act in the future. Importantly, its model can be translated into human-friendly and useful explanations.

12.5.2 Flexibility

Learning classifier systems have always been regarded as flexible systems, meaning that the rule base can update itself in response to a changing environment. For example, application of EpiCS (an LCS variant) to epidemiological surveillance has demonstrated the effectiveness of LCSs in both anticipating and describing a quickly evolving phenomenon, such as the occurrence of a certain epidemic disease. In Computational Economics [32, 41], LCSs have been effectively used to model the behaviour of buyers and sellers in artificial stock markets where agents must be able to adapt to rapidly changing market situations, and to develop successful strategies against competitors.

LCSs evolve a population of classifiers that collectively model the problem space. A single classifier in LCS models a distinct part of the problem (e.g. a niche). The classifiers within a population cooperate to cover the whole problem domain. Adaptation occurs in LCSs when new rules are created in response to environmental needs (e.g. new environmental inputs). This incremental, cooperative, and niche-based learning mechanism allow LCS to adapt effectively to a changing environment by maintaining and developing the different parts of the solution separately.

LCSs are able to accommodate different types of dependent or independent variables (e.g. binary, discrete, or continuous valued attributes) owing to their representation flexibility. Moreover, LCS is a stochastic learner; its discovery and learning capabilities are provided by a genetic algorithm (GA) [42]. GA is a stochastic search algorithm based on the mechanisms of natural selection and population genetics. The stochastic learning nature of LCS is advantageous in highly complex problems, where deterministic learning becomes intractable. The combination of representation flexibility and stochastic learning allows applications of LCS to many domains with multiple types of input and feedback. It also allows LCSs to leverage multiple sources of information simultaneously to efficiently search the solution space for promising solutions.

The combination of mechanisms used by the LCS enables the system to adapt to changes while interacting with the environment. In doing so the system is balancing exploration (acquisition of new knowledge, rules, evidences) with exploitation (the efficient use of learned knowledge to influence/control the environment).

12.5.3 Multi-Agent Coordination

Distributed artificial intelligence (DAI) [2], a generalisation of multi-agent systems, contributes the smartness required for the development of distributed autonomous systems that can tackle complex problems through some sort of coordination scheme. In a multi-agent system, AI agents work concurrently within a certain environment without any global coherent knowledge about that environment. To attain their own local goals, these agents may still need to coordinate their operations with each other. Therefore, they may benefit from having information about what others are doing or intending to do and offering them information about what they are doing. Researchers in the field of DAI have developed a variety of agent coordination schemes that in most cases rely on explicit or implicit sharing of information between agents. However, several studies suggest that the less an agent depends on shared information, the better it can adapt to dynamic and complex environments [33, 34]. Moreover, sharing information between agents in a real-environment has other shortcoming such as communication delays and failure of key agents (e.g. agents which control agent communication) and may not scale well with increasing number of agents. It is therefore desirable to have a coordination mechanism that imposes little cognitive burden on agents and does not suffer from these deficiencies.

Learning classifier systems have characteristics that render them more suitable for building multi-agent systems in such non-stationary problem domains compared to other types of machine learning. The rule-based mapping between the environmental perceptions and agent actions in LCSs can be used by multiple agents to learn coordination strategies without having to rely on explicit information sharing.

LCS based multi-agent systems were shown to perform well in developing effective coordination schemes in different applications. Bull et al. [3] used a Pittsburgh-style LCS [38] for the control of a quadrupedal robot, where each leg is represented by a separate LCS. Carse et al. [10] used a fuzzy Pittsburgh-style LCS multi-agent system for telecommunications network routing. Michigan-style LCS has also been used in multi-agent environments. For example, Dorigo and Schnep [17] used a hierarchical multi-agent system based on LCS to control an autonomous robot where agents at the lower level of the system learn simple behaviours whereas agents at higher levels learn to coordinate the activities of the lower level agents. Seredynski et al. [36] used a LCS multi-agent system to examine the use of local reward sharing in a simple iterated game. In robot navigation problems, XCS-based multi-agent system learned to solve the problem concurrently [33, 34]. Each agent learns its own environment independently, while global coordination between multiple agents emerges without explicit information sharing. Irvan et al. [23] used a multi-agent LCS architecture in which a shared memory concept was used to coordinate between multiple XCS-based agents. Seredynski [35] developed a parallel and distributed system based on LCS with game theoretic modelling to in-game playing simulation. Each agent in the game was represented by a LCS. The work demonstrated the strength of LCS in designing cooperative multi-agent systems. The proposed model showed that agents were capable of self-organising themselves and successfully evolved group-level behaviour. Shao et al. [37] used adapted XCS model to learn swarm-level behaviour for robots navigation.

As shown in the above studies, LCS provides an intuitive and clear way to build coordination schemes in a homogeneous multi-agent setting. In combination with human interpretability offered by the use of production rules, this allows the investigation of heterogeneous agents' interactions (including human agents).

12.6 Learning Classifier System Model for Shepherding

In this section, we investigate the performance of XCS-based model in learning shepherding behaviour in a simple sheep-dog herding scenario. The aim of this section is to demonstrate how XCS can capture shepherding behaviour into a set of simple IF THEN rules.

12.6.1 Sheep-Dog Herding Problem

We consider the shepherding problem, whereby an agent is learning how to guide a group of sheep to reach (within 10 m) a pasture. The shepherding problem consists of a paddock of size $L = 150$ m. A complex behaviour embodied as a set of state-response rules is learned by XCS-based shepherd model. The behaviour is learned in a simulation environment.

In this task, one agent is acting as the shepherd and a number of other agents act as sheep. The task is for the shepherd to guide the sheep into a pasture within a limited number of movements. The sheep react to the nearby shepherd as in the standard shepherding model defined by Strömbom [39] by moving away from it, otherwise the sheep moves in a random walk. Figure 12.5 shows the task environment with a paddock of length 100 m $\times$ 100 m. The shepherd is shown in red while the sheep are in blue, the global centre of mass of the sheep is shown in green. The shepherd must learn to control its own heading to get the sheep to move into pasture located at the bottom left corner of the paddock.

12.6.2 XCS Classifier Representation

Each classifier in the implemented XCS model (simply a rule) consists of conditions that match against the current sensors of the shepherd, and an action that suggests a heading for the next shepherd move. The condition part of the rule consists of discrete form of three variables: (1) the distance between the shepherd and the nearest sheep D_{cs}; (2) the distance between the shepherd and the target pasture D_T; and (3) the angle θ_{sh} between the two vectors connecting the shepherd and the target and the shepherd and global centre of mass (GCM) respectively as shown in Fig. 12.6. The action part represents the direction of the next move of the shepherd. The shepherd is allowed to move into eight different directions as shown in Fig. 12.7.

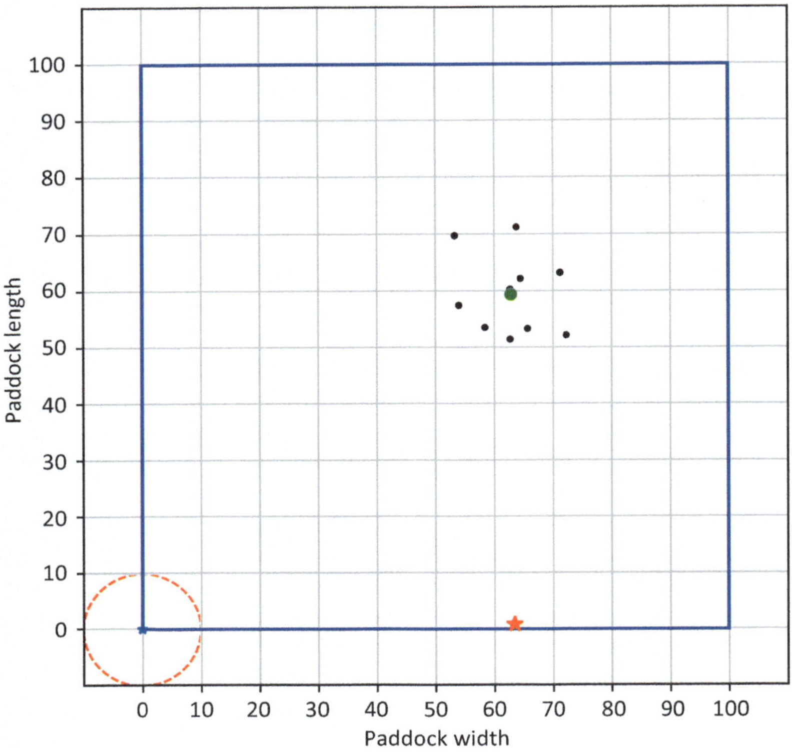

Fig. 12.5 Shepherding environment

12.6.3 *Experimental Setup*

The experiment was performed following the standard XCS settings used in the literature [43]. Each experiment consists of a number of trials of solving the sheep-dog herding problem. Each trial is either a learning trial or a test trial and is a separate problem that the system needs to solve. Each trial consists of the sheep and the shepherd being placed in randomly chosen positions and orientations within the paddock while the pasture is always positioned at location (0, 0). The shepherd then moves under control of the system until either it drives the sheep to the pasture location ($GCM - Target \leq 10$) or has taken $S = 1000$ steps, at which point the epoch (trial) ended. The experiments alternated between learning (exploration) and testing (exploitation) trials. In a learning trial, the system selects actions randomly from those represented in the match set. In a test trial, the system always selects the action with the highest prediction. The learning performance is computed as the average number of steps needed to guide the sheep to reach (within 10 m) the pasture location ($GCM - Target \leq 10$) in the testing trials. 3000 training and 3000 testing

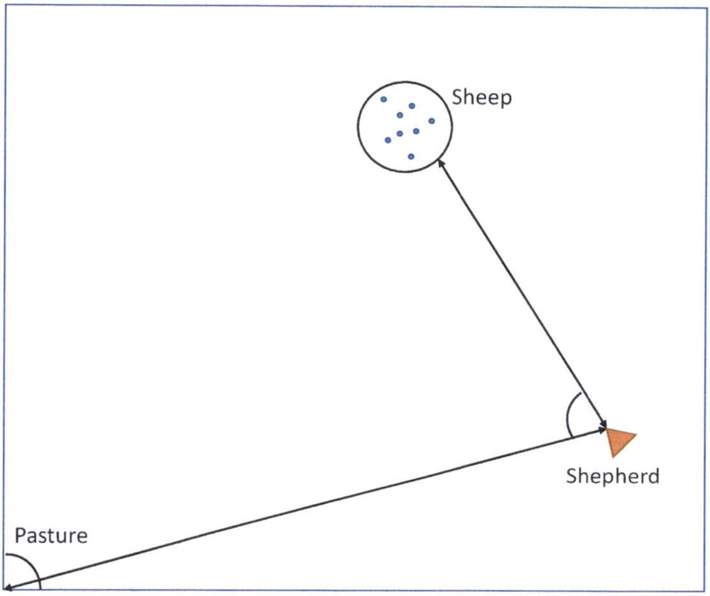

Fig. 12.6 Shepherd variables

Fig. 12.7 Shepherd movement directions

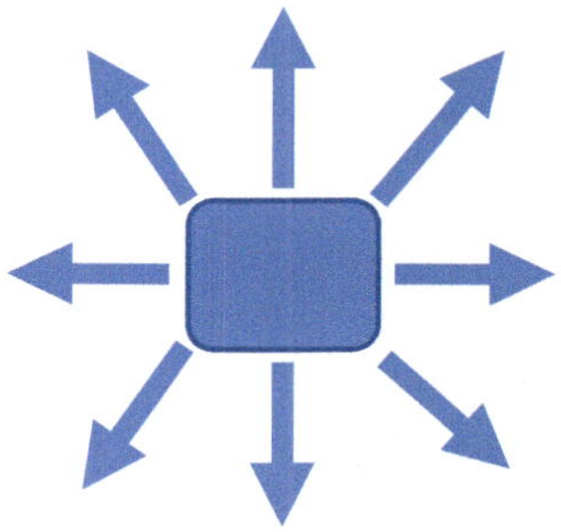

trials are used in this experiment. All the statistics reported are averaged over 10 runs. The shepherding simulation is initialised according to the parameters defined in Table 12.1. The standard Strömbom' s method is used in our experiment as a baseline. Strömbom's method is run 10 times using the same experimental setup and the performance is compared with XCS-based method.

XCS parameters are set following the standard XCS settings used in the literature [43] as follows: the population size $N = 6000$ classifiers, $P\# = 0.5$, $\beta = 0.2$, $\gamma = 0.7$, $\nu = 0.01$, $\theta_{mna} = 9$,$\theta_{ga} = 25$, $\epsilon_0 = 1$, $\theta_{del} = 20$, GA subsumption is on, action set subsumption is off. On contrary to the traditional rewarding approach of XCS in the literature, the proposed model is designed with multiple rewarding schemes that activate according to the current goal in the problem. For the first goal (reaching a driving position) the agent receives a reward

Table 12.1 Parameters for shepherding simulation

Parameter	Description	Typical values
L	Paddock length	100 m
Sheep parameters		
N	Total number of sheep agents	10
n	Number of nearest neighbours	6
r_s	Shepherd detection distance	65 m
r_a	Agent to agent interaction distance	2 m
ρ_a	Relative strength of repulsion from other agents	2
c	Relative strength of attraction to the n nearest neighbours	1.05
ρ_s	Relative strength of repulsion from the shepherd	1
h	Relative strength of proceeding in the previous direction	0.5
e	Relative strength of angular noise	0.3
d	Agent displacement per time step	1 m/ts
p	Probability of moving per time step while grazing	0.05
Shepherd parameters		
ds	Shepherd displacement per time step	1.5 m/ts
e	Relative strength of angular noise	0.3

$R = 100$. For the second goal (successfully guiding the sheep to the pasture), the agent receives a reward $R = 1000$, otherwise the agent receives reward $R = 0$. The driving position is a state with the angle $\theta_{sh} <= 15$ deg and the distance between the shepherd and the closest sheep is within (10–50)% of the shepherd detection distance r_s.

12.6.4 Results

Figure 12.8 shows the success rate of the XCS-based shepherd over learning iterations. A success is counted when the shepherd guides the sheep to the pasture area within the specified number of steps (1000 steps), otherwise the trial is counted as a failure. Experimental results show that XCS reached 96% success rate on average after 1700 learning iterations. Through a set of random interactions with the environment, the shepherding agent learned a hierarchical policy to first approach the sheep and then guide them to the pasture area. For every state in the herding environment and current sub-goal, the appropriate actions are determined and the agent moves towards the target position.

In addition to success rate, Fig. 12.9 reports the performance of XCS in the shepherding environment. The figure shows the average learning performance during the exploitation mode over 10 runs. The error bars represent the standard deviation of the number of steps taken to reach the goal. We can see XCS convergence trend after a certain number of learning iterations where the system performance plateaus

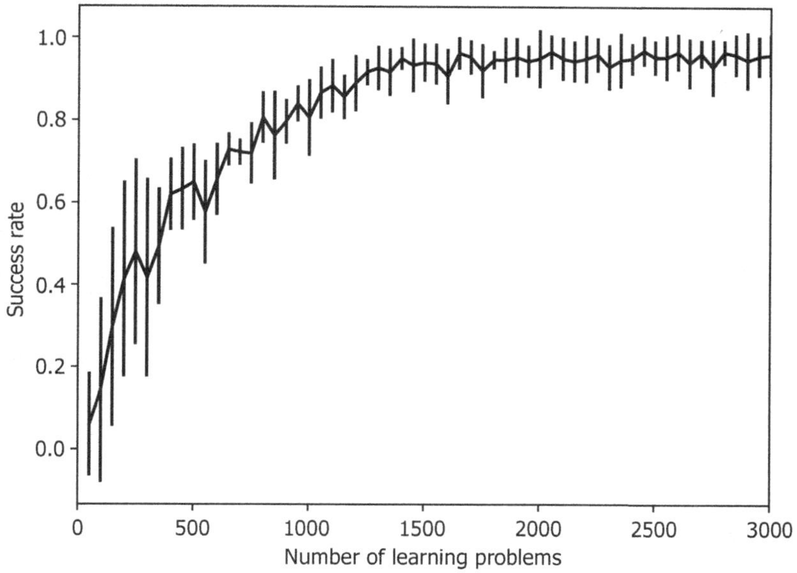

Fig. 12.8 Shepherding success rate

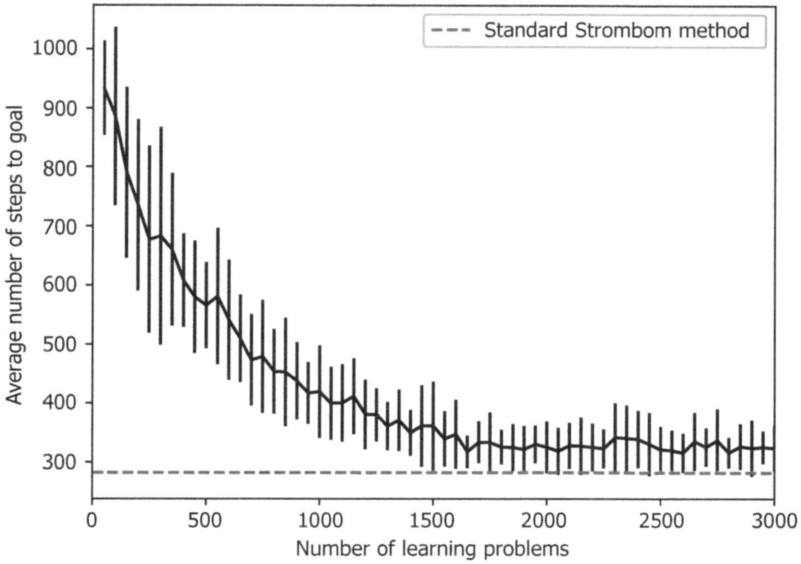

Fig. 12.9 Performance of XCS in shepherding problem

after 1700 learning iterations. The results of several simulations show that the proposed method is capable of achieving comparable performance to the standard Strömbom's method (shown in dashed red line).

Table 12.2 shows sample classifiers (rules) generated by the XCS model and used to guide the sheep-dog in the simulations. The fitness of each rule reflects the confidence in the rule's accuracy. The experience of a rule reflects the number of learning steps that the rule participated in and its accuracy was evaluated. The rule set generated at the end of each training cycle captures the sheep-dog behaviour in different situations. For example, the second and the fourth rules show that whenever the shepherd gets too close to any of the sheep it then stops to avoid splitting the sheep group regardless of the values of the distance to target or the shepherd angle. The fitness and experience of classifiers play an important role in deciding model decision where conflicting classifiers exist (classifiers with the same conditions but with different actions such as Classifiers 5 and 6).

12.7 Summary and Future Work

The aim of this chapter was to shed light on the capabilities of rule-based artificial intelligence to address HAT challenges. Literature from human autonomy interaction, cognitive engineering, and adaptive systems were leveraged to identify significant challenges in the design and development of successful human swarm teaming. Different challenges from the literature were discussed for human-robot teaming and how these challenges impact systems comprised of swarm of robots. We then proposed a taxonomy for these challenges in HAT based on artificial intelligence design perspectives. Three groups of challenges were proposed: transparency; flexibility; and coordination. In the second part of the chapter, we presented an overview of rule learning systems in general and focused on LCS framework. We also discussed the characteristics of LCS that enable it to address the highlighted challenges for HAT. A case study was presented and discussed where a LCS based model was designed and tested in a sheep-dog herding simulation.

Future work will aim at scaling the proposed model to more complex multi-agent scenario where multiple LCS-based shepherd interact and coordinate together to herd the sheep. For this complex task it may be desirable to co-evolve multiple behaviours of multiple cooperating XCS-based agents.

Table 12.2 Examples of classifiers (rules) generated by XCS-based shepherding model

Id	Condition	Action	Fitness	Experience
1	Shepherd to nearest sheep distance is between (14,28) **AND** shepherd to target distance is between (60,84) **AND** angle of shepherd to target and shepherd to GCM is between (36,64)	Move down left	0.999	3536
2	Shepherd to nearest sheep distance is between (0,14) **AND** shepherd to target distance is between (126,140) **AND** angle of shepherd to target and shepherd to GCM is between (36,64)	Stop	0.999	3400
3	Shepherd to nearest sheep distance is between (42,56) **AND** shepherd to target distance is between (126,140) **AND** angle of shepherd to target and shepherd to GCM is between (288,324)	Move up right	0.999	4336
4	Shepherd to nearest sheep distance is between (0,14) **AND** shepherd to target distance is between (126,140) **AND** angle of shepherd to target and shepherd to GCM is between (288,324)	Stop	0.999	4320
5	Shepherd to nearest sheep distance is between (28,42) **AND** shepherd to target distance is between (84,98) **AND** angle of shepherd to target and shepherd to GCM is between (108,144)	Move up left	0.999	3961
6	Shepherd to nearest sheep distance is between (28,42) **AND** shepherd to target distance is between (84,98) **AND** angle of shepherd to target and shepherd to GCM is between (108,144)	Move up right	0.561	2498
7	Shepherd to nearest sheep distance is between (70,84) **AND** shepherd to target distance is between (56,70) **AND** angle of shepherd to target and shepherd to GCM is between (108,144)	Move right	0.999	4561
8	Shepherd to nearest sheep distance is between (14,28) **AND** shepherd to target distance is between (84,98) **AND** angle of shepherd to target and shepherd to GCM is between (0,36)	Move down left	0.999	6581
9	Shepherd to nearest sheep distance is between (14,28) **AND** shepherd to target distance is between (126,140) **AND** angle of shepherd to target and shepherd to GCM is between (324,360)	Move up	0.999	3521
10	Shepherd to nearest sheep distance is between (14,28) **AND** shepherd to target distance is between (126,140) **AND** angle of shepherd to target and shepherd to GCM is between (324,360)	Move down	0.999	3217

References

1. Bohren, J., Rusu, R.B., Jones, E.G., Marder-Eppstein, E., Pantofaru, C., Wise, M., Mösenlechner, L., Meeussen, W., Holzer, S.: Towards autonomous robotic butlers: Lessons learned with the PR2. In: 2011 IEEE International Conference on Robotics and Automation, pp. 5568–5575. IEEE, Piscataway (2011)
2. Bond, A.H., Gasser, L.: Readings in Distributed Artificial Intelligence. Morgan Kaufmann, Burlington (2014)
3. Bull, L., Fogarty, T.C., Snaith, M.: Evolution in multi-agent systems: Evolving communicating classifier systems for gait in a quadrupedal robot. In: Proceedings of the 6th International Conference on Genetic Algorithms, pp. 382–388. Morgan Kaufmann, Burlington (1995)
4. Butz, M.: Rule-Based Evolutionary Online Learning Systems: Learning Bound, Classification, and Prediction (2004)
5. Butz, M.V.: Rule-Based Evolutionary Online Learning Systems. Springer, Berlin (2006)
6. Butz, M.V., Goldberg, D.E.: Bounding the population size in XCS to ensure reproductive opportunities. In: Genetic and Evolutionary Computation Conference, pp. 1844–1856. Springer, Berlin (2003)
7. Butz, M.V., Pelikan, M.: Analyzing the evolutionary pressures in XCS. In: Proceedings of the Third Genetic and Evolutionary Computation Conference (GECCO-2001), vol. 935, p. 942. Citeseer (2001)
8. Butz, M.V., Wilson, S.W.: An algorithmic description of XCS. In: International Workshop on Learning Classifier Systems, pp. 253–272. Springer, Berlin (2000)
9. Cao, Y., Ireson, N., Bull, L., Miles, R.: Design of a traffic junction controller using classifier system and fuzzy logic. In: International Conference on Computational Intelligence, pp. 342–353. Springer, Berlin (1999)
10. Carse, B., Fogarty, T.C., Munro, A.: Adaptive distributed routing using evolutionary fuzzy control. In: Proceedings of the 6th International Conference on Genetic Algorithms, pp. 389–396. Morgan Kaufmann, Burlington (1995)
11. Carse, B., Fogarty, T.C., Munro, A.: Evolving temporal fuzzy rule-bases for distributed routing control in telecommunication networks. Genetic Algorithms Soft Comput. Stud. Fuzziness Soft Comput. **8**, 467–488 (1996)
12. Chen, J.Y., Procci, K., Boyce, M., Wright, J., Garcia, A., Barnes, M.: Situation awareness-based agent transparency. Technical Report, Army Research Lab Aberdeen Proving Ground MD Human Research and Engineering ... (2014)
13. Debie, E., Shafi, K.: Implications of the curse of dimensionality for supervised learning classifier systems: theoretical and empirical analyses. Pattern Anal. Appl. **22**(2), 519–536 (2019). https://doi.org/10.1007/s10044-017-0649-0
14. Debie, E., Shafi, K., Lokan, C., Merrick, K.: Investigating differential evolution based rule discovery in learning classifier systems. In: 2013 IEEE Symposium on Differential Evolution (SDE), pp. 77–84. IEEE, Piscataway (2013)
15. Debie, E., Shafi, K., Merrick, K., Lokan, C.: An online evolutionary rule learning algorithm with incremental attribute discretization. In: 2014 IEEE Congress on Evolutionary Computation (CEC), pp. 1116–1123. IEEE, Piscataway (2014)
16. Dorigo, M., Colombetti, M.: Robot Shaping: An Experiment in Behavior Engineering. MIT Press, Cambridge (1998)
17. Dorigo, M., Schnepf, U.: Genetics-based machine learning and behavior-based robotics: a new synthesis. IEEE Trans. Syst. Man Cybern. **23**(1), 141–154 (1993)
18. Fox, J., Glasspool, D., Grecu, D., Modgil, S., South, M., Patkar, V.: Argumentation-based inference and decision making–a medical perspective. IEEE Intell. Syst. **22**(6), 34–41 (2007)
19. Fredivianus, N., Prothmann, H., Schmeck, H.: XCS revisited: a novel discovery component for the extended classifier system. In: Asia-Pacific Conference on Simulated Evolution and Learning, pp. 289–298. Springer, Berlin (2010)

20. Hayes, B., Shah, J.A.: Improving robot controller transparency through autonomous policy explanation. In: 2017 12th ACM/IEEE International Conference on Human-Robot Interaction (HRI), pp. 303–312. IEEE, Piscataway (2017)
21. Holland, J.: Progress in theoretical biology. Adaptation **4**, 264–293 (1976)
22. Holmes, J.H., Lanzi, P.L., Stolzmann, W., Wilson, S.W.: Learning classifier systems: New models, successful applications. Inf. Process. Lett. **82**(1), 23–30 (2002)
23. Irvan, M., Yamada, T., Terano, T.: Is XCS approach good for organizational-learning oriented classifier systems? In: 24th Annual Conference, 2010, pp. 1C52–1C52 (2010)
24. Klien, G., Woods, D.D., Bradshaw, J.M., Hoffman, R.R., Feltovich, P.J.: Ten challenges for making automation a "team player" in joint human-agent activity. IEEE Intell. Syst. **19**(6), 91–95 (2004)
25. Kulesza, T., Burnett, M., Wong, W.K., Stumpf, S.: Principles of explanatory debugging to personalize interactive machine learning. In: Proceedings of the 20th International Conference on Intelligent User Interfaces, pp. 126–137. ACM, New York (2015)
26. McDermotta, P.L., Walkera, K.E., Dominguez, C.O., Nelsonb, A., Kasdaglis, N.: Quenching the thirst for human-machine teaming guidance: Helping military systems acquisition leverage cognitive engineering research. In: 13th International Conference on Naturalistic Decision Making, pp. 236–240 (2017)
27. Mercado, J.E., Rupp, M.A., Chen, J.Y., Barnes, M.J., Barber, D., Procci, K.: Intelligent agent transparency in human–agent teaming for multi-UXV management. Hum. Factors **58**(3), 401–415 (2016)
28. Miller, T.: Explanation in artificial intelligence: insights from the social sciences. Artif. Intell. **267**, 1–38 (2019)
29. Orriols-Puig, A., Llorà, X., Goldberg, D.E.: How xcs deals with rarities in domains with continuous attributes. In: Proceedings of the 12th Annual Conference on Genetic and Evolutionary Computation, pp. 1023–1030. ACM, New York (2010)
30. Sabzehzar, A., Shan, W., Panahi, M.S., Saremi, O.: An improved eXtended classifier system for the real-time-input real-time-output (XCSRR) stability control of a biped robot. Procedia Comput. Sci. **61**, 492–499 (2015)
31. Saxon, S., Barry, A.: XCS and the monk's problems. In: International Workshop on Learning Classifier Systems, pp. 223–242. Springer, Berlin (1999)
32. Schulenburg, S., Ross, P.: Strength and money: An LCS approach to increasing returns. In: International Workshop on Learning Classifier Systems, pp. 114–137. Springer, Berlin (2000)
33. Sen, S., Sekaran, M.: Multiagent coordination with learning classifier systems. In: International Joint Conference on Artificial Intelligence, pp. 218–233. Springer, Berlin (1995)
34. Sen, S., Sekaran, M., Hale, J., et al.: Learning to coordinate without sharing information. In: Association for the Advancement of Artificial Intelligence, vol. 94, pp. 426–431 (1994)
35. Seredyński, F.: Task scheduling with use of classifier systems. In: AISB International Workshop on Evolutionary Computing, pp. 287–306. Springer, Berlin (1997)
36. Serendynski, F., Cichosz, P., Klebus, G.P.: Learning classifier systems in multi-agent environments. In: First International Conference on Genetic Algorithms in Engineering Systems: Innovations and Applications, pp. 287–292. IET, London (1995)
37. Shao, J., Lin, H., Zhang, K.: Swarm robots reinforcement learning convergence accuracy-based learning classifier systems with gradient descent (XCS-GD). Neur. Comput. Appl. **25**(2), 263–268 (2014)
38. Smith, S.F.: A learning system based on genetic adaptive algorithms. Ph.D. Thesis, University of Pittsburgh (1980)
39. Strömbom, D., Mann, R.P., Wilson, A.M., Hailes, S., Morton, A.J., Sumpter, D.J.T., King, A.J.: Solving the shepherding problem: heuristics for herding autonomous, interacting agents. J. R. Soc. Interf. **11**(100) (2014). https://browzine.com/articles/52614503
40. Studley, M.E.: Learning classifier systems for multi-objective robot control. Ph.D. Thesis, University of the West of England, Bristol (2006)
41. Tesfatsion, L.: Introduction to the JEDC special issue on agent-based computational economics. J. Econ. Dyn. Control **25**, 281–293 (2000)
42. Whitley, D.: A genetic algorithm tutorial. Stat. Comput. **4**(2), 65–85 (1994)

43. Wilson, S.W.: Classifier fitness based on accuracy. Evol. Comput. **3**(2), 149–175 (1995)
44. Wilson, S.W.: Get real! XCS with continuous-valued inputs. In: International Workshop on Learning Classifier Systems, pp. 209–219. Springer, Berlin (1999)
45. Zhong, J., Hu, X., Zhang, J., Gu, M.: Comparison of performance between different selection strategies on simple genetic algorithms. In: International Conference on Computational Intelligence for Modelling, Control and Automation and International Conference on Intelligent Agents, Web Technologies and Internet Commerce (CIMCA-IAWTIC'06), vol. 2, pp. 1115–1121. IEEE, Piscataway (2005)

Chapter 13
Human Performance Operating Picture for Shepherding a Swarm of Autonomous Vehicles

Raul Fernandez Rojas, Essam Debie, Justin Fidock, Michael Barlow, Kathryn Kasmarik, Sreenatha Anavatti, Matthew Garratt, and Hussein A. Abbass

Due to many factors that range from ethical considerations and accountability to technological imperfection in autonomous systems, humans will continue to be an integral part of any meaningful autonomous system. While shepherding offers a technological concept that allows a human to operate a significantly larger number of autonomous systems that a human can handle in today's environment, it is important to realise that a significant amount of accidents today are due to human error. The scalability promise that shepherding offers comes with possible challenges including those arising from the cognitive load imposed on human operators and the need to smoothly integrate the human, as a biological autonomous system, with the wider multi-agent autonomous system of future operating environments. In this chapter, we bring together the dimensions of this complex problem. We present carefully selected factors to cover human performance, especially for cognitively demanding tasks and situation awareness, and how these factors contribute to trust in the system. We then present the Human Factors Operating Picture (H-FOP), which offers a real-time situation awareness picture on human performance in this complex environment. We conclude with the concept of operation for integrating H-FOP with

R. F. Rojas (✉)
Faculty of Science and Technology, University of Canberra, Canberra, ACT, Australia
e-mail: raul.fernandezrojas@canberra.edu.au; r.fernandezrojas@adfa.edu.au

E. Debie · M. Barlow · K. Kasmarik · S. Anavatti · M. Garratt · H. A. Abbass
School of Engineering and Information Technology, University of New South Wales, Canberra, ACT, Australia
e-mail: e.debie@adfa.edu.au; m.barlow@adfa.edu.au; k.merrick@adfa.edu.au; s.anavatti@adfa.edu.au; m.garratt@adfa.edu.au; h.abbass@adfa.edu.au

J. Fidock
Defence Science and Technology Group, Edinburgh, SA, Australia
e-mail: Justin.Fidock@dst.defence.gov.au

H. A. Abbass, R. A. Hunjet (eds.), *Shepherding UxVs for Human-Swarm Teaming*, Unmanned System Technologies,
https://doi.org/10.1007/978-3-030-60898-9_13

the human-swarm teaming problem, with a focus on the reliance of shepherding as the swarm guidance and control method.

13.1 Introduction

Despite significant advances in autonomous systems and swarm robotics, human supervision is still an important part in the operational success of these systems. Robots can perform many complicated tasks and have the physical abilities to operate in extreme conditions (e.g., deep sea, collapsed buildings, planet exploration). While robots can look after the dirty, dull, and dangerous, humans need to offer ethical guidance, accountability functions, and cognitive governance for mission assurance.

Operating in complex environments such as search and rescue after a devastating hurricane or an earthquake is a non-trivial endeavour. The complexity imposed by managing many robots in these environments calls for the design of artificial intelligence (AI) software-defined agents (AISDA) to assist the human. These AISDA need to assist the human in the guidance and control of multiple robots, including swarms. Different types of AISDA could include those responsible to provide the human with key information needed to understand the distributed context of these complex situations [29], and interfaces for concepts such as shepherding to allow the human to operate simultaneously on a large number of robots using a few control points. Other forms of AISDA may include those offering real-time situation awareness picture on the cognitive health of the humans operating these systems. This last type is the context of this chapter. We present and discuss the Human-Factors Operating Picture (H-FOP), whereby AISDA are able to assess in real-time the cognitive demands on a human operator while interfacing with other agents such as a shepherding agent for swarm guidance and control.

This chapter is organised as follows. We first present on human performance and how models of human performance can improve the design of Human-Swarm teams. We do not attempt to provide a comprehensive review of human performance methods, but instead elicit key aspects of human performance and possible implications when shepherding a swarm. We then expand this discussion in four dimensions: task- and system-related metrics, cognitive workload metrics, situation awareness, and trust. We conclude with the H-FOP and a swarm shepherding example.

13.2 Human Performance

Human performance refers to the study and understanding of the types of human factors and ergonomics issues that can be addressed to constrain, enhance, and refine models of the interaction between humans and a system [57]. This information can be used for continuous and effective monitoring of human performance in diverse settings. In general, human performance can be seen as the successful accomplishment of a task by a single human operator or a group of human operators. However, understanding human factors is not only considering the physical and psychological abilities of each person (individual differences play a factor) but also other external factors (e.g., the environment, or other team members) that affect how people interact with different machine systems.

In the Human-Swarm domain, human performance refers to the human factors associated with the interaction between the human and the robotic swarm that affect the system-level performance to achieve common goals or tasks. In this domain, tasks can have diverse objectives, from performing a task manually alongside the swarm to a high-level supervisory control of an automated swarm system [33]. Regardless of the complexity of the task, human factors can assist in the identification of a metric of human performance for the task at hand. Often, in the quest to improve Human-Swarm systems, the main focus of attention is the technological part and, unfortunately, leaves the human element behind [36]. The human side of the system has been relegated to a secondary level of importance. Therefore, in order to improve the whole system, more efforts should be also put on measuring and improving human performance.

One objective of human factors is the use of human performance models to evaluate different design principles and alternatives in the interaction between humans and technological systems [16]. Humans play an important role in any human-machine systems, and identifying the human limitations and skills that might impact (positively or negatively) the performance on such systems, at an earlier stage of the design process, is imperative for many reasons. From the design and technological perspective, it is more appropriate to consider the human capabilities and work around different design ideas prior to the construction of the system, rather than trying to modify or adapt a system after it has been developed. This human performance approach is, in the long run, less expensive, less time consuming, and more efficient.

In addition, in the teleoperation of a robotic swarm, the human operator is constrained with the information provided through a terminal or console (computer-based workstations), when in many cases, the human is in a different geographical location to that of the operational place of the swarm. Through this terminal, the human launches, controls, monitors, and coordinates all the systems and sensors for the operational requirements of the swarm. In today's environment, multiple humans are in charge of a single UAV. As autonomy increases, a single human operator will be required to supervise a whole swarm. The human might not only be in charge of the manoeuvres to securely navigate the swarm but also in charge of taking complex decisions to accomplish the mission, all at the same time.

Research has also demonstrated that the operation of teleoperated robots is affected by other factors such as information lag, loss of visual cues, loss of orientation, attentional switches, depth perception, and motion effects [12, 14]; factors that can severely affect human performance and increase human errors.

Another challenge is the type of role that the human plays in the team. As the role of the human operator has transitioned from a manual controller in early teleoperated robots to one of a supervisory controller and decision-maker, to a more active cooperative player in Human-Swarm teams, factors affecting the human performance have also evolved [96]. Although the use of more automation in the teleoperation of robots and swarm control has been designed to ease their operations and to alleviate human factors such as workload, one of the major problems in this domain is the increased amount of automation used in recent applications. Research has shown that increased levels of automation negatively impact the human performance and also produce new problems[76]. These new problems include failure to check instruments, vigilance decrement, overconfidence to follow standard protocols, automation-induced complacency, and increased response time to detect problems [96]. Therefore, these failures increase the risk of human performance and overall team's performance decline.

From these examples, the needs to measure and predict human performance to mitigate these changes, and improve and maintain the overall Human-Swarm team performance, are apparent.

13.3 Performance Measures in Human-Swarm Teams

Many metrics have been proposed for human performance evaluation in various industries and adopted in the human-autonomy, human-robot, and Human-Swarm domain with some adaptations [45]. Other complementary measuring techniques continue to be implemented for each specific application by developing theoretical and empirical models to overcome the limitations with the main measures [34]. However, specific needs and characteristics that arise from the interaction with swarms might limit the adoption from other domains, problems such as complex movement, communication, leadership behaviours, greater uncertainty, visualization overload, and single control points such as the sheepherding control in this book [37]. These collective behaviours are not examined in classic human-robot interaction.

Human performance metrics aim to measure the success of an operator (or a group of operators) to accomplish a task. In the literature, metrics in human-swarm interaction come in multiple types [45]. We will present in this section an overview of Task- and system-related measures, human-related measures (subjective and objective), workload and trust metrics.

13.3.1 Task- and System-Related Measures

Task- and system-related measures give an indication of an operator's success in the assigned task by measuring different dimensions related to the completion of the task. For instance, Gawron [33] grouped well-known performance measures into six categories; although the related metrics in each category can be implemented in a Human-Swarm team, these metrics represent an overview and can be generalised for many domains and applications. These metrics are:

- **Accuracy** measures the level of correctness in the completion of a task, error is also used to measure the departure from an expected value; some related measures in this category are: percent correct, probability of correct detections, root-mean square, number of errors, and percent errors.
- **Time** measures how long a task takes to be completed and it reflects the operator's ability to complete the task, respond to changes, and supervise the swarm [37]; measures in this category are: detection time, movement time, search time, time to complete, and settling time.
- **Task battery** refers to a group of two or more tasks implemented sequentially or in parallel, and it serves to assess the effects of each task on different human's abilities. Metrics in this category are adopted from other domains, such as the Multiple Attribute Task Battery (MATB) or the Shared Aviation Task Battery (SAT-B) [23].
- **Domain-specific measures** assess the operator's skills to achieve a group of related tasks, and identify effects on the use of different control configurations. Similar to the task battery measures, these measures are adopted from other domains. These metrics are: Control Input Activity to measure activity for the aileron and elevator as a measure of flight-path control [15], Pilot Performance Index to measure performance (e.g., take off, climb, en route, decent, initial and final approach) for air transport missions [97], or the Driving Parameters Test to measure driver behaviour (e.g., speed, lateral distance, steering angle), system performance (e.g., time to completion, brake errors), and attentional behaviour (e.g., traffic lights, head movements) [8].
- **Critical incident** metrics are used to assess behavioural performance during activities that have a critical significance. These are generally applied to find worst-case scenario performance and to identify top priorities and actions in difficult situations [31].
- **Team performance** metrics are employed to assess the aptitude of a human team to accomplish a task, and they can serve as a configuration tool to adapt different coordination methods in particular scenarios and user preferences [84]. Team performance metrics from other highly complex domains can be applied in Human-Swarm systems, such as aviation crews, medical teams, or emergency response units [94].

13.3.2 Human-Related Measures

Metrics to assess Human Performance are based on human factors (or attributes) that affect (positively or negatively) the interaction between humans and swarms. These metrics mostly originated in human factors, human-computer interaction, human-robot interaction, and air-traffic control operations [37]. In this chapter we focus our attention on: workload, situation awareness, and trust due to their well-known effect on performance and the large research interest in other domains to use technological tools to gain objective measures of these human factors.

Human-related metrics can be divided into: (1) Subjective measures based on the operator's opinions, answers to questionnaires, or interviews. Other sources of subjective metrics are obtained by naturalistic observations and evaluations made by domain experts. (2) Objective measures are generally based on experimental methods used to collect physiological and/or behavioural information by a single sensor or a combination of different types of sensors (e.g., heart rate, skin temperature, facial expressions) simultaneously. Objective measures make it possible to obtain reliable information about physical stressors such as workload or situation awareness.

13.3.3 Workload

Mental (or cognitive) workload refers to the relationship between the amount of mental demands that a task imposes on a human operator and the cognitive resources available to be supplied by the human operator to complete the task [81]. When there is an increase in the intensity of difficulty of a task, there is also an increase in cognitive workload and a decrease in cognitive resources as more mental resources are needed. Research has shown that when an operator is under high cognitive workload, the cognitive workload approaches the operator's cognitive capacity, suboptimal decisions and human errors are expected. However, low workload can also lead to errors in a Human-Swarm team, since operators can become tired or distracted and might miss important information to maintain operational success.

Humans have a limited amount of resources (both physically and mentally); therefore, optimising these resources towards specific sets of tasks will produce better results. Understanding these human limitations is a challenge within work contexts due to many factors such as: individual differences (gender, age, ethnicity, etc.), motivational level or effort, previous experience, different strategies to solve problems. For example, the level of difficulty to accomplish a task might be seen differently by two operators; operator A might see the task difficult at first, but then find a good strategy to solve the task, while operator B might find the task extremely difficult and fail to complete the task. As human (or operator) resources are limited, there is a problem when a task demands more resources that are currently available [65]. Therefore, the need to carefully design tasks, that can be successfully accomplished by the users, requires more attention.

Explaining cognitive workload in terms of the task demands and the available resources to meet these demands has been explored in previous models of cognitive workload. Wickens [107] used attentional resources to explain workload as a function of the available resources to compensate task demands; in this demand-resource model, when demands approach its maximum, experienced operators either adapt their strategy to compensate or their performance declines [116]. Megaw [69] called this type of workload model Limited Resource Model, one of the assumptions of this model is that the human's processing capacity is limited and it is shared among other physiological activities such as perception, comprehension, and response.

Figure 13.1 presents the relationship between the task demands (horizontal axis), the operator's resources to complete the task and performance (vertical axis). In this figure it is possible to observe that, when the task demands (or difficulty of a task) reach a point of high (or low) mental resource supply (dash-dotted line), a decrement in performance (solid line) and increase of errors are expected. In under load conditions, the operator losses alertness and attention (disengagement), while in overload conditions, the operator faces higher task complexity, multitasking, or less time for suitable information processing. One possibility to overcome these two conditions is the investment of additional resources by the operator [35], however, this could be a good strategy in cases of underload, but could have an adverse effect in cases of overload.

There is an optimal range of cognitive load (dotted line), which is directly related with best performance. This optimal range can be seen between the two transition points (red lines). These red lines divide the two regions of demand-resource space, the area on the left is also called the "reserve capacity" area, and the region on the right is called the "overload region" [116]. These two regions are of interest to human factors researchers due to their implication for real-time prediction and

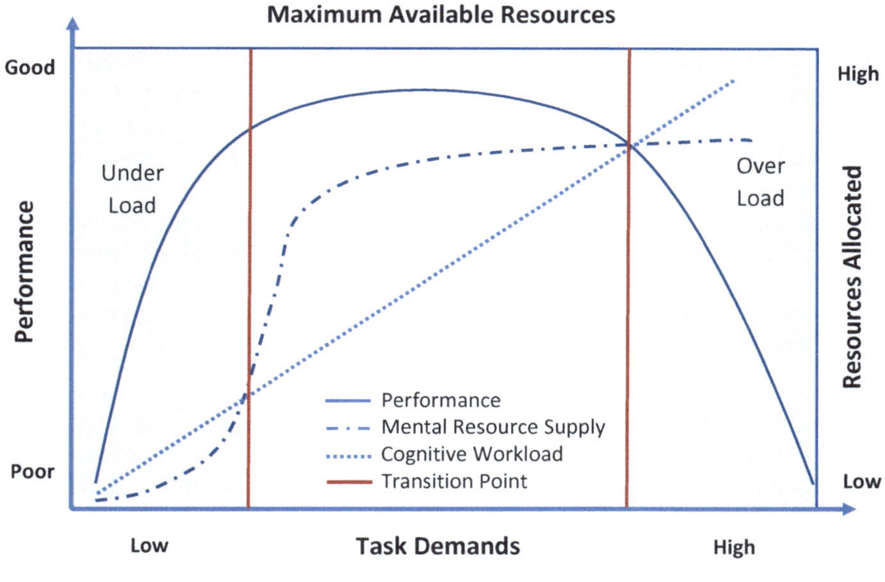

Fig. 13.1 Workload model, showing the relationship between performance and task demands

assessment of cognitive workload. For example, ergonomists and designers are interested in predicting when performance drops as a consequence of demand exceeding supply, and in design strategies to overcome situations when overload or underload occurs. For instance, reducing the resource demands of the task (e.g., reducing the task load or automating subtasks), increasing operator's training, distributing the task load among the team members, or reducing multitasking to increase sequential tasks. Therefore, understanding and modelling these transitions (red lines) will help develop better solutions to maintain good performance longer and safer systems.

There are different approaches to perform subjective measures of cognitive workload, including rating scales, questionnaires, and interviews. Many studies have shown that subjective measures are reliable, highly sensitive to different levels of workload, non-intrusive, easy to apply, can be used in many domains and applications, highly tolerable by the operator [34, 40, 92]. There are many techniques in the literature and here we present the most common ones.

- The Subjective Workload Assessment Technique (SWAT) is a rating scale proposed by Reid and Nygren [85]. SWAT classifies an operator's experiences into three dimensions (Time Load, Mental Effort, and Psychological Stress), each with three categorical levels (Low, Moderate, and High). These dimensions are combined to produce a rating scale of cognitive workload. SWAT is considered to be the most common technique reported in the literature [9]. This technique was originally designed to measure the workload associated with aircraft cockpit and crew-station environments.
- Another technique to assess operator's mental workload is the NASA Task Load Index (NASA-TLX) [38]. This method uses six dimensions: mental demand, physical demand, temporal demand, performance, frustration level, and effort, each with 10 or 20-points scale. An overall rating is then calculated as the weighted mean of all six ratings. One of the limitations of NASA-TLX is the lack of continuous measurement during task performance. Participants typically answer survey questions after task is completed and they may be unable to recall workload experienced during a trial.
- Other rating scales techniques proposed in the literature include the Modified Cooper-Harper Scale [109], and Subjective Workload Dominance (SWORD) [104]. Other subjective measures replace the need for asking operators to perform and rate the real task, with expert opinions to rate task performance based upon prior experience with other comparable systems [111].

Although, subjective measures are considered the gold standard due to ease of use, objective measures are the most desirable because they are a direct indicator of the operator's performance [96]. Cognitive workload is a multidimensional construct that exhibits the interaction between the operator's cognitive skills (e.g., information-processing behaviour, strategies, training, experience, and effort) and involuntary activities in the nervous, cardiovascular, muscular, and respiratory system [66]. Objective measures allow real-time assessment of the operator's workload, by measuring these distinctive physiological changes, which can be examined to make inferences about the cognitive workload. Based on this description, objective measures can be classified into: neurophysiological, physiological, and behavioural measures.

- Neurophysiological indicators of mental workload include neuroimaging methods such as electroencephalography (EEG) and functional near-infrared spectroscopy (fNIRS). EEG measures the brain's electrical activity and examination of distinctive patterns in brain activity indicate different levels of mental workload. EEG studies in workload have reported sensitivity to different intensities of workload in various tasks, such as: air traffic control [1], car driving [21], flight simulation [5], and shepherding [87]. Despite its proved sensitivity to workload, the number of sensors and their set up might restrict the operators' capacity to move and to complete their activities, which limits the use of EEG in some applications. fNIRS, on the other hand, uses near-infrared light to measure oxygenated (OxyHb) and deoxygenated (DeoxyHb) haemoglobin concentrations in the cerebral cortex [89, 90]. fNIRS serves as an indication of the amount of mental effort employed in a particular cortical region in response to the task at hand[88]. Previous studies have reported that in activities of high cognitive workload there is a decrease in DeoxyHb and an increase in OxyHb [83].
- Physiological indicators are linked to changes in the activity of the autonomic nervous system (ANS) [66]. Mental stressors affect involuntary activities such as those induced by cardiac and respiratory activities [102]. For example, under high workload conditions, an increase of heart rate is expected to supply the body with higher levels of oxygenated blood. In these conditions, a technology such as electrocardiography (ECG) can be used to measure the electrical activity generated by the heart and provide an indication of changes in the ANS. Metrics such as heart rate variability, inter-beat-interval, and blood pressure have shown sensitivity to increased cognitive effort [24]. Another useful technology is electrodermal activity (EDA), also known as galvanic skin response (GSR); in this method variations in electrical conductance of the skin are measured by the use of sensors attached to the skin. In general, skin conductance shows a positive correlation with increase of workload [77]. Electrooculography (EOG) is another technique to measure cognitive workload, this technique measures the electrical potential field between the cornea and the retina to record eye movement. Sensors are attached above and below each eye, and eye movements are recorded as changes in the electric field [73]. Pupillometry is another technique to measure eye activity, this method measures the pupil's response to different visual stimuli. Previous studies have reported that blink duration declines when task demands increase and during periods of high cognitive load [7].
- Behavioural information is another approach for assessing cognitive workload. Behavioural metrics including keystroke dynamics, mouse tracking, and body positioning can be used as effective metrics for assessing cognitive workload. In human-computer interaction (HCI), user attention and engagement can be measured by activities such as clicking, scrolling, or keyboard typing [61]. For example, keystroke features (i.e., use of backspace, delete, end, and arrow keys)

have shown changes in typing patterns to be affected by increased cognitive workload during mental multiplication and memory, which in turn affected task performance [105]. Body posture is another useful behavioural measure, seated posture was shown to correlate with task engagement and boredom[24]. Increased pressure on the seat was associated with increased engagement, while increased pressure on the back of the chair was associated with boredom.

13.3.4 Situation Awareness

Situational awareness (SA) is a term used to describe the recognition of the elements in the surrounding environment in a given situation, understanding the current meaning of these elements, prediction of the behaviour of these elements in the future, and then using this information to act [26]. In her seminal work, Mica Endsley [25] provided a situational awareness framework that decomposes the situational awareness into three levels: (1) Perception of elements in current situation, to gather all the information that is currently available; (2) comprehension of current situation, to synthesise raw information into meaningful patterns/clues, and to understand and identify the importance of the current situation; and (3) projection of future states, where cumulative understanding of the elements in the surrounding environment is used to predict the status and dynamics of these elements in the future. For instance, pilots need to know their aircraft status at all times, the operational environment in which they are flying, and the relationship between both components (e.g., strong winds created turbulence). Therefore, SA has been considered as a critical component in decision-making [33].

Early research on SA took place in the military and aviation domains [49, 51]. SA has been widely adopted by researchers in other fields and domains including sport, medicine [6], and human-machine interactive systems (i.e., teleoperation systems). SA has become a critical research component in almost any area that involves humans performing tasks in complex environments. Considering that the operator's tasks in most complex systems (e.g., nuclear power plants, battlefield, engine room of a ship) can be summarised as a series of cognitive activities such as monitoring, detecting, understanding, diagnosing, decision-making, planning, and implementing, the operator's tasks can be significantly influenced by the operators' SA [34]. Therefore, correct situation awareness is accepted as one of the most significant additions to safe operation in complex systems.

Loss of SA has been correlated with poor human performance [47]. In teleoperated systems, one of the major factors that causes loss of SA is data transmission difficulties [82]. Data transmission failures exhibit a common set of problems including information latency (timing is too late to be effective), and information loss (content is completely lost, not consistently complete or inaccurate) [60]. Information loss, in particular, has been shown to have a degrading impact on SA. In aviation, for instance, Thornton [100] reported that the number of communications requesting clarification made by aircraft crew members was positively correlated with the number of committed errors. In surgery management, Helmreich and Davies [39] observed that communication breakdowns that resulted in loss of awareness of the patient's condition resulted in failure to note adverse changes in patient status.

Another critical factor that contributes to the loss of situation awareness is the use of new technologies in the operator's operatorial environment. Although, it is expected that operators can be better aware of situation with the use of new technologies, it might have an adverse effect and deteriorate SA of the operators [34]. For example, a study in the implications of crew performance issues in advanced control rooms of nuclear plants found that the use of technology has critical implications for plan safety and affects the role of the operators in the plant [79]. With the use of technology, information is presented in a compact form (e.g., processed and integrated into a higher level of information) and operator's awareness can be affected due to different interpretation of information. Another problem is the use of poorly designed displays, these can mislead or confuse the operator, make important information difficult to find, and even operators can lose their ability to use well-learned skills to recognise critical information [79]. Thus, the need of an integrated validation system to ensure the appropriate use of technology to support operators' situation awareness and overall performance.

Another important component in the composition of SA is the perception of time and the temporal dynamics associated with events. Time is an important part in the comprehension of SA, it refers to the ability to identify time constraints for an event to occur, or when an event will have an impact on the operator's goals and tasks; but time also is a critical part of planning ahead and knowing when an action, response, or other activity needs to be taken [27]. Time is also important in the dynamics and natural occurrence of real-life situations, and how this constant change (and the rate of change) is important for planning future situations. As the situation is constantly evolving, the operator's SA must also be constantly changing or there is an inadequate or outdated perception of SA. In highly dynamic environments, this demands the operators to adapt their cognitive strategies for maintaining SA [27].

A well-accepted SA model presented by Endsley [26] is shown in Fig. 13.2. In particular, this model of situation awareness is based on information-processing theory [108]. In this model, it is possible to observe a number of factors that influence SA, including the individual and task/system factors.

Individual factors refer to the operators' own ability to remain attentive and retain information in working memory, which affects their perception of situation awareness. When the operator has learned a new skill through experience or training, by storing this new information in long-term memory in the form of schemata and mental models [17]; this skill can facilitate the integration, comprehension, and the projection of such information in a more effective way. Thus, this process frees working memory and efficiently assigns more resources for perceptual processes. In addition, prototypical or repetitive situations are incorporated into mental models to facilitate easy retrieval of actions from long-term memory, which facilitates rapid decision-making [17]. In addition, situation awareness is affected by the operator's

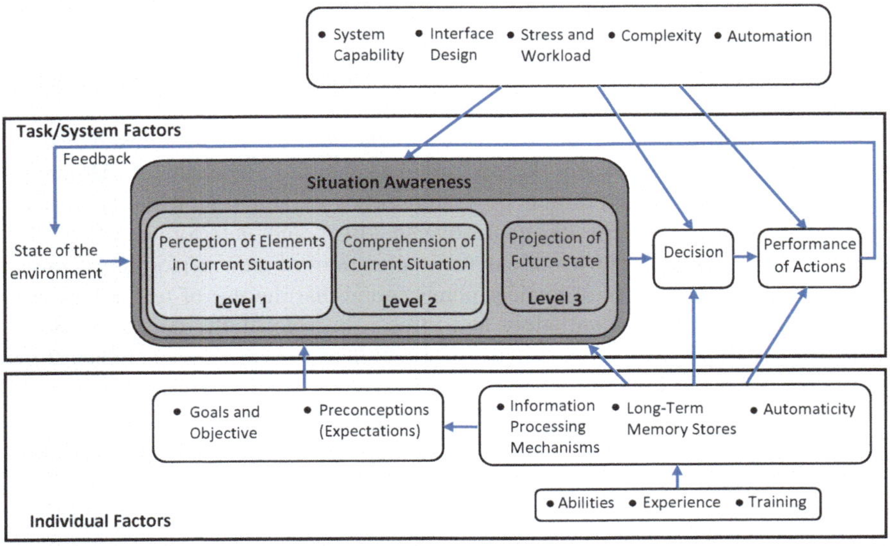

Fig. 13.2 Model of factors affecting Situation Awareness, based on [25]

goals and expectations, which influence how attention and perceptual processes are deployed.

Task-, system-, and environment-related factors play a role in the perception of situation awareness. For instance, workload affects the operators' SA since in high-cognitive tasks the volume of information received by the operators might impede their ability to perceive SA; thus, incomplete perception and lack of information integration will create erroneous SA. Another challenge is the complexity of the system to be operated. In this particular case, developing good strategies and/or internal mental models to aid in the integration of information is imperative; however, it needs extensive training and might be beyond the abilities of the operator [17]. Automation of tasks is another factor that alters the perception of SA. Although, automation facilitates the operation of a task and helps in the reduction of cognitive workload, research has also shown that the use of automation hinders the operators' ability to detect errors, and subsequently decision-making, to overcome problems [53]. Examining these factors for their potential impact in the perception of SA needs to be taken in consideration.

Subjective metrics and questionaries have been developed for the assessment of SA. These methods are well-accepted in the evaluation of design concepts due to their easy and low-cost implementation. Subjective metrics are also easily implemented in less controlled and real-world applications [17]. These measures allow the collection of the operator's perception of SA that can be evaluated against reality. These types of assessments can be administered during interruptions in a simulation or post-tests. However, some of the limitations are that the participants tend to overgeneralise or rationalise their answers [75]. Here, we present some of the most popular techniques in the literature:

- Situational awareness global assessment technique (SAGAT) was developed by Endsley [26] to measure individual situational awareness. In human-machine systems, it is used to evaluate system design to ensure that the system in question supports the operator's SA requirements. SAGAT is used during a simulation of the system. It works by freezing the simulation at randomly selected times so that the operator can answer questions about their perception of the situation at that time. The questions examine the three levels of SA (perception, comprehension, and projection). Subjects' answers are compared with the correct answers that are automatically collected in the computer database.
- Situational awareness rating technique (SART) is another well-known technique, designed by Taylor [99]. This technique is administered post-trial and was originated to assess pilots' SA. Participants need to rate ten dimensions on a seven-point rating scale, with 1 as the lowest rating and 7 as the highest rating. The ten dimensions to assess the operator's SA are: familiarity of the situation, focussing of attention, information quantity, information quality, instability of the situation, concentration of attention, complexity of the situation, variability of the situation, arousal, and spare mental capacity.
- Situational awareness-linked instances adapted to novel tasks (SALIANT), this technique measures SA in team settings. This method has five stages: (1) identify behavioural indicators of team SA, 24 behaviour indicators are identified from the literature; (2) develop scenarios, the scenarios are used to demonstrate team SA behaviours; (3) define specific and acceptable responses; (4) write a script that can be used to evaluate consistently all teams; and (5) create an observation form to describe the scenarios and responses from each team. One advantage of this method is that it identifies the absence or presence of behaviours which serves as indication of level of SA.
- Situation awareness subjective workload dominance (SA-SWORD) is a modified version of SWORD, this technique was developed by Vidulich and Hughes [103]. This technique is focused on the assessment of SA as perceived by the use of displays. Operators evaluate and compare their perception of situation awareness obtained by different monitors. The tasks studied by this method include: flash detection, colour identification, and location.

Some of the limitations from the SA subjective metrics can be overcome by objective measures. For instance, they do not require to stop the simulation, thus, they do not interfere with the primary task. In addition, objective metrics can provide a continuous and real-time indication on the operator's SA. However, one of the disadvantages of objective measures is that they are more expensive to apply, and more difficult to interpret and analyse than subjective metrics. Objective indicators of SA are: behavioural and neurophysiological methods.

- Behavioural indicators, some information about SA can be gained from examining operator's behaviour on specific tasks. For instance, eye fixation has been used in the nuclear industry to measure the operators' SA, this technique is known as visual indicator of situation awareness (VISA) [34]. Eye blinks have been used as an objective measure of SA in pilots during simulated air-to-ground combat fight missions [110]. Other behavioural indices include time to answer a question and corresponding reaction as observed from verbalizations and appropriateness during particular situations [17].

- Neurophysiological measures rely on the relationship between the operator's cognitive activity and the changes in their central and peripheral nervous system during specific stimulus or actions. Electroencephalography (EEG) is a neuroimaging technique used in previous attempts to measure SA objectively. For instance, Catherwood et al. [10] used EEG to map brain activity during loss of situation awareness. In another study, Yeo et al. [115] used EEG to monitor SA in an air traffic controller (ATC) task, the EEG model is based on the response time (latency) from the subjective questionnaire. Fernandez Rojas et al. [28] used EEG to measure SA in presence of delay and dropout of information in a teleoperated system, their findings highlighted EEG power bands that can be used for the objective assessment of SA. Although, neurophysiological measures might help infer whether received information is correctly processed and assessed by an operator to maintain good SA, it might also help determine whether particular stimulus are important in the perception of SA [110].

13.3.5 Trust

Trust is a fundamental attribute in determining the viability of complex swarm systems and its impact on the performance of Human-Swarm teams. In general, human-robot and Human-Swarm systems bring new opportunities to diverse areas of our society. For instance, the use of robotic assembly lines in the manufacturing industry, auto drivers in the design of self-driving cars, and autopilot systems in the trajectory control of airplanes. However, these examples of autonomous systems also bring new challenges in how humans' trust in autonomous partners affect the overall team performance. Trust can define the way people interact with technology, this has been defined as trust in automation [42]. Research has shown that automation is severely restricted and becomes impractical if untrusted by humans, in particular in applications where high levels of cooperation are needed [2].

Trust overlaps across multiple research areas and domains. It is a social concept in a human-human interaction, which is based on the expectation of trustworthiness among agents. For instance, on the one hand the truster assigns a task to the trustee and accepts the risk that the trustee could be untrustworthy; on the other hand, the trustee takes the task and commits to be trustworthy [106]. Similar to the social domain, trust in a human-autonomy scenario can improve the effectiveness and efficiency of the interaction in complex situations. If autonomy is appreciated as trustworthy by the human, the human could improve its reliance on the autonomous system and delegate to it more complex tasks. From these scenarios, the concept of trust among humans, and between autonomous systems and humans is very similar, where in both scenarios, human trust in team members (humans and/or autonomous systems) significantly affects system performance.

The importance of trust in the relationship between the human operator and the swarm impacts (positively or negatively) team performance [43, 44]. With the increasing level of complexity in swarm robotics and other autonomous systems, the role of the human operator has changed from direct manual control to supervisory control [58]. In their role as supervisory controllers, human operators can decide how to interact with the swarm at different degrees of control. However, the amount of interaction between the operator and the swarm might play a factor on the performance of the overall team. Inappropriate dependence (or over trust) on supervisory control might lead to situations in which the operators face abnormal behaviours that the system is not designed for. While, excessive manual control (or distrust) leads to an increase of cognitive workload, which also affects the overall performance of the team.

Another factor that influences human performance is the various levels of trust in automation. For instance, the use of faulty equipment (e.g., uncalibrated sensors) leads to disuse of the technology and creates low levels of trust. On the other hand, high levels of trust in unreliable systems, as in the previous example, leads to overreliance and failure to corroborate the received information and the source of such information; this unjustified assumption of perfect automation leads to poor performance, this has been referred as the complacency issue [81]. However, complacency is not the sole reason for failing to monitor the autonomous system. Parasuraman [80] showed that monitoring of the information sources used by automation was also inadequate when operators had other actions to complete and automation reliability was stable.

Marsh and Dibben [67] proposed three main layers of trust in human-autonomy interaction, these are: dispositional trust, situational trust, and learned trust. This model of trust is presented in Fig. 13.3.

Dispositional trust refers to the operator's tendency to trust automation, independent of the operational situation or specific system. Hoff et al. [42] identified four sources of variability in dispositional trust: culture, age, gender, and personality. Research has shown that people with high levels of dispositional trust in automation tend to trust reliable systems, but their level of trust is considerably affected when errors occur [42]. Overall, individual differences significantly influence the disposition of individuals to trust automated systems. Therefore, taking into consideration individual differences and tendencies of various operators might increase the acceptability of the autonomous system.

Situational trust refers to the variability of trust depending on the situational context. There are two main causes of variability in this type of trust: the environment (external) and the context-dependent characteristics of the human operator (internal) [42]. External variability refers to the environmental factors that can shift the dynamics of human-automation trust. The operator's trust can be influenced by the complexity of the task, the potential risk and benefits associated with the

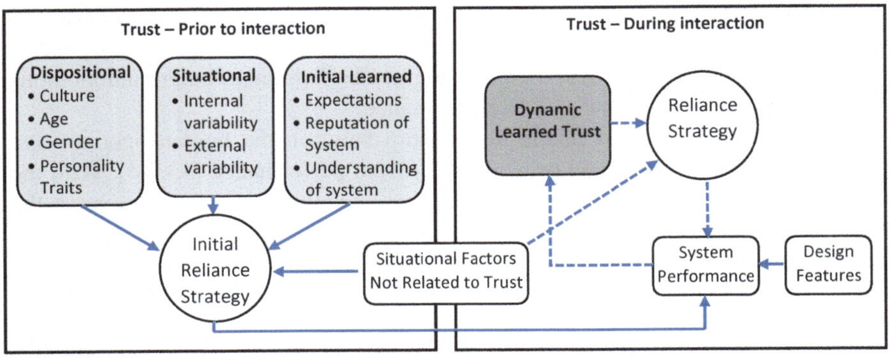

Fig. 13.3 Factors that affect trust in automation. The dotted lines represent factors that can change during the interaction. Model based on [42]

use of automation, the type of system, the organizational settings, or the framing of the task. These factors can have a positive or negative effect depending on the specific situational characteristics of the task at hand. For instance, trust can be affected if performance variability is observed when the same system is used in similar tasks; or the operator's reliance on automation can decrease when greater risk is involved in the task [62]. On the other hand, internal variability refers to the operator's characteristics that depend on context; the operator's trust can be affected by self-confidence, level of expertise, mood, or attentional ability [42]. For example, when the operator's self-confidence is high, it is expected that automation is utilised less frequently; subject matter expertise also affects trust in automation, less experienced operators tend to rely more on automation than more experienced operators [93]; another factor that affects the trust formation process is the operator's mood, operators with positive moods exhibit higher initial trust in automation [98].

Learned trust concerns with the previous interactions with automation and the lessons learned from these experiences that can affect the trust of novel systems. This layer of trust is highly influenced by the performance of automation. In real-time situations, the operator's trust might change during the course of an interaction, due to differences in automation's performance; these effects can be described as initial and dynamic trust [42]. Initial learned trust symbolises the trust before interacting with the automated system, this can alter the trust formation process and resulting decisions. Initial learned trust can also be seen as pre-existing knowledge such as system's reputation, expectations, and opinions and knowledge about the purpose of the automated systems. On the other hand, dynamic learned trust refers to the perceived trust during the interaction. When the performance of the automation fluctuates during the interaction it impacts the operator's trust; thus, the operator's reliance strategy might change, which in turn can affect the system performance (completing the cycle). In addition, design features can affect trust in automation, including: the ease of use, the arrangement and aesthetic of its interface (appearance), communication style (voice, text, visual), the logic exhibited by the automated system (transparency), and the level of control the operator has over the automation.

Finally, the model of trust also presents situational factors not related to trust that affect human-autonomy interaction, in particular with the interdependent relationship among trust, reliance strategy, and performance [42]; these factors are: effort needed to engage a system, time constraints, the operator's situational awareness, and the alternatives to using automation. The dotted arrows in Fig. 13.3 represent factors that can change during the interaction of the system.

Some methods from the fields of economics, organizational and applied psychology have been proposed to measure trust. In the design of measures of trust, the role of trust needs to be understood under different perspectives, such as [106]: through an analysis of what social scientists and psychologists have found in their research on trust; by developing game-based models to represent trust during different interactions; and through the design of computational models of trust to explicitly embed them in decision-making models for human-autonomy interaction.

Initial research in human-autonomy trust has used human-human trust models from social psychology research as a measure of trust. These previous studies have tried to explain the user's development of trust in an intelligent system by adapting instruments of interpersonal measurement or making their own measures [64]. For instance, Larzelere and Houston [56] used questionnaires to measure trust in terms of benevolence and honesty between partners. Kramer [54] investigated trust within organizations, where an individual indicated their degree of trust towards themselves and other people in the same organization. Other studies have shown that these terms might be dynamic and changing over time as the relationships evolve [46]. For example, Rempel et al. [86] provided a hierarchical model of trust and suggested that some factors of trust might develop with time and after more emotional investment.

On the other hand, in studies of trust using human-machine systems, trust has been investigated by using trust questionnaires. For example, Singh et al. [95] presented a rating scale to measure individuals' potential for complacency, by examining attitudes towards different automated devices. In another study, Muir and Moray [72] explored operator's trust in automated systems in a supervisory control scenario using rating scales to assess individuals' attitudes towards reliability and trustworthiness of the autonomous systems. Although, these studies are based on applied psychology and reflected (in part) human-human interaction, these identified some social attributes that influence trust.

Here, we present some subjective measures of trust in human-autonomy systems found in the literature:

- The general trust scale: it is a questionnaire developed by Toshio Yamagishi [114]. It is a 6-item questionnaire is based on Yamagishi's trust scale [113] and uses general statements to measure individual's attitudes about honesty and trustworthiness of others. Although, this scale has been designed as a measure of trust and distrust in psychological studies, it has also been adapted to measure trust in automation [13].

- Trust in close relationships scale: this metric was designed by Rempel et al. [86] to gauge levels of trust in a partner. This model defines trust in hierarchical levels, with the trust at a given stage formulated as the result of the previous stage. According to this model, predictability prevails early in the relationship, and succeeds by dependability and faith [101]. This model suggests that individuals will identify the predictability of an autonomous system by assessing the consistency and accuracy of its behaviour. This judgment of predictability is based on the actual system's predictability, the operator's skills to estimate the behaviour of the autonomous system, and the stability of the operational environment [101].
- Trust in hybrid inspection systems: this trust questionnaire was design by Master et al. [68]. This questionnaire helps measure the effects of human trust and determines how it develops over time in a hybrid inspection system given different types of errors (e.g., misses and false alarms). This metric explores four dimensions of trust: competence, predictability, reliability, and faith. This questionnaire has been used to identify the impact of the levels of trust an operator has in an automated system [101].
- Interpersonal trust scale: this method was design by Rotter [91]. It aims to differentiate people on their propensity to trust others [59] and that this propensity evolves through the number of interactions with other social agents (e.g., co-workers, teachers). The operator's propensity to trust autonomy can directly affect the use of that autonomous system [70]. Thus, research suggests that due to the human component in human-autonomy teams, a trust scale to measure the operator's propensity to trust should be included [112].

As subjective measures require individuals to report their level of trust, these methods face some limitations. For instance, not all participants can accurately describe or define specific trust attitudes or might not be keen to share their true attitude [3]. In addition, as trust can evolve over time depending on the situation, subjective measures do not accurately represent this change, since they represent the individuals' trust at a fixed time interval; thus, subjective measures of trust might lack reliability. To overcome the limitations with subjective measures, some objective measures have been proposed.

- Behavioural measures: these techniques aim to identify unconscious behavioural patterns produced by individuals during their interaction with machines. For example, DeSteno et al. [20] identified a group of non-verbal behaviours in an interaction between a human and a teleoperated robot that can serve as an indication of trust towards the robot [32]. Four non-verbal cues were indicative of untrustworthy behaviour, these are: face touching, hands touching, arms crossing, and leaning backwards. Another behavioural metric used in the literature is proxemics, this method is used in the study of human use of space. Joose et al. [48] measured the distance maintained by the human with respect to the robot during the interaction. Proxemics metrics not only help obtain the individual's perceived safety during a physical interaction with a robot but also the acceptance of autonomous systems [32].

- Neurophysiological measures include methods such as EEG, functional near-infrared spectroscopy (fNIRS), and functional magnetic resonance imaging (fMRI). These methods try to identify neural correlates of different trust-related behaviours. For instance, fNIRS was used to measure human suspicion during the interaction with autonomy, and how this mental state (before making a judgement) affects the decision to trust or distrust information [41]. EEG has been used to study the neural response of human trust at various situations, capability-based trust [22], trust decisions based of initial impressions [4], trust and distrust using word elicitation [78], and trustworthiness in artificial agents during an investment game [106]. Similar to fNIRS, fMIR measures changes associated with blood flow and neural activation; fMRI presents a much higher resolution than EEG and fNIRS, however, fMRI is not portable and much more expensive (both to purchase and maintain). A well-known study on trust using fMRI by Krueger et al. [55], this technique was used to examine the functional brain activity in a trust game during human-human interaction; this study identifies specific brain areas that help maintain conditional and unconditional trust in a partnership.
- Physiological measures have been used to measure physiological reactions to trust. Electrodermal activity (EA) has been used to measure the skin conductance with respect to trust, research has shown that increased electrodermal activity is associated with loss of trust [74]. Electrocardiogram (ECG) measures the electrical activity generated by the heart, ECG activity has shown positive correlation with shared perceptions of technology trustworthiness among members of teams [71]. Heart rate variability is well-known for polygraph tests, it measures the time interval between heartbeats; this type of metric has been utilised to estimate the level of trust in natural dialogs of real-world scenarios in human-robot-human interaction [52].

13.4 Human Performance Operating Picture for Swarm Shepherding

So far, we have discussed different methodologies to measure human performance by means of assessing both task-related (Sect. 13.3.1) and human-related metrics (Sect. 13.3.2). However, their usage and implementation in Human-Swarm teams for a shepherding task have not yet been discussed. In this section, we discuss how both types of metrics can be used and integrated, and the considerations and constraints for their implementation, by providing an overview of our current work. Finally, we also provide an operational framework to design a trusted Human-Swarm team based on our current research.

Human-Swarm teaming for shepherding requires effective coordination of effort between the human operator and the swarm system to achieve mission success. For example, disaster response operations require teams to plan and complete distributed tasks in a timely manner to mitigate losses of human lives [30]. This process of managing resources in an efficient way has been a subject of research in the literature, and diverse multi-agent task allocation approaches have been proposed to solve such type of tasks [11, 18].

This process is defined as adaptive automation, it refers to the dynamic distribution of control of system functions or tasks between the human and the swarm over time with the purpose of optimizing overall team performance [50]. This distribution of functions requires an allocation authority (or agent) that can take the decision to give a task or function to either the human or the swarm; this authority can be seen as a specific function that controls the operation of both the human and the swarm. To decide who is responsible for a given set of tasks, the allocation agent needs to take in consideration changes in the states of both the physical (or virtual) environment and the human. In other words, the allocation logic is based on contextual information from the operational environment and human mental states and/or behaviour to coordinate tasks among the agents.

During this allocation strategy, the operators change their operational approach in how they cooperate with the swarm. In the literature there are two main approaches for how a human operator can work together with an autonomous partner to achieve a common goal. One way in which this cooperation can be achieved is by placing the human operator in-the-loop as an active controller of the swarm in either the same operational location or in teleoperated control. Another way to achieve this is by placing the human on-the-loop, in which the role of the human takes a secondary role as a supervisor and intervenes only when is needed.

Based on this function allocation strategy, Fig. 13.4 presents a close-loop system for Human-Swarm teaming. This model exhibits three main components (or agents):

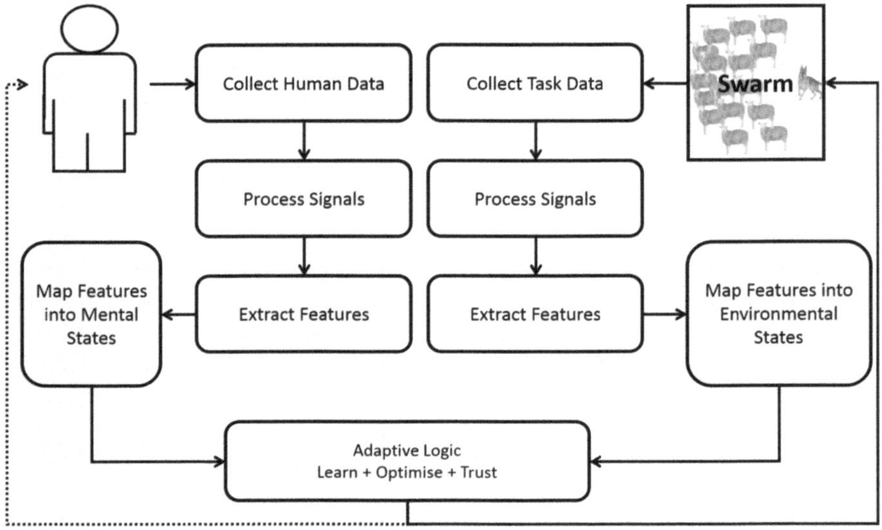

Fig. 13.4 Closed loop Human-Swarm Team. This framework shows how an adaptive allocation logic can improve team performance

the human, the swarm, and the adaptive allocation logic (AAL). In this example, the AAL (an AI agent) receives information about the states of both the human in terms of mental states (e.g., workload, situation awareness, trust) and the shepherd and the swarm (e.g., errors, risk level, time for completion). Then, it triggers the allocation of a task to a given agent based on their state. For instance, if the operator is working in-the-loop (manual control) to guide the swarm and the operator's cognitive state is perceived as overloaded, then the AAL might decide to give autonomous control to the swarm and maintain the operator on-the-loop to supervise the swarm until the operator's state is perceived as normal. The dotted lines represent recommendations sent to the operator and the dashed lines are commands given to the shepherd.

From this image, it is evident that the AAL serves as supervisor and has control over the human operator and the swarm. This overall control allows the AAL to intervene in the team's actions to redirect resources, reallocate tasks, or adjust tasks efforts. However, to achieve this dynamic allocation of tasks and functions an appropriate Human-Swarm modelling and analysis framework is needed. This framework should evaluate Human-Swarm performance based on the impact of the task reallocation on the operator's cognitive state (trust, situation awareness, and workload) and also swarm performance. In order to achieve this high level of adaptation some important design considerations should be considered.

13.4.1 Design Considerations

The objective of human performance metrics is not only to provide evidence of the operator's performance but also the overall team performance, including the autonomous agents in the team. Taking into consideration both the individual performance and the team performance is an important feature that a framework should include to achieve high level of adaptation. Modelling human performance using cognitive states and behaviour is a complex task that involves many factors. Therefore, building reliable human performance metrics that can facilitate high-level AAL needs to be considered within the framework. Some considerations for reliable human performance assessment are [63]:

- Human performance metrics should not interfere with the operators' operations to complete their tasks. Thus, subjective metrics that depend on stopping the task (e.g., freezing a simulation) to collect data should not be considered for realistic applications.
- Realistic scenarios require continuous measures that can provide constant monitoring of the human's performance. This continuous monitoring needs to be stable and with a sample resolution that allows the AAL to intervene in timely manner.
- Reliable human performance metrics need to detect transitions between different states (e.g., from normal cognitive load to overload), so the AAL triggers correct measures depending on the operator's state.

- The metrics need to be able to provide reliable data in diverse body-postures and orientations, and allow the operator to move freely with maintaining the quality of the data.

However, other factors should be considered; for example, the understanding of task requirements, environmental factors, or the needs to achieve individual goals and team goals. Some other factors to be considered are:

- Contextual information from the environment should be considered. Knowing geographical, environmental, or sociological factors in the operational environment, that could affect the team performance, need to be considered. Thus, contextual information should be used when planning and decision-making are done.
- Decision-making needs to integrate individual goals and team goals to assign tasks between the human and the swarm.
- Data integration from different metrics (e.g., EEG, heart rate, or contextual data from the environment) needs to be carefully designed to allow efficient intervention when needed.
- Transparency in decision-making by the AAL improves trust in the Human-Swarm team.
- Knowing individual and team limitations/strengths should be considered in planning and decision-making.

In the following section, we present the Human Factors Operating Picture (H-FOP) framework that considers these desired requirements to offer real-time indicators of human performance for effective Human-Swarm teams.

13.4.2 Elements of H-FOP Design

Elements of H-FOP, a proprietary technology from our group, are presented in Fig. 13.5. This framework was developed to fuse different cognitive metrics and to offer a better data representation on human performance in a real-time shepherding task. H-FOP has two main components, the H-FOP-Visualisation and Reporting Toolbox (RVT) (H-FOP-RVT) and the H-FOP Artificial Intelligent (AI) agent (H-FOP-AI).

The H-FOP-AI extracts cognitive indicators from the human operator in real-time to provide decision-makers a complete description of the operator and the task. This component is subdivided into two parts: (1) the Analytics AI, which offers analytic capabilities to explore the large stream of data received from the different sensors on humans and the task; and (2) the Control AI, which provides control capabilities with the ability to alter the environment or the user interface.

The H-FOP-VRT component obtains important information from the human and the task. This visualization tool reports information in an easy-to-understand and compact way to be used by an agent acting as a decision-maker to support

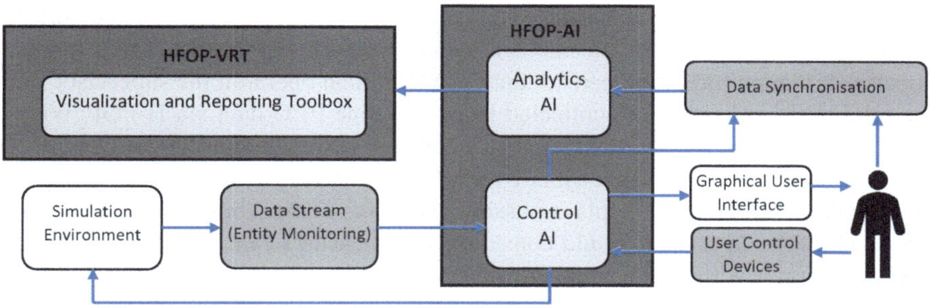

Fig. 13.5 Human factors operational picture (H-FOP)

decisions based on models of operator's behavioural and cognitive states and task performance. This tool also adds transparency to the decision-making process and provides evidence of specific situations that need attention.

There are also other components that help capture information from control devices (e.g., joystick, mouse, keyboard) that the operator uses to interact with the simulation environment, the data stream component that serves to stream data out of the simulation environment, and the data synchronisation layer that collects and synchronises different modalities [19] of sensor data (e.g., EEG, GSR, heart rate). In a realistic scenario, the simulation (virtual) environment can be replaced by the (physical) swarm in which the operator can behave as a shepherd in order to control the swarm.

13.4.3 Shepherding a Group of Robots for Effective Human-Swarm Teaming

We conclude this chapter by bringing together the previous dimensions of human performance into the shepherding problem. The advantage that shepherding offers for swarm guidance is that it puts the guidance of many agents in the hands of a few shepherds, or even a single shepherd. The human operator can guide the swarm using this single control point. This makes it easy to transfer most of the current literature that has been focusing on a human operating a single robot to the human-swarm teaming literature with ease.

To illustrate the concept, we will assume that the operational environment corresponds to an urban setting and with specific hazard zones that need to be avoided. For this example, the swarm (a group of UGVs) might be transporting medical supplies in a disaster area. The human side of the team controls and distribute commands to the ground robots by means of control devices and receives sensor information on the user interface to maintain situation awareness.

Performance metrics from both the human operator and the task can be visualised on H-FOP-VRT. In this tool, it is possible to observe the operator's cognitive and behavioural state. These information can be used by another member of the human team to trigger adjustment mechanisms when needed. For instance, if a human supervisor is monitoring the performance of the human operator, the supervisor can read the workload indicators collected from real-time EEG data via H-FOP. When H-FOP detects changes in the operator's cognitive state, which might indicate that the operator's cognitive capacity is overloaded. The system then could trigger a recommender system that displays possible measures to overcome the situation and that the human supervisor should consider; such measures can either substitute the current operator with another operator to control the swarm, or enable an AI agent to assist the operator to guide and control the swarm.

In case an AI agent is authorised to help the operator, the AI agent needs to balance the division of tasks to neither overload nor underload the human operator. This is done by the AI agent since important information about the operator's cognitive states is received from H-FOP to make decisions on the distribution of tasks. The human operator and the AI agent can continuously work together as shepherds to guide the swarm in challenging situations, while the human supervisor maintains full situation awareness on the shepherds and the system performance. Once the level of workload can be maintained by the operator and the situation is less challenging, the supervisor can remove the AI agent to give full control to the operator to shepherd the swarm.

To achieve this dynamic interaction between team members and efficient use of their resources, a multimodal assessment of human performance in dynamic environments is needed [19]. This multimodal integration to measure human performance should include sensors that comply with the operational settings and technical requirements, which can provide reliable data that can be transformed into actionable information. Advances in human sciences (e.g., psychology, neuroscience), engineering, and computer science (e.g., artificial intelligence and machine learning) will allow this vision.

13.5 Conclusions

Shepherding offers one or a few control points that allow a human to manage a swarm. The reduction of the complexity required to manage a large swarm in one or more control points is achieved through appropriate artificial intelligence to increase level of autonomy by the shepherds. As a new concept and environment for the human, many unknowns exist on the level of cognitive workload the environment is imposing on the human operator. Fortunately, the vast cognitive engineering and human factor literature could be transferred to this domain, where the human-shepherd interaction is still a single human and a single robot interaction problem, assuming there is a single shepherd. Nevertheless, the human needs to maintain situation awareness on the swarm, which is a distributed system.

In this chapter, we attempted to review the vast literature cognitive engineering literature and our previous work to present a single point of reference for readers interested in human-swarm teaming. While the literature is vast, we selected a subset that is most relevant to the human-shepherding context.

The chapter covered aspects of subjective and objective metrics for task performance, cognitive workload, situation awareness, and trust. The Human-Factors Operating Picture (H-FOP) was then presented to offer a real-time situation awareness picture on human performance. Such a picture could be used for real-time system-level adaptation mechanisms that allow the system to tune the autonomy of the shepherds based on human-cognitive performance.

Achieving effective Human-Swarm teaming requires effective coordination of efforts between the human operator and the swarm system. To achieve this dynamic behaviour between team members, it will be necessary to obtain a multidisciplinary approach to assess human performance and obtain actionable information that can be used to trigger remedies to coordinate team's efforts. To achieve this vision, teaming with other autonomous agents (e.g., AI or machine) will be necessary to leverage Human-Swarm capabilities.

For future work, it is important to use multiple data modalities to enhance human performance assessment. This approach would increase: (1) sensor validation to handle data imperfection due to the quality of sensors data, (2) sensor robustness to handle sensor failure and also accommodate individual variability, (3) sensor interaction to provide complimentary measures to each other, while at the same time, one modality could be used to constrain another modality.

References

1. Abbass, H.A., Tang, J., Amin, R., Ellejmi, M., Kirby, S.: Augmented cognition using real-time EEG-based adaptive strategies for air traffic control. In: Proceedings of the Human Factors and Ergonomics Society Annual Meeting, vol. 58, pp. 230–234. SAGE Publications Sage CA, Los Angeles (2014)
2. Abbass, H.A., Petraki, E., Merrick, K., Harvey, J., Barlow, M.: Trusted autonomy and cognitive cyber symbiosis: open challenges. Cognit. Comput. **8**(3), 385–408 (2016). https://doi.org/10.1007/s12559-015-9365-5
3. Bindewald, J.M., Rusnock, C.F., Miller, M.E.: Measuring human trust behavior in human-machine teams. In: International Conference on Applied Human Factors and Ergonomics, pp. 47–58. Springer, Berlin (2017)
4. Blais, C., Ellis, D.M., Wingert, K.M., Cohen, A.B., Brewer, G.A.: Alpha suppression over parietal electrode sites predicts decisions to trust. Soc. Neurosci. **14**(2), 226–235 (2019)
5. Blanco, J.A., Johnson, M.K., Jaquess, K.J., Oh, H., Lo, L.C., Gentili, R.J., Hatfield, B.D.: Quantifying cognitive workload in simulated flight using passive, dry EEG measurements. IEEE Trans. Cogn. Develop. Syst. **10**(2), 373–383 (2016)
6. Bogner, M.S.: Human Error in Medicine. CRC Press, Boca Raton (2018)
7. Borghini, G., Astolfi, L., Vecchiato, G., Mattia, D., Babiloni, F.: Measuring neurophysiological signals in aircraft pilots and car drivers for the assessment of mental workload, fatigue and drowsiness. Neurosci. Biobehav. Rev. **44**, 58–75 (2014)
8. Brookhuis, K., Waard, D.D., Mulder, B.: Measuring driving performance by car-following in traffic. Ergonomics **37**(3), 427–434 (1994)

9. Cain, B.: A review of the mental workload literature. Technical Report, Defence Research And Development Toronto (Canada) (2007)
10. Catherwood, D., Edgar, G.K., Nikolla, D., Alford, C., Brookes, D., Baker, S., White, S.: Mapping brain activity during loss of situation awareness: an EEG investigation of a basis for top-down influence on perception. Hum. Factors **56**(8), 1428–1452 (2014)
11. Chapman, A.C., Micillo, R.A., Kota, R., Jennings, N.R.: Decentralised dynamic task allocation: A practical game: Theoretic approach. In: Proceedings of The 8th International Conference on Autonomous Agents and Multiagent Systems, vol. 2, pp. 915–922. International Foundation for Autonomous Agents and Multiagent Systems (2009)
12. Chen, J.Y., Haas, E.C., Barnes, M.J.: Human performance issues and user interface design for teleoperated robots. IEEE Trans. Syst. Man Cybern. Part C (Appl. Rev.) **37**(6), 1231–1245 (2007)
13. Chien, S.Y., Sycara, K., Liu, J.S., Kumru, A.: Relation between trust attitudes toward automation, hofstede's cultural dimensions, and big five personality traits. In: Proceedings of the Human Factors and Ergonomics Society Annual Meeting, vol. 60, pp. 841–845. SAGE Publications Sage CA, Los Angeles (2016)
14. Cooke, N.J.: Human factors of remotely operated vehicles. In: Proceedings of the Human Factors and Ergonomics Society Annual Meeting, vol. 50, pp. 166–169. SAGE Publications Sage CA, Los Angeles (2006)
15. Corwin, W.H., Sandry-Garza, D.L., Biferno, M.H., Boucek Jr, G.P., Logan, A.L.: Assessment of crew workload measurement methods, techniques and procedures. Volume 1. Process, methods and results. Technical Report, Douglas Aircraft Co Long Beach CA (1989)
16. Council, N.R., et al.: Quantitative Modeling of Human Performance in Complex, Dynamic Systems. National Academies Press, Washington (1990)
17. Council, N.R., et al.: Tactical Display for Soldiers: Human Factors Considerations. National Academies Press, Washington (1997)
18. Das, G.P., McGinnity, T.M., Coleman, S.A., Behera, L.: A distributed task allocation algorithm for a multi-robot system in healthcare facilities. J. Intell. Rob. Syst. **80**(1), 33–58 (2015)
19. Debie, E., Rojas, R.F., Fidock, J., Barlow, M., Kasmarik, K., Anavatti, S., Garratt, M., Abbass, H.A.: Multimodal fusion for objective assessment of cognitive workload: a review. IEEE Trans. Cybern. 1–4 (2019)
20. DeSteno, D., Breazeal, C., Frank, R.H., Pizarro, D., Baumann, J., Dickens, L., Lee, J.J.: Detecting the trustworthiness of novel partners in economic exchange. Psychol. Sci. **23**(12), 1549–1556 (2012)
21. Dijksterhuis, C., de Waard, D., Brookhuis, K., Mulder, B., de Jong, R.: Classifying visuomotor workload in a driving simulator using subject specific spatial brain patterns. Front. Neurosci. **7**, 149 (2013)
22. Dong, S.Y., Kim, B.K., Lee, K., Lee, S.Y.: A preliminary study on human trust measurements by EEG for human-machine interactions. In: Proceedings of the 3rd International Conference on Human-Agent Interaction, pp. 265–268. ACM, New York (2015)
23. Dorneich, M.C., Passinger, B., Hamblin, C., Keinrath, C., Vašek, J., Whitlow, S.D., Beekhuyzen, M.: Evaluation of the display of cognitive state feedback to drive adaptive task sharing. Front. Neurosci. **11**, 144 (2017)
24. Elkin-Frankston, S., Bracken, B.K., Irvin, S., Jenkins, M.: Are behavioral measures useful for detecting cognitive workload during human-computer interaction? In: Advances in The Human Side of Service Engineering, pp. 127–137. Springer, Berlin (2017)
25. Endsley, M.R.: Situation awareness global assessment technique (SAGAT). In: Proceedings of the IEEE 1988 National Aerospace and Electronics Conference, pp. 789–795. IEEE, Piscataway (1988)

26. Endsley, M.R.: Toward a theory of situation awareness in dynamic systems. Human Fact. **37**(1), 32–64 (1995). 06296. https://doi.org/10.1518/001872095779049543
27. Endsley, M.R., Garland, D.J.: Situation awareness analysis and measurement. CRC Press, Boca Raton (2000)
28. Fernandez Rojas, R., Debie, E., Fidock, J., Barlow, M., Kasmarik, K., Anavatti, S., Garratt, M., Abbass, H.A.: Encephalographic assessment of situation awareness in teleoperation of human-swarm teaming. In: Proceedings of The 26th International Conference on Neural Information Processing 2019. Springer, Berlin (2019)
29. Fernandez-Rojas, R., Perry, A., Singh, H., Campbell, B., Elsayed, S., Hunjet, R., Abbass, H.A.: Contextual awareness in human-advanced-vehicle systems: a survey. IEEE Access **7**, 33304–33328 (2019)
30. Fischer, J.E., Greenhalgh, C., Jiang, W., Ramchurn, S.D., Wu, F., Rodden, T.: In-the-loop or on-the-loop? Interactional arrangements to support team coordination with a planning agent. In: Concurrency and Computation: Practice and Experience, p. e4082 (2017)
31. Flanagan, J.C.: The critical incident technique. Psychol. Bullet. **51**(4), 327 (1954)
32. Gaudiello, I., Zibetti, E., Lefort, S., Chetouani, M., Ivaldi, S.: Trust as indicator of robot functional and social acceptance. an experimental study on user conformation to iCub answers. Comput. Human Behav. **61**, 633–655 (2016)
33. Gawron, V.J.: Human performance, workload, and situational awareness measures handbook. CRC Press, Boca Raton (2008)
34. Ha, J.S., Seong, P.H., Lee, M.S., Hong, J.H.: Development of human performance measures for human factors validation in the advanced MCR of APR-1400. IEEE Trans. Nuclear Sci. **54**(6), 2687–2700 (2007)
35. Hancock, P.A., Meshkati, N.: Human Mental Workload. North-Holland, Amsterdam (1988)
36. Harbour, J.L.: Lean Human Performance Improvement. Productivity Press, Boca Raton (2014)
37. Harriott, C.E., Seiffert, A.E., Hayes, S.T., Adams, J.A.: Biologically-inspired human-swarm interaction metrics. In: Proceedings of the Human Factors and Ergonomics Society Annual Meeting, vol. 58, pp. 1471–1475. SAGE Publications Sage CA, Los Angeles (2014)
38. Hart, S.G., Staveland, L.E.: Development of NASA-TLX (Task load index): results of empirical and theoretical research. Adv. Psychol. **52**, 139–183 (1988). 07598
39. Helmreich, R.L., Davies, J.M.: 3 Human factors in the operating room: Interpersonal determinants of safety, efficiency and morale. Baillière's Clin. Anaesthesiol. **10**(2), 277–295 (1996). 00147. https://doi.org/10.1016/S0950-3501(96)80017-1
40. Hill, S.G., Iavecchia, H.P., Byers, J.C., Bittner Jr, A.C., Zaklade, A.L., Christ, R.E.: Comparison of four subjective workload rating scales. Hum. Factors **34**(4), 429–439 (1992)
41. Hirshfield, L., Bobko, P., Barelka, A., Sommer, N., Velipasalar, S.: Toward interfaces that help users identify misinformation online: using fnirs to measure suspicion. Augment. Hum. Res. **4**(1), 1 (2019)
42. Hoff, K.A., Bashir, M.: Trust in automation: Integrating empirical evidence on factors that influence trust. Hum. Factors **57**(3), 407–434 (2015)
43. Hussein, A., Elsawah, S., Abbass, H.A.: Trust mediating reliability–reliance relationship in supervisory control of human–swarm interactions. Hum. Factors (2019) https://doi.org/10.1177/0018720819879273
44. Hussein, A., Elsawah, S., Abbass, H.A.: The reliability and transparency bases of trust in human-swarm interaction: principles and implications. Ergonomics **63**, 1–19 (2020)
45. Hussein, A., Ghignone, L., Nguyen, T., Salimi, N., Nguyen, H., Wang, M., Abbass, H.A.: Towards bi-directional communication in human-swarm teaming: A survey (2018). Preprint arXiv:1803.03093
46. Jian, J.Y., Bisantz, A.M., Drury, C.G.: Foundations for an empirically determined scale of trust in automated systems. Int. J. Cognit. Ergon. **4**(1), 53–71 (2000)
47. Jones, D.G., Endsley, M.R.: Sources of situation awareness errors in aviation. Aviat. Space Environ. Med. **67**, 507–512 (1996). 00423

48. Joosse, M., Sardar, A., Lohse, M., Evers, V.: BEHAVE-II: The revised set of measures to assess users' attitudinal and behavioral responses to a social robot. Int. J. Soc. Rob. **5**(3), 379–388 (2013)
49. Kaber, D.B., Riley, J.M., Zhou, R., Draper, J.: Effects of visual interface design, and control mode and latency on performance, telepresence and workload in a teleoperation task. In: Proceedings of the Human Factors and Ergonomics Society Annual Meeting, vol. 44, pp. 503–506. SAGE Publications Sage CA, Los Angeles (2000)
50. Kaber, D.B., Wright, M.C., Prinzel III, L.J., Clamann, M.P.: Adaptive automation of human-machine system information-processing functions. Hum. Factors **47**(4), 730–741 (2005)
51. Keller, J., Lebiere, C., Shay, C.R., Latorella, K.: Cockpit system situational awareness modeling tool. Human Perform. Situation Awareness Autom. Current Res. Trends HPSAA II **2**, 66–71 (2004). 00010
52. Khalid, H.M., Shiung, L.W., Nooralishahi, P., Rasool, Z., Helander, M.G., Kiong, L.C., Ai-vyrn, C.: Exploring psycho-physiological correlates to trust: Implications for human-robot-human interaction. In: Proceedings of the Human Factors and Ergonomics Society Annual Meeting, vol. 60, pp. 697–701. SAGE Publications Sage CA, Los Angeles (2016)
53. Kolling, A., Walker, P., Chakraborty, N., Sycara, K., Lewis, M.: Human interaction with robot swarms: A survey. IEEE Trans. Human-Mach. Syst. **46**(1), 9–26 (2015)
54. Kramer, R.M.: Trust and distrust in organizations: emerging perspectives, enduring questions. Ann. Rev. Psychol. **50**(1), 569–598 (1999)
55. Krueger, F., McCabe, K., Moll, J., Kriegeskorte, N., Zahn, R., Strenziok, M., Heinecke, A., Grafman, J.: Neural correlates of trust. Proc. Natl. Acad. Sci. **104**(50), 20084–20089 (2007)
56. Larzelere, R.E., Huston, T.L.: The dyadic trust scale: toward understanding interpersonal trust in close relationships. J. Marriage Family **42**, 595–604 (1980)
57. Laughery, K., Lebiere, C., Archer, S.: Modeling human performance in complex systems. In: Handbook of Human Factors and Ergonomics, pp. 965–996 (2006)
58. Lee, J., Moray, N.: Trust, control strategies and allocation of function in human-machine systems. Ergonomics **35**(10), 1243–1270 (1992)
59. Lee, J.D., See, K.A.: Trust in automation: designing for appropriate reliance. Hum. Factors **46**(1), 50–80 (2004)
60. Lingard, L., Espin, S., Whyte, S., Regehr, G., Baker, G.R., Reznick, R., Bohnen, J., Orser, B., Doran, D., Grober, E.: Communication failures in the operating room: An observational classification of recurrent types and effects. BMJ Qual. Saf. **13**(5), 330–334 (2004). 01001
61. Longo, L.: Human-computer interaction and human mental workload: Assessing cognitive engagement in the world wide web. In: IFIP Conference on Human-Computer Interaction, pp. 402–405. Springer, Berlin (2011)
62. Lyons, J.B., Stokes, C.K.: Human–human reliance in the context of automation. Hum. Factors **54**(1), 112–121 (2012)
63. Ma-Wyatt, A., Fidock, J., Abbass, H.: Quantifying and predicting human performance for effective human-autonomy teaming. In: Proceedings of The 2018 International Conference on Science and Innovation for Land Power, pp. 915–922. Industry Defence and Security Australia Limited, Victoria (2018)
64. Madsen, M., Gregor, S.: Measuring human-computer trust. In: 11th Australasian Conference on Information Systems, vol. 53, pp. 6–8. Citeseer (2000)
65. Maior, H.A., Pike, M., Wilson, M.L., Sharples, S.: Continuous detection of workload overload: An fnirs approach. In: Contemporary Ergonomics and Human Factors 2014: Proceedings of the international conference on Ergonomics & Human Factors 2014, Southampton, 7–10 April 2014, p. 450. CRC Press, Boca Raton (2014)
66. Marieb, E.N., Hoehn, K.: Human Anatomy & Physiology. Pearson Education, London (2007)
67. Marsh, S., Dibben, M.R.: The role of trust in information science and technology. Ann. Rev. Inf. Sci. Technol. **37**(1), 465–498 (2003)
68. Master, R., Gramopadhye, A., Melloy, B., Bingham, J., Jiang, X.: A questionnaire for measuring trust in hybrid inspection systems. In: Proceedings of the Industrial Engineering Research Conference, Dallas (2000)

69. Megaw, T.: The definition and measurement of mental workload. In: Evaluation of Human Work, pp. 530–556, 3rd edn. CRC Press, Boca Raton (2005)
70. Merritt, S.M., Heimbaugh, H., LaChapell, J., Lee, D.: I trust it, but i don't know why: effects of implicit attitudes toward automation on trust in an automated system. Hum. Factors **55**(3), 520–534 (2013)
71. Montague, E., Xu, J., Chiou, E.: Shared experiences of technology and trust: an experimental study of physiological compliance between active and passive users in technology-mediated collaborative encounters. IEEE Trans. Human-Mach. Syst. **44**(5), 614–624 (2014)
72. Muir, B.M., Moray, N.: Trust in automation. Part II: experimental studies of trust and human intervention in a process control simulation. Ergonomics **39**(3), 429–460 (1996)
73. Müller, J.A., Wendt, D., Kollmeier, B., Brand, T.: Comparing eye tracking with electrooculography for measuring individual sentence comprehension duration. PloS One **11**(10), e0164,627 (2016)
74. Mutz, D.C., Reeves, B.: The new videomalaise: effects of televised incivility on political trust. Am. Political Sci. Rev. **99**(1), 1–15 (2005)
75. Nisbett, R.E., Wilson, T.D.: Telling more than we can know: verbal reports on mental processes. Psychol. Rev. **84**(3), 231 (1977)
76. Norman, D.A.: The 'problem' with automation: inappropriate feedback and interaction, not 'over-automation'. Philos. Trans. R. Soc. Lond. B Biol. Sci. **327**(1241), 585–593 (1990)
77. Nourbakhsh, N., Wang, Y., Chen, F., Calvo, R.A.: Using galvanic skin response for cognitive load measurement in arithmetic and reading tasks. In: Proceedings of the 24th Australian Computer-Human Interaction Conference, pp. 420–423. ACM, New York (2012)
78. Oh, S., Seong, Y., Yi, S.: Preliminary study on neurological measure of human trust in autonomous systems. In: IIE Annual Conference. Proceedings, pp. 1066–1072. Institute of Industrial and Systems Engineers (IISE), Georgia (2017)
79. O'Hara, J.M., Hall, R.E.: Advanced control rooms and crew performance issues: implications for human reliability. IEEE Trans. Nuclear Sci. **39**(4), 919–923 (1992)
80. Parasuraman, R., Molloy, R., Singh, I.L.: Performance consequences of automation-induced'complacency'. Int. J. Aviation Psychol. **3**(1), 1–23 (1993)
81. Parasuraman, R., Sheridan, T.B., Wickens, C.D.: Situation awareness, mental workload, and trust in automation: viable, empirically supported cognitive engineering constructs. J. Cogn. Eng. Decis. Mak. **2**(2), 140–160 (2008)
82. Parush, A., Kramer, C., Foster-Hunt, T., Momtahan, K., Hunter, A., Sohmer, B.: Communication and team situation awareness in the OR: implications for augmentative information display. J. Biomed. Inf. **44**(3), 477–485 (2011). 00093
83. Peck, E.M.M., Yuksel, B.F., Ottley, A., Jacob, R.J., Chang, R.: Using fnirs brain sensing to evaluate information visualization interfaces. In: Proceedings of the SIGCHI Conference on Human Factors in Computing Systems, pp. 473–482. ACM, Berlin (2013)
84. Polvichai, J.: Modeling team performance for coordination configurations of large multi-agent teams using stochastic neural networks. Ph.D. Thesis, University of Pittsburgh (2008)
85. Reid, G.B., Nygren, T.E.: The subjective workload assessment technique: a scaling procedure for measuring mental workload. Adv. Psychol. **52**, 185–218 (1988). 00635
86. Rempel, J.K., Holmes, J.G., Zanna, M.P.: Trust in close relationships. J. Pers. Soc. Psychol. **49**(1), 95 (1985)
87. Rojas, R.F., Debie, E., Fidock, J., Barlow, M., Kasmarik, K., Anavatti, S., Garratt, M., Abbass, H.: Electroencephalographic workload indicators during teleoperation of an unmanned aerial vehicle shepherding a swarm of unmanned ground vehicles in contested environments. Front. Neurosci. **14** (2020)
88. Rojas, R.F., Huang, X., Ou, K.L.: Region of interest detection and evaluation in functional near infrared spectroscopy. J. Near Infrared Spec. **24**(4), 317–326 (2016)
89. Rojas, R.F., Huang, X., Ou, K.L.: Toward a functional near-infrared spectroscopy-based monitoring of pain assessment for nonverbal patients. J. Biomed. Opt. **22**(10), 106,013 (2017)
90. Rojas, R.F., Huang, X., Ou, K.L.: A machine learning approach for the identification of a biomarker of human pain using fnirs. Sci. Rep. **9**(1), 5645 (2019)

91. Rotter, J.B.: A new scale for the measurement of interpersonal trust 1. J. Personal. **35**(4), 651–665 (1967)
92. Rubio, S., Díaz, E., Martín, J., Puente, J.M.: Evaluation of subjective mental workload: A comparison of SWAT, NASA-TLX, and workload profile methods. Appl. Psychol. **53**(1), 61–86 (2004)
93. Sanchez, J., Rogers, W.A., Fisk, A.D., Rovira, E.: Understanding reliance on automation: effects of error type, error distribution, age and experience. Theor. Issues Ergon. Sci. **15**(2), 134–160 (2014)
94. Shah, J., Kim, B., Nikolaidis, S.: Human-inspired techniques for human-machine team planning. In: 2012 AAAI Fall Symposium Series (2012)
95. Singh, I.L., Molloy, R., Parasuraman, R.: Automation-induced "complacency": development of the complacency-potential rating scale. Int. J. Aviat. Psychol. **3**(2), 111–122 (1993)
96. Spravka, J.J., Moisio, D.A., Payton, M.G.: Unmanned air vehicles: A new age in human factors evaluations. Technical Report, Air Force Flight Test Center Edwards AFB CA (2003)
97. Stein, E.S.: The measurement of pilot performance: A master-journeyman approach. Technical Report, Federal Aviation Administration Technical Center Atlantic City NJ (1984)
98. Stokes, C.K., Lyons, J.B., Littlejohn, K., Natarian, J., Case, E., Speranza, N.: Accounting for the human in cyberspace: Effects of mood on trust in automation. In: 2010 International Symposium on Collaborative Technologies and Systems, pp. 180–187. IEEE, Piscataway (2010)
99. Taylor, R.: Situational awareness rating technique (SART): The development of a tool for aircrew systems design. In: Situational Awareness, pp. 111–128. Routledge, Abingdon (2017)
100. Thornton, R.C.: The effects of automation and task difficulty on crew coordination, workload, and performance. Doctor of Philosophy (PhD), Dissertation, Psychology, Old Dominion University (1992). 00019
101. Uggirala, A., Gramopadhye, A.K., Melloy, B.J., Toler, J.E.: Measurement of trust in complex and dynamic systems using a quantitative approach. Int. J. Ind. Ergon. **34**(3), 175–186 (2004)
102. Veltman, J., Gaillard, A.: Physiological indices of workload in a simulated flight task. Biol. Psychol. **42**(3), 323–342 (1996)
103. Vidulich, M.A., Hughes, E.R.: Testing a subjective metric of situation awareness. In: Proceedings of the Human Factors Society Annual Meeting, vol. 35, pp. 1307–1311. SAGE Publications Sage CA, Los Angeles (1991)
104. Vidulich, M., Tsang, P.: Techniques of subjective workload assessment: a comparison of SWAT and the NASA-Bipolar methods. Ergonomics **29**(11), 1385–1398 (1986). 00067
105. Vizer, L.M., Zhou, L., Sears, A.: Automated stress detection using keystroke and linguistic features: an exploratory study. Int. J. Human-Comput. Stud. **67**(10), 870–886 (2009)
106. Wang, M., Hussein, A., Rojas, R.F., Shafi, K., Abbass, H.A.: Eeg-based neural correlates of trust in human-autonomy interaction. In: 2018 IEEE Symposium Series on Computational Intelligence (SSCI), pp. 350–357. IEEE, Piscataway (2018)
107. Wickens, C.D.: Multiple resources and performance prediction. Theor. Issues Ergon. Sci. **3**(2), 159–177 (2002)
108. Wickens, C.D., Hollands, J.G., Banbury, S., Parasuraman, R.: Engineering Psychology and Human Performance. Psychology Press, London (2015)
109. Wierwille, W.W., Casali, J.G.: A Validated Rating Scale for Global Mental Workload Measurement Applications. pp. 129–133. Sage Publications Sage CA, Los Angeles (1983). 00247
110. Wilson, G.F.: Strategies for psychophysiological assessment of situation awareness. In: Situation Awareness Analysis and Measurement, pp. 175–188. CRC Press, Boca Raton (2000)
111. Xie, B., Salvendy, G.: Review and reappraisal of modelling and predicting mental workload in single- and multi-task environments. Work & Stress **14**(1), 74–99 (2000). https://doi.org/10.1080/026783700417249. 00142
112. Yagoda, R.E., Gillan, D.J.: You want me to trust a ROBOT? The development of a human–robot interaction trust scale. Int. J. Soc. Rob. **4**(3), 235–248 (2012)

113. Yamagishi, T.: The provision of a sanctioning system as a public good. J. Pers. Soc. Psychol. **51**(1), 110 (1986)
114. Yamagishi, T., Yamagishi, M.: Trust and commitment in the united states and japan. Motiv. Emot. **18**(2), 129–166 (1994)
115. Yeo, L.G., Sun, H., Liu, Y., Trapsilawati, F., Sourina, O., Chen, C.H., Mueller-Wittig, W., Ang, W.T.: Mobile EEG-based situation awareness recognition for air traffic controllers. In: 2017 IEEE International Conference on Systems, Man, and Cybernetics (SMC), pp. 3030–3035. IEEE, Piscataway (2017)
116. Young, M.S., Brookhuis, K.A., Wickens, C.D., Hancock, P.A.: State of science: mental workload in ergonomics. Ergonomics **58**(1), 1–17 (2015)

Index

H. A. Abbass, R. A. Hunjet (eds.), *Shepherding UxVs for Human-Swarm Teaming*, Unmanned System Technologies,
https://doi.org/10.1007/978-3-030-60898-9

The manufacturer's authorised representative in the EU is Springer Nature Customer Service Centre GmbH, Europaplatz 3, 69115 Heidelberg, Germany. If you have any concerns regarding our products, please contact ProductSafety@springernature.com

Printed and bound by CPI Group (UK) Ltd, Croydon, CR0 4YY
30/09/2026
02225694-0012